Methods for the Study of Deep-Sea Sediments, Their Functioning and Biodiversity

Methods for the Study of Deep-Sea Sediments, Their Functioning and Biodiversity

Edited by

Roberto Danovaro

CRC Press
Taylor & Francis Group
Boca Raton London New York

CRC Press is an imprint of the
Taylor & Francis Group, an **informa** business

CRC Press
Taylor & Francis Group
6000 Broken Sound Parkway NW, Suite 300
Boca Raton, FL 33487-2742

International Standard Book Number: 978-1-4398-1137-5 (Hardback)

Library of Congress Cataloging-in-Publication Data

Danovaro, Roberto.
Methods for the study of deep-sea sediments, their functioning and biodiversity / author, Roberto Danovaro.
p. cm.
"A CRC title."
Includes bibliographical references and index.
ISBN 978-1-4398-1137-5 (alk. paper)
1. Benthic animals--Research--Methodology. 2. Benthos--Research--Methodology. 3. Marine sediments--Research--Methodology. 4. Marine biology--Methodology. I. Title.

QL121.D316 2010
551.46--dc22 2009032044

Visit the Taylor & Francis Web site at
http://www.taylorandfrancis.com

and the CRC Press Web site at
http://www.crcpress.com

To all women who devote their enthusiasm for deep-sea life to research and, in particular, to Cinzia who had the idea to write this book.

Contents

SECTION III Deep-Sea Benthic Life

SECTION IV Deep-Sea Benthic Diversity

SECTION V Deep-Sea Benthic Functioning

SECTION VI Suggested Readings for the Study of Deep-Sea Biodiversity and Ecosystem Functioning

Foreword

The concept of deep-sea biology as a distinct discipline was formulated with the *Challenger* expedition of 1872 to 1876. Although organisms from the deep sea were known before this time, the *Challenger* expedition was the first systematic attempt to determine if and what animals lived in the deep sea. Its many volumes are a testament to systematics and taxonomy, but at the same time they address questions about the then unknown composition of the deep-sea bed and the waters of the ocean interior. Biodiversity remained a fundamental theme of deep-sea biological research up until the *Galathea* expedition of 1950 to 1952, progress being interrupted by war and fiscal constraints. This so-called heroic age of deep sea exploration was carried out from vessels that ranged from sailing ships, through steam to diesel-powered, although the sampling technology remained the same, being composed of relatively coarse-meshed nets in trawls and baited traps. Biodiversity was considered low.

In the 1960s, with the advent of more sophisticated sampling devices that collected small bodied animals from the deep-sea, biodiversity was shown to be high, although the driving laboratory techniques were a microscope and the specialist's brain. However, sampling was haphazard to some extent, and it was only with the development of submersible technology in the late 1960s that scientists could enter the deep sea and sample at the submeter scale. Submersibles also gave the opportunity of establishing hypothesis-driven experiments in the deep sea, but it was with the discovery of hydrothermal vents in the 1977 that such technologies really came into their own, as deep-sea scientists could assess biodiversity and ecosystem function at these sporadic sites on a submeter scale. At the same time ship-deployed quantitative samplers were also being developed that allowed sampling for macro- and meiofauna and their use in vitro experimentation in pressure chambers aboard ships.

However, deep-sea animals are not independent of their environment, and it was in the 1990s that the relationship between biodiversity and deep-sea ecosystem function started to become established. Increasingly, sophisticated techniques allowed a wide variety of microorganisms to be collected and cultured, and the relationship between organic carbon input to the deep-sea, microbial activity, and faunal diversity and function started to become resolved. As with all aspects of biology, a quantum leap forward occurred with the use of molecular techniques, and these are now as routinely used in deep-sea biology as they are in any biological laboratory.

Volumes on deep-sea biology in the past have addressed pattern and process on a variety of scales, but in most cases there has been little attention given to the techniques by which we, as deep-sea biologists, must wrest our knowledge of this apparently remote and challenging environment. In many cases the techniques used have been unique to the situation under study. But to enable comparison of biodiversity and ecosystem function to be made among different deep-sea regions it is vital to establish protocols that all can use. Roberto Danovaro brings a wealth of practical experience to this volume, as well as being one of the world's leading deep-sea ecologists. This book of techniques will prove to be invaluable to deep-sea biologists, from those well established to the new postgraduate starting out on a scientific exploration and analysis of the most remote, potentially the most biodiverse, and the most exciting of ecosystems on planet Earth.

Paul Tyler
Professor of Deep-Sea Biology
National Oceanography Centre
University of Southampton
United Kingdom

Premise

The deep sea is the largest environment on the planet, the least well known, and one of the least studied. It contains extremely large, continuous habitats such as the millions of km^2 of abyssal plains and the 65,000-km-long mid-oceanic ridge system. At the same time, it encloses relatively small (hundreds of km^2 to only a few m^2), localized geological features such as canyons, seamounts, deep-water coral reefs, hydrothermal vents and fluid seepages on mud volcanoes, and pockmarks or faults that support unique microbial and faunal communities. What little we know about deep-sea ecosystems supports the hypothesis that more species occur in the deep sea than anywhere else on Earth. As much as 90% of species collected in a typical abyssal sediment sample are new to science.

The multidisciplinary ESF-EUROCORES Programme EuroDEEP was launched in June 2007 aiming at the exploration and identification of the different deep-sea habitats, assessing both the abiotic and biotic processes that sustain and maintain deep-sea communities. EuroDEEP is a program for deep-sea biology and ecology that strongly depends and requires collaboration between taxonomists, microbiologists, ecologists, physical and chemical oceanographers, and geologists.

Education and outreach is a very important aspect of the program since the resulting scientific data are a prerequisite for the sustainable use and the development of management and conservation options aiming at the sustainable use of marine resources that will benefit society as a whole. EuroDEEP acknowledges, therefore, this extremely valuable initiative by Prof. Danovaro, partner in the BIOFUN Project in EuroDEEP, and supports this book as a basis for the formation of a new generation of deep-sea scientists.

Dr. Inge Jonckheere
EuroDEEP Programme Coordinator—ESF

Preface

This book has been written to provide practical information for scientists, postdoctoral, graduate, and undergraduate students approaching the study of deep-sea habitats, their characteristics, functioning, and biodiversity. It is complementary to the texts and manuals already available for the study of the marine benthos, a key textbook being, for instance, *Methods for the Study of Marine Benthos* edited by A. Eleftheriou and A. McIntyre. This book includes all relevant protocols and details for the design and analysis of benthos and for the characterization of sediments. If you do not own it, please acquire it! Your library certainly needs such a comprehensive view of the marine benthos. The main difference between the present book and others already available is that in this book there are detailed protocols dealing with the analysis of benthic viruses (or virio-benthos), benthic prokaryotes, including Bacteria and Archaea, protozoa, foraminifera, meiofauna, macrofauna, and megafauna. The quantitative study of the biotic components includes analyses of their biodiversity, protocols for investigating some aspects of functional biodiversity, and some key variables commonly used to describe ecosystem functioning.

In the last few years there has been growing interest in the integration of biodiversity (taxonomy and autoecology) with ecosystem functioning (sinecology and interdisciplinary). It is my opinion, however, that this book is new within the panorama of the available textbooks, in the sense that it can be considered as the first book attempting to link research on biodiversity with a standardized approach to the study of the functioning of the marine ecosystems. I hope this book is recognized for the significant, if not exhaustive, contribution it makes to the expansion and widening of research in the field of biodiversity and ecosystem functioning in different habitats and ecosystems types.

This is only the beginning, a starting point to be further implemented by scientists affiliated to that network of excellence called Marine Biodiversity and Ecosystems Functioning (MARBEF), which has been able to bring together for the first time marine benthonologists, biogeochemists, and taxonomists with the perspective of obtaining a better understanding, not only of the number of new species in different marine systems, but also on how biodiversity interacts with the functioning of these ecosystems.

This book does not address the issues of the ecology of deep-sea organisms, nor does it explain the functioning of deep-sea ecosystems. These topics are covered in the "bible" of deep-sea biologists by Paul Tyler and John Gage, *Deep-Sea Biology: A Natural History of Organisms at the Deep-Sea Floor* (Cambridge University Press). I recommend that readers refer to this textbook and to another book by Peter Herring, *The Biology of the Deep Ocean*, for help in coping with the gaps in knowledge of the biology and ecology of deep-sea ecosystems.

Here, particular attention has been paid to the description of the analytical protocols and, when needed, additional references have been provided to the reader.

The book is organized as follows: After a brief introduction, there is a description of the sampling procedures and/or the sample treatment and then the laboratory protocols have been described in detail, providing information also on instrument setting and/or the solutions used. All materials have been reported in detail and the procedures described step by step, with additional notes on crucial steps or on the possible difficulties arising from the analysis. The calculation formulas were reported for each specific variable, along with a careful evaluation of the conversion coefficients involved. In certain cases, also, information has been added to check the procedure. The final section of references report the source of different methodologies. In addition, in each chapter there is a synthesis of the work and a visual scheme given. This scheme of the protocol can be copied or brought to the laboratory as a support during the laboratory activities and for tracking each laboratory step. Pictures, schemes, and/or diagrams are reported for all methods. Several pictures of the organisms or of the components investigated have been included to provide visual support to the scientists working in the laboratory.

In this manual for the first time, all living components of the benthos have been given attention. This is, to my opinion, important as in the future a better understanding of the structure and functioning of the marine ecosystems, and in particular of deep-sea ecosystems, cannot be achieved without also considering the smallest components of the benthos, which dominate numerically and account for most of the total biomass. These components also play a key role in deep-sea ecosystem functioning, thus closing, conceptually, the loop of the rational of this book.

I thank in advance all scientists and readers who would like to provide inputs, suggestions, and information for the updating and improvement of this manual, so that it can become an even more useful tool for the community of scientists working in the field of deep-sea biodiversity and ecosystem functioning.

Concluding, this manual is really a first attempt. The idea of this book was suggested by my wife Cinzia, in order to provide practical support and a guideline to protocols for scientists working in my laboratory. But this manual represents also the most updated methodological support for the study of deep-sea benthos that has been developed in the last 20 years. Several methods of specific research fields—in particular, the benthic microbiology—are evolving rapidly, and certainly this book will need updates in the coming years.

Finally, this manual represents a contribution toward achieving objectives in understanding the increasingly relevant topics dealing with deep-sea sciences. The need for improving our knowledge of the deep-sea ecosystems has also been highlighted by the launch of several European Union projects such as the HERMES project (Hotspot Ecosystem Research on the Margins of European Seas), followed by the HERMIONE project (Hotspot Ecosystem Research and Man's Impact on European Seas), and the ESF (European Science Foundation) Eurodeep project BIOFUN (BIOdiversity and Ecosystem FUNctioning Contrasting Southern European Deep-Sea Environments). These large-scale, pan-European projects were born with the specific objective of understanding the link and interactions between the biosphere and geosphere of the deep sea. I do hope that this is just the beginning of the effort that we all have to make to better know the function of the deep sea and the almost unexplored ecosystems that are hosted in the permanently dark portion of the biosphere.

Roberto Danovaro

Contributors

Roberto Danovaro contributed to all chapters and is at the Dipartimento di Scienze del Mare, Università Politecnica delle Marche, Ancona, Italy.

The following contributors (in alphabetical order) are at the Dipartimento di Scienze del Mare, Università Politecnica delle Marche: **Silvia Bianchelli** (co-author of Chapter 17), **Lucia Bongiorni** (co-author of Chapters: 15, 40), **Cinzia Corinaldesi** (co-author of Chapters: 11, 14, 31, 33), **Antonio Dell'Anno** (co-author of Chapters: 5, 6, 7, 8, 9 10, 29, 32, 37), **Cristina Gambi** (co-author of Chapters: 17, 24, 26, 38, 41, 42), **Gian Marco Luna** (co-author of Chapters: 20, 21, 22, 23), **Mirko Magagnini** (co-author of Chapters: 11, 33), **Massimiliano Molari** (co-author of Chapters: 13, 34, 37), **Antonio Pusceddu** (co-author of Chapters: 1, 2, 3, 4), **Daniela Zeppilli** (co-author of Chapter 17).

Emmanuelle Geslin is co-author of Chapter 16 and is at the University of Angers, Angers, France.

Leonardo Langone and **Stefano Miserocchi** are co-authors of Chapter 29 and are at the Istituto di Scienze Marine – Consiglio Nazionale Delle Ricerche, Bologna, Italy.

Elena Manini is co-author of Chapters 12, 13, 30, 34, 35, 36, 41 and is at the Istituto di Scienze Marine – Consiglio Nazionale Delle Ricerche, Ancona, Italy.

Caterina Morigi is co-author of Chapter 16 and is at the Geological Survey of Denmark and Greenland, Copenhagen, Denmark.

Joan B. Company and **Eva Ramirez-Llodra** are co-authors of Chapters 19, 25, 39 and are at the Institut de Ciències del Mar, Consejo Superior de Investigaciones, Barcelona, Spain.

Giulia Pelosi, Marianna Mea and **Paolo Paroncini** are acknowledged for their contribution to the editing and graphics of the protocols. They are at the Dipartimento di Scienze del Mare, Università Politecnica delle Marche, Ancona, Italy.

Introduction

The oceans cover approximately 70% of the world's total surface. The oceanic water masses extend down to a depth of more than 11 km, with an average depth of 3,850 m, and in several regions of the world a thick layer of sediments (up to 10 km) has accumulated over the crust of the mantel. The overall volume of the oceans represents more than 95% of the biosphere. Light can penetrate only in the uppermost layer of the water column. Typically, less than 1% of the light reaching the surface penetrates down to 150–200 m. At deeper depths the light is insufficient to sustain the photosynthetic primary production, and no light is available beneath 1000 m depth. The permanently dark portion of the globe represents, therefore, more than 65% of the world's surface and more than 90% of the biosphere. Deep-sea ecosystems collectively determine the biogeochemical processes that regulate the Earth system.

Despite the huge dimensions of the dark portion of the global biosphere, so far less than roughly 5% of the deep-sea ecosystems have been explored. Deep-sea ecosystems are remote, subjected to high pressure and accessible only by means of high tech submersibles and/or sampling gear (Snelgrove 1999; Ellingsen and Gray, 2002). It has been estimated that the total amount of sediment collected at a depth higher than 3000 m is equivalent to approximately a couple of football pitches. Therefore, less than 0.001% of the deep-sea floor has been investigated in detail. It is therefore evident that our view of the deep-sea ecosystems is extremely limited.

The discovery of new species and life forms hidden in the ocean interior and of the new ecosystems lying on the deep seafloor represents one of the last frontiers of the scientific research. At the same time, the investigation of the deep-sea habitats requires highly sophisticated technologies, and huge costs are associated with ship time and the instrumentation required to investigate the deep sea. This makes investigations in deep-sea environments extremely expensive, with costs often comparable or higher than extraterrestrial explorations. This is certainly one reason that can explain the delay of deep-sea research when compared with other fields of science. At the same time the exploration of the deep sea also represents an enormous opportunity for new discoveries.

During the last decade, central issues in ecological sciences have been (1) the identification of factors determining and controlling biodiversity (Gaston 2000), (2) the comprehension of how many species live within a set of habitats (Vandermeulen et al. 2001), and (3) the relationship between biodiversity and ecosystem functioning (Loreau et al. 2001). Although these issues are valid for both terrestrial and marine ecosystems, there is still a huge gap between studies of biological diversity carried out in terrestrial systems and those on marine environments (Rex et al. 1993, 2000; Ellingsen and Gray 2002). The gap of knowledge between terrestrial and marine biodiversity limits our ability to define a general theory of biodiversity patterns and regulating factors.

Deep-sea ecosystems, lacking primary photosynthetic production, largely depend for their energy budget on allochthonous inputs from organic matter, but a clear and global picture of deep-sea metabolism and the pathways of energy fueling this huge biome are not yet available.

The knowledge of deep-sea biodiversity is also limited. We do not know yet how many undiscovered species are hosted by the deepest parts of the oceans nor do we know exactly the patterns and drivers of deep-sea biodiversity. Depth-related patterns of biodiversity have been suggested to be unimodal, with a peak at intermediate depths (between 1000 and 2500 m depth) and lower values at shallower depths and in abyssal systems (Rex 1981; Rex et al. 1997; Etter and Grassle 1992; Gray et al. 1997). However, an increasing number of studies indicates that such patterns are not universal (Stuart et al. 2001; Levin and Gage 1998).

If spatial patterns of biodiversity in the deep sea are far from being clarified, even more uncertain is our comprehension of the mechanisms driving deep-sea biodiversity. Among them, those more frequently invoked are (a) sediment grain size and substrate heterogeneity (Etter and Grassle 1992); (b) productivity, organic content, and/or microbial features (Jahnke 1996; Lampitt and Antia 1998); (c) food resources (Levin and Gage 1998); (d) oxygen availability (Diaz and Rosemberg 1995); (e) current regimes (Lambshead et al. 2001), and (f) catastrophic disturbances (Levin et al. 2001). However, all these factors are subject to strong debate and criticism because they are often site specific and constrained by local (or regional) conditions (Levin et al. 2001; Ellingsen and Gray 2002).

In the last two decades enormous progress has been made towards comprehension of deep-sea biodiversity and ecosystem functioning. Thanks to the collective effort of the European Union and the collaborative projects between the European Union, United States, and Japan, among others, we are entering a new era for the investigation of deep-sea systems. The information acquired in the last years is negating most if not all of the hypotheses and paradigms advanced in the past. Deep-sea ecosystems, for instance, are not as stable and homogeneous as we believed. Hotspot deep-sea ecosystems, for instance, are found much more widely distributed than previously thought. Our knowledge is therefore rapidly expanding from stable slopes and deep canyons to cold seeps and mud volcanoes along continental margins, from seamounts to unstable slopes, from hydrothermal vents to permanent anoxic systems. Biodiversity in the deep sea is not only high, but apparently higher than in coastal areas. The adaptation of deep-sea organisms to extreme conditions is accompanied by a wide use of symbiotic relationships. The same applies for the energy-area theory of latitudinal biodiversity gradients. Hypotheses on factors causing deep-sea dwarfisms or gigantisms have been shown to be wrong or unconvincing. We have hypothesized that deep-sea organisms are starving with low metabolic rates due to low temperatures, but increasing evidence suggests that this is not the case. We have for a long time assumed that the deep sea is deprived of food, and now we have begun to understand that the amount of available food is not so low and its quality can be higher than on the continental shelves. The calculations we are making on deep-sea metabolism are not balanced and let's hypothesize that a large fraction of the total biomass is sustained by the "dark energy" (e.g., methane and other carbon hydrates, gases, and elements) or other sources. We have assumed that deep-sea systems are stable and homogeneous; conversely, we see now that they are frequently characterized by abrupt changes and are extremely heterogeneous at all spatial scales, suggesting that geosphere and biosphere are intimately linked. We have estimated that deep-sea ecosystems are not influenced by climate change or anthropogenic impacts (or if they do respond, the response could take decades), whereas we understand now that they can be variously threatened by anthropogenic impacts and react promptly to episodic "signals" from the surface.

To improve our knowledge of deep-sea ecosystems we need financial support, adequate technologies, and standardized protocols. This book aims at contributing to this latter need.

REFERENCES

Diaz, R. J. and R. Rosemberg. 1995. Marine benthos hypoxia: A review of its ecological effects and the behavioural responses of benthic macrofauna. *Oceanography and Marine Biology Annual Review* 33: 245–303.

Ellingsen, K. E. and J. S. Gray. 2002. Spatial patterns of benthic diversity: Is there a latitudinal gradient along the Norwegian continental shelf?. *Journal of Animal Ecology* 71: 373–389.

Etter, R. J. and J. F. Grassle. 1992. Patterns of species diversity in the deep sea as a function of sediment particle size diversity. *Nature* 369: 576–578.

Gaston, K. J. 2000. Global patterns in biodiversity. *Nature* 405: 220–227.

Gray J. S., G. C. B. Poore, K. I. Ugland, R. S. Wilson, F. Olsgard, and Ø. Johannessen. 1997. Coastal and deep-sea benthic diversities compared. *Marine Ecology Progress Series* 159: 97–103.

Jahnke, R. 1996. The global ocean flux of particulate organic carbon, areal distribution, and magnitude. *Global Biogeochemistry Cycles* 10: 71–88.

Lambshead, P. J. D., J. Tietjen, A. Glover, T. Ferrero, D. Thistle, and A. J. Gooday. 2001. The impact of large-scale natural physical disturbance on the diversity of deep-sea North Atlantic nematodes. *Marine Ecology Progress Series* 214: 121–126.

Lampitt, R. and R. Antia. 1997. Particle flux in the deep-seas: Regional characteristics and temporal variability. *Deep-Sea Research Part I* 44: 1377–1383.

Levin, L. A. and J. D. Gage. 1998. Relationships between oxygen, organic matter, and the diversity of bathyal macrofauna. *Deep-Sea Research I* 45: 129–163.

Levin, L. A., R. J. Etter, M. A. Rex et al. 2001. Environmental influences on regional deep-sea species diversity. *Annual Review of Ecology and Systematics* 32: 51–93.

Loreau, M., S. Naeen, P. Inchausti et al. 2001. Biodiversity and ecosystem functioning: Current knowledge and future challenges. *Science* 294: 804–808.

Rex, M. A.. 1981. Community structure in the deep-sea benthos. *Annual Review of Ecology and Systematics* 12: 331–353.

Rex, M. A., C. T. Stuart, R. R. Hessler et al. 1993. Global-scale latitudinal patterns of species diversity in the deep-sea benthos. *Nature* 365: 636–639.

Rex, M. A., R. J. Etter, and C. T. Stuart. 1997. Large scale patterns of species diversity in the deep-sea benthos. In *Marine biodiversity: Patterns and processes*, ed. R. F. G. Ormond, J. D. Gage, and M. V. Angel, 94–121. Cambridge: Cambridge Univ. Press.

Rex, M. A., C. T. Stuart, and G. Coyne. 2000. Latitudinal gradients of species richness in the deep-sea benthos of the North Atlantic. *Proceeding of the National Academy of Sciences of the United States of America* 97: 4082–4085.

Snelgrove, P. V. R. 1999. Getting to the bottom of marine biodiversity: sedimentary habitats. *BioScience* 49: 129–138.

Stuart, C. T., M. A. Rex, and R. J. Etter. 2001. Large spatial and temporal patterns of deep-sea benthic species diversity. In *Ecosystems of the world: Ecosystems of deep oceans*, ed. P. A. Tyler, 295–311. Amsterdam: Elsevier.

Vandermeulen, M. A., A. J. Hudson, and S. M. Scheiner. 2001. Three evolutionary hypotheses for the hump-shaped productivity-diversity curve. *Evolutionary Ecology Research* 3: 379–392.

Section I

Deep-Sea Environmental Variables

1 Total Organic Matter and Water Contents, Grain Size, Bulk Density, Porosity, and Redox Potential of Sediments

CONTENTS

1.1 INTRODUCTION

The distribution of benthic fauna is the result of a complex array of either biotic or abiotic factors. The multifactorial nature of the interactions between benthic fauna and the physical–chemical characteristics of their habitats makes it rather difficult to ascertain which among the various interacting abiotic factors mainly shapes the composition, spatial distribution, and temporal variability of the benthic assemblages; therefore, in principle, many abiotic measurements should be carried out to get the best picture. While this aspect has been increasingly faced by experimental coastal ecology (e.g., Dunson and Travis 1991; Chapman and Underwood 1995), unfortunately, it has almost been neglected in the ocean interior where the remote and unfavorable conditions for sampling seriously limit the synoptic determinations of biotic and abiotic features of the deep sea and/or the experimental appraisal of cause–effect relationships (McClain and Rex 2001). Many of the measurements made on deep-sea sediment samples are often conducted on deck once the sediment is retrieved, and corrections for the decompression effects are generally almost unknown. In spite of the possible biases, many easy physical–chemical measurements can be carried out on retrieved sediments. Among those mostly reported in the deep-sea literature are the total organic matter and water content of the sediment, the grain size and bulk density, and the values of the redox potential. In this chapter we report the most common procedures for the measurement of all of the aforementioned variables.

1.2 FIELD PROCEDURES

Marine sediment collection for the assessment of grain size and total organic matter must be carried out avoiding any manipulation of the sample that could affect the vertical stratification of the sediment or a relevant alteration of the *in situ* physical–chemical properties of the surface layer (e.g., resuspension, warming, desiccation). A ship and special equipment are requested for sediment sampling at deeper depths. Common devices are grabs, box corers, and multiple corers (Figure 1.1). Nonetheless, the use of box corers or multiple corers is preferable. Multiple corers, moreover, are equipped with transparent Plexiglas® (Rohm & Haas Company, Philadelphia, PA) liners that permit analysis of sample quality and integrity immediately after retrieval and measurement of the depth of the redox potential discontinuity (RPD) layer. The reduced velocity of sediment penetration of multiple corer liners allows collection of a totally undisturbed sediment sample.

Once the sediment core is extruded from the liner, whenever studies of the vertical distribution of either the grain size or total organic matter are required, the core can be sliced into the layers needed using a plastic (inert) knife. The thickness of the sediment layers must be chosen depending on the sampling strategy adopted. Generally, the top 5 cm of the corer are sliced into 0.1–1.0 cm thick layers, whereas, deeper in the core, layer thickness can be extended to 2–5 cm. Sediment layers for the subsequent analysis of total organic matter contents are then immediately homogenized and stored frozen at −20°C until analysis (generally within 3 months from collection). For grain size analysis, sediment samples can be stored at 4°C until processing.

1.3 LABORATORY PROCEDURES: TOTAL ORGANIC MATTER AND WATER CONTENT OF SEDIMENTS

The total organic matter content of marine sediments is determined by ignition loss, according to Parker (1983).

1.3.1 Equipment and Instruments

Stainless steel spatula
Pipette (20–200 μL)
Analytical balance (± 0.1 mg)
Oven (60°C–450°C)
Vacuum glass vessel

1.3.2 Materials

Aluminum foils resistant to high temperature (>600°C)

1.3.3 Procedure

1. Homogenize a wet sediment sample corresponding to 50–100 mg dry weight and place it in an aluminum cup (diameter 1–3 cm and height 0.5–1.5 cm), previously weighed (±1 mg), and made from commercial aluminum foils (check that the foil is resistant to 600°C).
2. Dry in the desiccator at 60°C for 24 h or until constant weight (dry sediment, DW).
3. Weigh the cups containing the dry sediment (±1 mg).
4. Place in a muffle furnace at 450°C for 4 h.
5. Weigh the cups containing the calcinated sediment (±1 mg).

Water content of marine sediments is assessed as the difference between the wet and dry sediment weights and expressed as a percentage of the sediment initial weight. The total organic matter

(a)

(b)

(c)

FIGURE 1.1 A large grab (a), a box corer (b), and a multiple corer (c).

content is obtained as the difference between the dry and calcinated sediment weights. This content is then normalized to the sediment DW and expressed as milligram organic matter per gram of DW, or as a percentage.

1.4 LABORATORY PROCEDURES: POROSITY AND BULK DENSITY

1.4.1 Equipment and Instruments

Analytical balance (±0.1 mg)

1.4.2 Procedure

1. A known volume of intact wet sediment is weighed (±0.1 mg) and desiccated at 60°C for 24 h to remove all of the interstitial water.
2. DW is reweighed, and the weight loss is assumed to include all of the interstitial water, which in turn is assumed to be equal to the interstitial volume of the known-volume sediment sample.
3. Sediment porosity (*n*) is then calculated as the (percentage) ratio between the interstitial volume and the (known) total sediment volume [NOTE 1].
4. The sediment bulk density is then defined as the ratio between the weight of the wet sediment and its (known) volume.
5. The bulk density of a saturated sediment [NOTE 1] can also be determined from the porosity and grain density (Hein 1991) according to the following formula:

$$BD_{sat} = n \cdot D_w + (1-n) \cdot D_g \tag{1.1}$$

where

BD_{sat} is the bulk density of saturated sediment.
n is the porosity.
D_w is the water density.
D_g is the density of the grains [NOTE 2].

1.5 LABORATORY PROCEDURES: GRAIN SIZE

Grain size, a fundamental property of marine sediments, linked to sediment transport and deposition (McLaren and Bowles 1985), affects sediment geotechnics (Hein 1991), and is one of the major factors controlling the distribution of the living benthic biota (Nybakken 1982).

There are several different approaches to the measurement of sediment grain size, based on different physical properties related with their linear dimension (e.g., diameter of the grains, sedimentation speed, effects of particles on the resistivity of an electrolyte, laser diffraction, and extinction time of a laser beam).

The grain size of the sediment is usually reported according to the Wentworth scale, which assumes that the dimensional scale of the diameter of grains is logarithmic and based on phi units (equal to the $\log_2$ of the diameter in millimeters) (Table 1.1).

1.5.1 Equipment and Instruments

Large-volume beaker (200–1000 mL)
Sieves (for particles, >63 μm)
Ultrasonic bath
RX sedigraph, or Coulter counter, or laser beam-based instrument

TABLE 1.1
Wentworth Scale for Classifying Sediments Based on Mean Grain Size

Sediment Size Description	Grain Diameter (phi units)
Cobbles and boulders	Less than −6
Very coarse pebbles	−6 to −5
Coarse pebbles	−5 to −4
Medium pebbles	−4 to −3
Fine pebbles	−3 to −2
Very fine pebbles	−2 to −1
Very coarse sand	−1 to 0
Coarse sand	0 to 1
Medium sand	1 to 2
Fine sand	2 to 3
Very fine sand	3 to 4
Coarse silt	4 to 5
Medium silt	5 to 6
Fine silt	6 to 7
Very fine silt	7 to 8
Clay	Grater than 8

Source: Data from Poppe, L. J. et al. 2000.

Analytical balance (±0.1 mg)
Oven (60°C)

1.5.2 Solutions

10% H_2O_2
0.6% Sodium hexametaphosphate

1.5.3 Procedure

Removal of organic matter:

1. Sediment samples are treated with an excess of 10% H_2O_2 for 24–48 h in a large-volume beaker (800–1000 mL) [NOTE 3].
2. Dry the sediment in the oven at 105°C for 24 h and weigh it (±0.1 mg).
3. The dry sediment is sieved on a series of standard sieves, meshes of which decrease at regular size intervals (typically of 0.5 phi) down to 4 phi (62.5 μm), mounted on an automatic shaker for 15 min.
4. Sediment fractions retained by each sieve are weighed and normalized to 100% of the initial total DW of the sediment sample.
5. Prepare a sediment suspension in water (see the instrument specifications).
6. Add a 0.6% solution of sodium hexametaphosphate to the dry sediment as a dispersant, and treat it in the ultrasonic bath for 10 min.
7. Leave the dispersant to react for 24 h, then sonify again for 1 min.
8. Analyze the sample using an RX sedigraph, a Coulter counter, or a laser-beam-based instrument.

1.5.4 Notes

1. When the interstitial volume is almost entirely occupied by water, sediments are defined as saturated, and the water content is equivalent to porosity.
2. Generally the value 2.65—related to the quartz density—is utilized.
3. The reaction of organics with water peroxide determines the release of gas and sediment bubbling so that, if the height of the beaker is insufficient, some sediment can outflow.

1.6 LABORATORY PROCEDURES: DEPTH OF REDOX POTENTIAL DISCONTINUITY (RPD)

Transparent Plexiglas is required for the measurement of the RPD layer depth, which is defined as the depth at which the sediment color turns from brown to grey black. This simple measurement is made in centimeters and is used to identify the depth at which most of the fauna is confined. In deep-sea sediments, it is uncommon to observe an RPD depth in the top 20 cm.

1.7 LABORATORY PROCEDURES: REDOX POTENTIAL

The values of the redox potential of the sediment provide a rough indication of the levels of oxygen available within the substrate. The measurement of the redox potential is crucial to the investigation of the microclimate interstitial conditions of the sediment as affected by burrowing organisms (either plants, algae, or dwelling organisms) or anthropogenic factors. The measurement of the redox potential is typically carried out by means of a potentiometric probe equipped with an indicator electrode (made of inert material such as platinum) and a reference electrode. Measured in millivolt, often reported as Eh (hydrogen standard electrode) the redox potential has a low-definition significance because of the multifactors interacting in producing it, and, as such, has only a semiquantitative empirical significance. The generally positive values of the redox potential are associated with well-oxygenated sediments, whereas highly negative values (<–200 mV) are typical of suboxic or anoxic conditions. Another empirical measurement of the oxygen levels present in the sediment is represented by the depth (millimeter or centimeter) of the RPD layer and is assumed to occur when the color of the sediment shifts from light to dark brown and black concomitantly with the occurrence of an anoxic layer deeper in the sediment (Figure 1.2).

FIGURE 1.2 Measure of the depth of the redox potential discontinuity (RPD) layer.

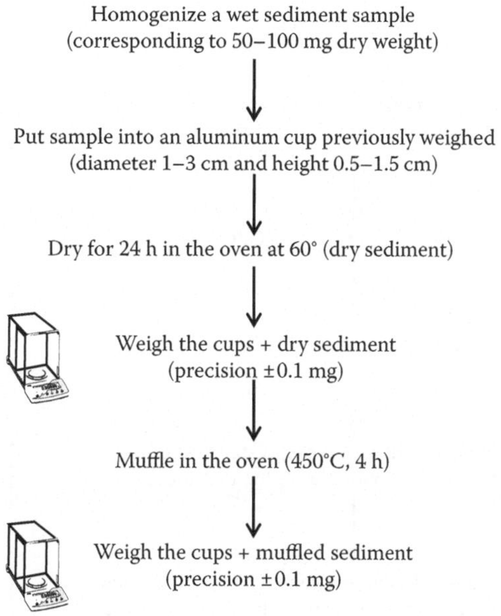

PROTOCOL 1.1 Quantification of total organic matter and the water content of sediments.

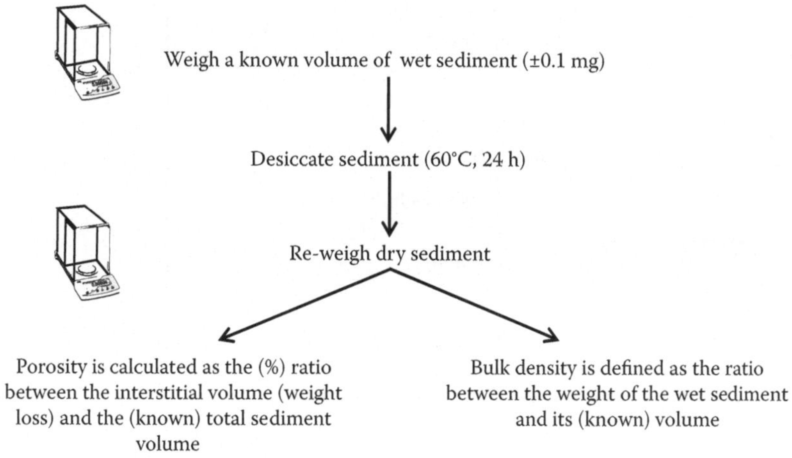

PROTOCOL 1.2 Determination of the porosity and bulk density of sediments.

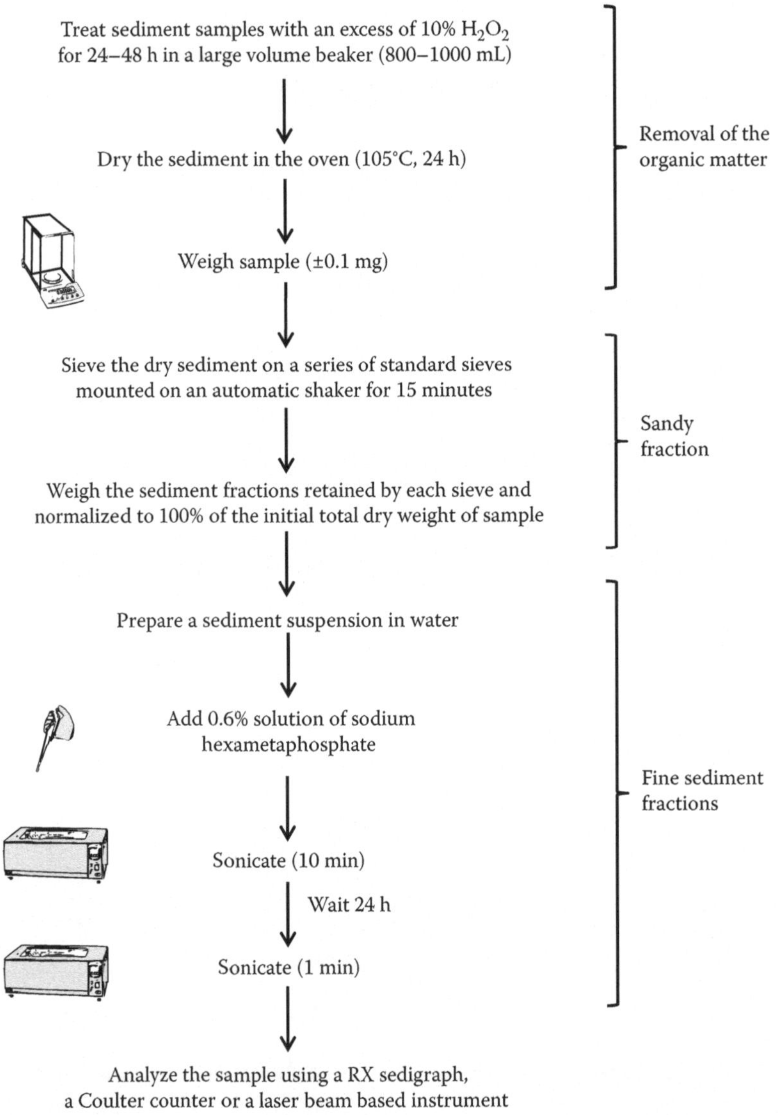

PROTOCOL 1.3 Determination of the grain size of sediments.

REFERENCES

Chapman, G. and A. J. Underwood. 1995. *Coastal Marine Ecology of Temperate Australia.* UNSW Press, Sydney, Australia.

Dunson, W. A. and J. Travis. 1991. The role of abiotic factors in community organization. *The American Naturalist* 138(5): 1067–1091.

Hein, F. J. 1991. The need for grain size analysis in marine geotechnical studies. In *Principles, methods, and application of particle size analysis.* ed. J. P. M. Syvitski, 346–362. Cambridge: Cambridge University. Press.

McLaren, P. and D. Bowles. 1985. The effect of sediment transport on grain size distributions. *Journal of Sedimentary Petrology* 55:457–470.

McClain, C. and M. Rex. 2001. The relationship between dissolved oxygen concentration and maximum size in deep-sea turrid gastropods: An application of quantile regression. *Marine Biology* 139: 681–685.

Nybakken, J. W. 1982. *Marine biology: An ecological approach.* Harper and Row, New York.

Parker, J. G. 1983. A comparison of methods used for the measurement of organic matter in sediments. *Chemistry and Ecology* 1: 201–210.

Poppe, L. J., A. H. Eliason, J. J. Fredericks, R. R. Rendigs, D. Blackwood, and C. F. Polloni. 2000. USGS east-coast sediment analysis—Procedures, database, and georeferenced displays. U.S. Geological Survey Open File Rep. 00-358.

2 Total Organic Carbon, Total Nitrogen, and Organic Phosphorus in Marine Sediments

CONTENTS

2.1 INTRODUCTION

The ocean floor is a depository for many organic molecules, among which are carbon (C), nitrogen (N), and phosphorus (P). The relative importance of these three organic pools depends on their origin, degradation rates, and utilization by benthic consumers. Investigations of the spatial and temporal distribution of total organic C (TOC) in marine sediments are important from a trophodynamic perspective (Magni 2003; Hyland et al. 2005). TOC contents in the sediment are generally a function of inputs from the water column, *in situ* production, and oxygen availability (e.g., respiration) (Graf 1992; Jahnke 1996; Hartnett et al. 1998; Hedges et al. 1999). Since N is utilized by consumers at a faster rate than C, C:N ratios higher than 20 are typically associated with refractory organic matter of terrestrial origin (Meyers 1994). Deviations from the Redfield ratio (C:N:P = 106:16:1) can thus provide insights into the origin and aging of sediment organic matter (Müller 1977).

Phosphorus is a critical element in the biosphere, limiting biological productivity and thus modulating the global carbon cycle and climate (Compton et al. 2000). Much of the phosphorus removed from seawater to sediments is in the form of P included into organic matter (Delaney 1998). Other P chemical forms include phosphorus associated with apatite and calcium carbonate minerals, and with hydrous ferric oxides (Ruttenberg and Berner 1993). Marine sediments are the main reservoir of organic P (on a global scale ca. 5.2×10^{18} mol; Van Cappellen and Ingall 1994) and, as such,

FIGURE 2.1 Sampling using a box corer. The box corer is subsampled with transparent Plexiglas liners.

the regeneration of phosphates from this pool is a key step in the regulation of P availability in the overlaying water column.

Total organic P pool in marine sediments is composed of a wide variety of compounds such as P monoesters, diesters, and polyphosphates. Among these, nucleic acids, phospholipids, and metabolic intermediates such as ATP undergo selective degradation, being intrinsically more labile than other P compounds (Dell'Anno and Danovaro 2005).

2.2 FIELD PROCEDURES

2.2.1 Sampling

Marine sediment collection for the assessment of TOC, total N, and organic P must be carried out avoiding any manipulation of the sample that could affect the vertical stratification of the sediment or a relevant alteration of the *in situ* physical–chemical properties of the surface layer (e.g., resuspension, warming, desiccation). The most common devices for sediment sampling are grabs, box corers, and multiple corers (Figure 2.1). The use of box corers or multiple corers is preferable in sandy or muddy sediments. Multiple corers, moreover, are equipped with transparent Plexiglas® liners that permit the analysis of sample quality and integrity immediately after retrieval. The reduced velocity of sediment penetration of multiple corer liners allows collection of a totally undisturbed sediment sample. The most recent models of box corers can collect high-quality sediment samples, and recent studies carried out on synoptically collected samples in calm sea conditions, using either a box corer or a multicorer, demonstrated that these two different sampling devices lead to similar results for organic matter determinations (Danovaro et al. 1998).

2.2.2 Sample Storage

Once the sediment core is extruded from the liner (Figure 2.2), whenever studies of the vertical distribution of organics is required, the core can be sliced into the layers needed using a plastic (inert) spatula. The thickness of the sediment layers must be chosen depending on the sampling strategy adopted. Generally, the top 5 cm of the corer are sliced into 0.1–1.0 cm thick layers, whereas, deeper in the core, layer thickness can be extended to 2–5 cm. Sediment layers are then

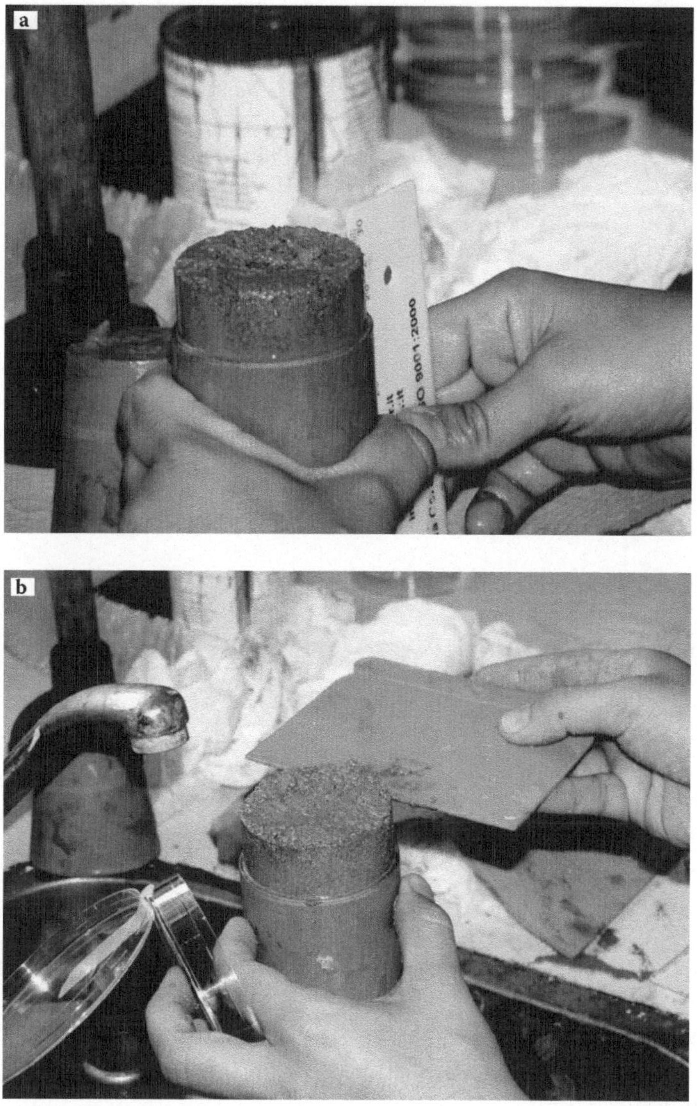

FIGURE 2.2 Slicing of the subcore in different layers.

Continued

immediately homogenized and stored frozen at −20°C until analysis (generally within 3 months from collection).

2.3 LABORATORY PROCEDURES: CHN DETERMINATION USING AUTOMATED CHN ANALYZERS

Several CHN analyzers can be used for the rapid determination of TOC and total N in marine sediments. These instruments allow the introduction of a sediment sample with oxygen and a carrier gas (typically helium) into a combustion column made of an oxidation catalyst. The resulting gas mixture (CO_2, NO_x, and H_2O) enter a reduction column, where NO_x is reduced to N_2, and O_2 in excess is eliminated. CO_2 and N_2 are then determined by thermal conductivity (Yamamuro and Kayanne 1995). In order to remove inorganic C, sediment samples must be treated with HCl for carbonate removal (Hedges and Stern 1984).

FIGURE 2.2 *Continued.*

2.3.1 Equipment and Instruments

Stainless steel spatula
Pipette (20–200 μL)
Analytical balance (±1 μg)
Oven (60°C)
Vacuum glass vessel
CHN analyzer

2.3.2 Materials and Solutions

Silver cups
1 N HCl

2.3.3 Procedure

1. Homogenize a wet sediment sample corresponding to 50–100 mg dry weight.
2. Dry overnight in the oven at 50°C.
3. A subsample of 1–10 mg dry sediment (annotated weight) is inserted into silver cups for CHN analyzer.
4. The dry sample is then treated with a few drops of organic free 1 N HCl (for carbonate-poor sediments).
5. In the case of carbonate-rich sediments (>wt 50% $CaCO_3$), the silver cups can be exposed to pure HCl vapors in a glass vacuum vessel for 24–48 h.
6. After acidification, samples are dried again for 1 h at 50°C and weighed.
7. Silver cups are crimped and transferred to the CHN for analysis.
8. Set combustion column temperature at 1020°C–1050°C, depending on the instrument.
9. Standards are prepared using silver cups containing preweighed acetanilide samples.
10. Blanks are made using empty silver cups.
11. Any CHN analyzer has its own software that allows the calculation of TOC and TN concentrations in each sample.

2.4 LABORATORY PROCEDURES: SPECTROPHOTOMETRIC DETERMINATION OF PHOSPHORUS POOLS

The protocol for the extraction and quantification of P pools in the sediments was developed by Aspila et al. (1976) and widely applied to marine sediments collected worldwide. This method is simple, relatively fast, and sufficiently accurate for determining synoptically inorganic and total phosphorous pools in marine sediments. The organic P concentrations are then calculated as the differences between the total and inorganic P contents. Before analyses sediment samples must be freeze dried or desiccated in an oven at 80°C–100°C for 24 h until a constant weight is yielded.

2.4.1 Equipment and Instruments

Stainless steel spatula
Pipettes (1.0–5.0 mL)
Analytical balance (±1 mg)
Oven (60°C–80°C)
Muffle furnace (up to 600°C)
Spectrophotometer
Shakers

2.4.2 Materials and Chemical Products

Graduate glass flask
Beaker
Graduate cylinders
Aluminum foils
Concentrated sulfuric acid (96% H_2SO_4)
Ammonium molybdate tetrahydrate [$(NH_4)_6Mo_7O_{24} \cdot 4H_2O$]
Antimony potassium tartrate [$K(SbO)C_6H_4O_6$]
Ascorbic acid [$C_6H_8O_6$]
Chloroform [$CHCl_3$]
Potassium dihydrogen phosphate [KH_2PO_4]
Concentrated hydrochloric acid (36% HCl)

2.4.3 Solutions and Preparation of Solutions

Solutions

a. 1 N HCl
b. Standard stock solution of phosphorous (6 μg at P mL^{-1})
c. 5 N H_2SO_4
d. 4% [$(NH_4)_6Mo_7O_{24} \cdot 4H_2O$] solution
e. 0.3% [$K(SbO)C_6H_4O_6$]
f. 1.8% [$C_6H_8O_6$] solution

Preparation

a. Fill 1 L graduate glass flask with about 500 mL of reagent-grade water. Add slowly 83.3 mL of concentrated HCl and then bring up to 1 L with reagent-grade water.
b. Dissolve 0.815 g KH_2PO_4 with ca. 300 mL of reagent-grade water. Transfer the solution into a 1 L graduate glass flask and add 1 mL of $CHCl_3$. Bring up to 1 L with reagent-grade water. Transfer the solution into a dark glass bottle. This solution is stable for several months.

c. Fill a 1 L graduate glass flask with about 500 mL of reagent-grade water. Add slowly 140 mL of concentrated H_2SO_4 and then bring up to 1 L with reagent-grade water.
d. Dissolve 40 g of $(NH_4)_6Mo_7O_{24} \cdot 4H_2O$ in 1 L of reagent-grade water.
e. Dissolve 3 g of $K(SbO)C_6H_4O_6$ in 1 L of reagent-grade water.
f. Dissolve 18 g of ascorbic acid in 1 L of reagent-grade water. This solution must be prepared fresh each time it is needed.

2.4.4 Procedure

1. Homogenize a wet sediment sample corresponding to 1–2.0 g dry weight.
2. Dry in an oven at 80°C for 24 h.
3. A subsample of 0.3–0.5 g dry sediment (annotated weight) is placed in an alumina crucible and treated in a muffle furnace at 550°C for 2 h.
4. After treatment in the muffle furnace, weigh the samples again.
5. Weigh another dry subsample of 0.3–0.5 g.
6. Place combusted and uncombusted samples separately into two 100 mL glass beakers.
7. Add 50 mL of 1 N HCl to each sample.
8. Stir the samples for 12 h at room temperature.
9. Transfer the samples into centrifuge tubes.
10. Centrifuge the samples for 3000 rpm for 10 min.
11. Filter the supernatant with Whatman GF/F filters (nominal pore size 0.45 μm).
12. Transfer 2 mL of filtrates into a 50 mL graduate glass flask.
13. Add about 10 mL of reagent-grade water.
14. Add 5 mL of mixed reagent*.
15. Bring the volume up to 50 mL.
16. Incubate for 10 min.
17. Transfer an aliquot of the sample into a glass cuvette.
18. Read absorbance (ABS) at 885 nm.

2.4.5 Preparation of Standard P Curves

Dilute a stock solution of P 1:100 with reagent-grade water (the final concentration is 0.06 μg at P mL^{-1}). From this, transfer 0.5, 1, 2, 4, and 8 mL into glass graduate flasks. Add 2 mL of 1 N HCl and then 5 mL of mixed reagent. Bring the volume up to 50 mL with reagent-grade water. Concentrations of these solutions are 0.012, 0.024, 0.048, 0.072, and 0.120 μg at P 50 mL^{-1}.

2.4.6 Calculations

Convert the ABS of samples into concentrations from the standard curve. The obtained concentrations of the samples are in microgram P-PO_4.

To obtain P concentrations in parts per million per gram (ppm g^{-1}), the following formula is used:

$$[Total\ \ P]\ g^{-1} = ([concentration\ measured\ in\ combusted\ samples] \cdot 25 \cdot 30.97)/g \tag{2.1}$$

* Mixed reagent must be prepared fresh before use by mixing the following reagents in the order given:
- 50 mL of 5 N H_2SO_4
- 15 mL of 4% $[(NH_4)_6Mo_7O_{24} \cdot 4H_2O]$ solution
- 30 mL of 1.8% $[C_6H_8O_6]$ solution
- 5.0 mL of 0.3% $[K(SbO)C_6H_4O_6]$ solution

$$[Inorganic\ \ P]\ g^{-1} = ([concentration\ measured\ in\ uncombusted\ samples] \cdot 25 \cdot 30.97) / g \quad (2.2)$$

where 30.97 is the atomic weight of P, and 25 is the dilution factor.

For the calculation of total P concentration, g is the weight of the combusted samples, whereas, for the calculation of inorganic P, g is the weight of the uncombusted samples.

2.5 NOTES

There are potential sources of error in organic P determinations using the analytical procedure developed by Aspila et al. (1976). The first is that the organic P pool is obtained by differences between total and inorganic P determinations. In sediments enriched in inorganic P (phosphorites), the calculated organic P concentration can be a small difference between two large numbers. Thus, small errors in the analytical values can lead to errors in the calculated organic P concentration. In most sediments analyzed so far, organic P concentrations are large enough relative to inorganic P concentrations so that the uncertainty in the calculated organic P concentration is generally low (<10%–15%). Another potential problem is partial hydrolysis of organic P by 1 N HCl in sediment samples previously not treated at 550°C.

Finally, there is the possibility that small amounts of inorganic P are mobilized during heating at 550°C (but not during HCl attack alone) and are thereby included in the calculated values for organic P.

Homogenize a wet sediment sample
(corresponding to 50–100 mg dry weight)

↓ Dry overnight in the oven at 50°C

Put a sub-sample (1–10 mg, annotate weight) into silver cup

↓

Add few drops of organic free 1N HCl

↓ In case of carbonate-rich sediments (>wt50% $CaCO_3$), the silver cups can be exposed to pure HCl vapors in a glass vacuum vessel for 24–48 hrs

Dry sample (50°C, 1 h)

↓

Weigh

↓

Close and transfer the cups to the CHN

↓

Set combustion column temperature at 1020–1050°C
(depending on the instrument)

↓

Prepare standards using silver cups containing
pre-weighted acetanilide samples

↓

Prepare blanks using empty silver cups

↓

Run the instrumental analysis
Any CHN analyzer has its own software that allows the
calculation of TOC and TN concentrations in each sample

PROTOCOL 2.1 CHN determination using automated CHN analyzers.

Homogenize a wet sediment sample
(corresponding to 1–2.0 g dry weight)

↓ Dry overnight in the oven (80°C, 24 h)

Put a sub-sample (0.3–0.5 g, annotate weight) in aluminum cup

↓

Transfer samples in a muffle (550°C, 2 h)

↓

Weigh the samples

↓

Transfer combusted samples in a glass beaker (100 mL)

↓

Put a sub-sample (0.3–0.5 g, annotate weight)
of un-combusted sediment in another glass beaker (100 mL)

↓

Add in each beaker 50 mL of 1N HCl

↓ Stir samples (room T, 12 h)

Centrifuge (3000 × g; 10 min)

↓

Filter supernatant with Whatman GF/F filters
(nominal pore size 0.45 μm)

↓

Transfer 2 mL of filtrates into a 50 mL graduated glass flask

↓

Add 10 mL of reagent grade water

↓

Add 5 mL of the mixed reagent

↓

Fill the volume at 50 mL

↓ Incubate for 10 min

Transfer an aliquot of sample in a glass cuvette

↓

Read ABS at 885 nm

PROTOCOL 2.2 Spectrophotometric determination of phosphorus pools.

REFERENCES

Aspila, K. I., H. Agemian, and A. S. Y. Chau. 1976. A semi-automated method for the determination of inorganic, organic and total phosphate in sediments. *Analist* 101: 187–197.

Compton, J., D. Mallinson, C. R. Glenn et al. 2000. Variations in the global phosphorus cycle. In *Marine authigenesis: From global to microbial*, ed. C. R. Glenn, L. Prévôt-Lucas, and J. Lucas, 21–33. Society for Sedimentary Geology Special publication no. 66.

Danovaro, R., D. Marrale, N. Della Croce, A. Dell'Anno, and M. Fabiano. 1998. Heterotrophic nanoflagellates, bacteria, and labile organic compounds in continental shelf and deep-sea sediments of the eastern Mediterranean. *Microbial Ecology* 35: 244–255.

Delaney, M. L. 1998. Phosphorus accumulation in marine sediments and the oceanic phosphorus cycle. *Global Biogeochemical Cycles* 12: 563–572.

Dell'Anno, A. and R. Danovaro. 2005. Extracellular DNA plays a key role in deep-sea ecosystem functioning. *Science* 309: 2179.

Graf, G. 1992. Benthic–pelagic coupling: A benthic view. *Oceanography and Marine Biology* 30: 149–190.

Hartnett, H. E., R. G. Keil, J. I. Hedges, and A. H. Devol. 1998. Influence of oxygen exposure time on organic carbon preservation in continental margin sediments. *Nature* 391: 572–574.

Hedges, J. I. and J. H. Stern. 1984. Carbon and nitrogen determinations of carbonate containing solids. *Limnology and Oceanography* 29: 6547–663.

Hedges, J. I., F. S. Hu, A. H. Devol, H. E. Hartnett, E. Tsamakis, and R. G. Keil. 1999. Sedimentary organic matter preservation: A test for selective degradation under oxic conditions. *American Journal of Science* 299: 529–555.

Hyland, J., L. Balthis, I. Karakassis et al. 2005. Organic carbon content of sediments as an indicator of stress in the marine benthos. *Marine Ecology Progress Series* 295: 91–103.

Jahnke, R. A. 1996. The global ocean flux of particulate organic carbon: Areal distribution and magnitude. *Global Biogeochemical Cycles* 10: 71–88.

Magni, P. 2003. Biological benthic tools as indicators of coastal marine ecosystems' health. *Chemistry and Ecology* 19: 363–372.

Meyers, P. A. 1994. Preservation of elemental and isotopic source identification of sedimentary organic matter. *Chemical Geology* 114: 289–302.

Müller, P. J. 1977. C/N ratios in Pacific deep-sea sediments: Effect of inorganic ammonium and organic nitrogen compounds sorbed by clays. *Geochimica et Cosmochimica Acta* 41: 765–776.

Ruttenberg, K. C. and R. A. Berner. 1993. Authigenic apatite formation and burial in sediments from non-upwelling, continental margin environments. *Geochimica et Cosmochimica Acta* 57: 991–1007.

Van Cappellen, P. and E. Ingall. 1994. Benthic phosphorus regeneration, net primary production, and ocean anoxia: A model of the coupled marine biogeochemical cycles of carbon and phosphorus. *Paleoceanography* 9: 677–692.

Yamamuro M. and H. Kayanne. 1995. Rapid direct determination of organic carbon and nitrogen in carbonate-bearing sediments with a Yanaco MT-5 CHN analyzer. *Limnology and Oceanography* 40: 1001–1005.

3 Bioavailable Organic Matter *Total and Enzymatically Hydrolyzable Proteins, Carbohydrates, and Lipids*

CONTENTS

3.1 INTRODUCTION

Organic matter in marine sediments is a mixture of compounds exhibiting different levels of bioavailability to consumers, ranging from labile (i.e., immediately digestible; Mayer et al. 1995) to refractory (recalcitrant to decomposition). Refractory compounds (such as humic and fulvic acids, structural carbohydrates, and "black" carbon) generally account for most of the sedimentary organic matter, and as they have very low degradation rates, they accumulate easily in marine sediments (Middelburg et al. 1999). In contrast, highly labile compounds, representing the fraction of organic matter that is rapidly digested by benthic consumers, are subjected to wider temporal and spatial changes (Mayer et al. 1995).

However, between these two opposite levels of bioavailability, there is a continuum of characteristics that passes through intermediate levels of bioavailability. Unfortunately, the assessment of the lability of sediment organic C is not an easy task and generally implies several operational assumptions.

The biopolymeric fraction of sediment organic C, measured as the sum of protein, carbohydrate, and lipid carbon, has often been reported as the fraction of total organic C potentially available to benthic consumers (Fabiano et al. 1995).

Since biopolymeric C may contain molecules that become available to consumers only after microbial reworking, its use as a proxy of labile organic C depends on its origin and molecular composition. The amount of labile (bioavailable) organic matter in marine sediments can be extrapolated according to mathematical models based on the exponential decrease of organic carbon with increasing depth in the sediments (Rice and Rhoads 1989). Such models, however, assume that different classes of organic compounds undergo degradation and utilization at similar rates (Danovaro et al. 2001).

More recent investigations reported that only a fraction (5%–30%) of biopolymeric organic carbon in marine sediments can actually be degraded enzymatically and thus, is readily available for heterotrophic nutrition (Dauwe and Middelburg 1998; Dauwe et al. 1999a, 1999b; Dell'Anno et al. 2000; Danovaro et al. 2001; Pusceddu et al. 2003).

3.2 FIELD PROCEDURES

Marine sediment collection for the assessment of quantity and biochemical composition of organic matter must be carried out by avoiding any manipulation of the sample, which could affect the vertical stratification of the sediment or a relevant alteration of the *in situ* physical–chemical properties of the surface layer (e.g., resuspension, warming, and desiccation).

For the assessment of the organic matter content of marine sediments, the most common devices for sediment sampling are grabs, box corers, and multicorers.

The use of grabs, although generally avoided because of the strong reworking of the top-surface sediment layers and the consequent alteration of substrate vertical stratification, can sometimes be necessary when the sediment grain size is close to that of gravel, or for particular substrates. Nonetheless, the use of box corers or multicorers is preferable in sandy or muddy sediments. Multicorers, moreover, are equipped with transparent Plexiglas® liners that permit analysis of sample quality and integrity immediately after retrieval (Figure 3.1).

Moreover, the reduced velocity of sediment penetration of multicorer liners allows the collection of a totally undisturbed sediment sample. However, the most recent models of box corers also allow the collection of high-quality sediment samples, and recent studies carried out on synoptically collected samples in calm sea conditions, using either a box corer or a multicorer, demonstrated that these two different sampling devices lead to similar results in terms of quantity and biochemical composition of sediment organic matter (Fabiano et al. 1995).

Once the sediment core is extruded from the liner, whenever studies of the vertical distribution of organics is required, the core can be sliced into the layers needed, using a plastic (inert) knife. The thickness of the sediment layers must be chosen depending on the sampling strategy adopted. Generally, the top 5 cm of the corer are sliced into 0.1–1.0 cm thick layers, whereas, deeper in the core, layer thickness can be extended to 2–5 cm. Sediment layers are then immediately homogenized and stored frozen at –20°C until analysis (generally within 3 months from collection).

(a)

(b)

FIGURE 3.1 Multicorer equipped with transparent Plexiglas liners, (a) before and (b) after sediment sampling.

3.3 LABORATORY PROCEDURES: TOTAL PROTEINS

The protein content of natural samples can be determined using different analytical protocols (Rice 1982; Mayer et al. 1986; Bradford 1976). Among these, the most frequently utilized are the Bradford (1976) and the Hartree (1972) protocols, the latter being a modification of the widely used Lowry et al. (1951) procedure, rearranged by Rice (1982). Experimental evidence in the laboratory has demonstrated that both these protocols provide similar results in terms of bovine serum albumin (BSA) equivalents (Berges et al. 1993). In the present volume, details are provided on the Hartree (1972) protocol after modifications by Rice (1982), but slightly modified. This colorimetric method allows the reaction of proteins with rameic tartrate and the Folin–Ciocalteau reactive in basic environment (pH 10). The reaction provides a stable blue coloration whose intensity is proportional to the protein concentrations in the reaction solution.

3.3.1 Equipment and Instruments

Stainless steel spatula
Graduated glass tubes (15 mL)
Pipette (1–5 mL, 20–200 μL)
Optical glass cuvette (1 cm optical length)
Centrifuge (800 × g)
Spectrophotometer (340>λ>750 nm)
Oven (60°C)
Analytical balance (±0.1 mg)

3.3.2 Solutions

Solution A—Dissolve 2 g of NaK-tartrate and 100 g of Na_2CO_3 anhydrous in 500 mL of 1 N NaOH (20 g in 500 mL of reagent-grade water); the solution, uncolored and transparent, is diluted to 1000 mL with reagent-grade water. Solution A should be stored at 4°C, and it is stable for 2–3 months, after which the Na_2CO_3 tends to precipitate. In this case, the solution cannot be further utilized.

Solution B—Dissolve 2 g of NaK-tartrate and 1 g of $CuSO_4$ in 90 mL of reagent-grade water; the slightly blue solution is diluted with 10 mL of 1 N NaOH. The final solution, of a transparent electric blue color, when stored at 4°C, is stable for several weeks.

Solution C—1 mL of the Folin–Ciocalteau reagent is diluted with 15 mL of reagent-grade water. The solution, of a transparent yellow color, is unstable and must be prepared a few minutes before being used.

BSA standard stock solution—Dissolve 1 mg of BSA in 1 mL of reagent-grade water each time a calibration curve is prepared. Dilute the stock solution according to the expected range of concentrations in the sample.

3.3.3 Procedure

1. An aliquot of wet or frozen sediment (0.1–2.0 g; ±0.1 mg) is placed in a 15 mL glass tube.
2. 1 mL of reagent-grade water is added and the sediment vigorously shaken using a vortex (1 min).
3. The tube is treated with ultrasound (in a sonifier bath, 50–100 W) for 3 min with 30 s of interval between each minute of sonication.
4. To each tube is added 0.9 mL of solution A and the tube shaken with a vortex for 1 min.
5. Tubes are placed in a hot-water bath (50°C) for 10 min.

6. To each tube is added 0.1 mL of Solution B and the tube is shaken with vortex (1 min) and left to react for 10 min at room temperature.
7. To each tube is added 3 mL of Solution C and the tube vigorously shaken with vortex (1 min).
8. Tubes are placed in a hot-water bath (50°C) for 10 min.
9. Tubes are then centrifuged (15 min at 800 × g).
10. After centrifugation, the supernatant is analyzed with the spectrophotometer in an optical glass cuvette (1 cm optical length) at absorbance of 650 nm against a blank of reagent-grade water.
11. After analysis, the supernatant is discarded, the sediment dried at 60°C for 24 h, and the tubes weighed (±0.1 mg).
12. A blank of reagents (1 mL of reagent-grade water) and at least three sediment replicates of the sample previously calcinated at 500°C (4 h) are processed in the same way.

3.3.4 Calculations

The protein content of the sample is calculated from calibration curves of standard solutions of BSA and analyzed following the procedure described. The detection limit of the method is about 2.5 μg of BSA per mL^{-1} of reacting solution. Accuracy allows variability among replicates of BSA solution to be kept within a variation coefficient (CV) lower than 7%. Protein concentration is then normalized to sediment dry weight and expressed as milligram of albumin equivalents per gram of dry sediment.

The total protein concentration in the sediment is calculated by Equations 3.1–3.3:

$$PRT_c\ (mg\ g^{-1}) = \left\{\left[\left(A_{650c} - A_{650blkr}\right)\cdot C\right] + K\right\}\cdot P_S^{-1}\cdot 10^{-3} \tag{3.1}$$

$$PRT_{blks}\ (mg\ g^{-1}) = \left\{\left[\left(A_{650c} - A_{650blkr}\right)\cdot C\right] + K\right\}\cdot P_{sc}^{-1}\cdot 10^{-3} \tag{3.2}$$

$$PRT_c\ (mg\ g^{-1}) = PRT_c - PRT_{blks} \tag{3.3}$$

where

A_{650c} = absorbance of the sample at 650 nm.
$A_{650blkr}$ = absorbance of the reagents blank (average of three replicates).
$A_{650blks}$ = absorbance of calcinated sediments (sediment blank) at 650 nm.
C = coefficient of the linear interpolation of absorbance (independent variable) and concentration (dependent variable, μg mL^{-1}) of the BSA standard solutions. C is normally around 300 if the linear interpolation is in the form CONCENTRATION = [ABSORBANCE • C] + K.
K = constant of the linear interpolation of absorbance (independent variable) and concentration (dependent variable, μg mL^{-1}) of the BSA standard solutions.
P_s = weight of the dry sediment after the analysis.
P_{sc} = weight of the calcinated sediment.
PRT_c = protein concentration in the sample.
PRT_{blks} = protein concentration in the calcinated sediment.

3.4 LABORATORY PROCEDURES: TOTAL CARBOHYDRATES

Concentrations of total carbohydrates were determined according to Dubois et al. (1956) and optimized for sediments by Gerchacov and Hatcher (1972) after minor modifications. This colorimetric assay is based on the reaction between sugars and phenol in the presence of concentrated sulfuric acid. The method is nonspecific and allows concentrations of total carbohydrates, cellulose included, to be determined. The method described here is generally preferred because molar absorbance for most sugars lies within a limited range and because anomalous colors are avoided. Moreover, the procedure allows higher reproducibility than the protocol described by Dubois et al. (1956).

3.4.1 Equipment and Instruments

Stainless steel spatula
Graduated glass tubes (15 mL)
Pipette (1–5 mL; 20–200 μL)
Optical glass cuvette (1 cm optical length)
Centrifuge (800 × g)
Spectrophotometer (340>λ>750 nm)
Oven (60°C)
Analytical balance (±0.1 mg)

3.4.2 Solutions

5% Phenol aqueous solution
Concentrated sulfuric acid [H_2SO_4]

3.4.3 Procedure

1. An aliquot of wet or frozen sediment (0.1–2 g) is placed into a 15 mL glass tube, and 1 mL of reagent-grade water is added.
2. The tube is vigorously shaken by a vortex (1 min).
3. The tube is treated with ultrasound (in a sonifier bath, 50–100 W) for 3 min with 30 s of interval between each minute of sonication.
4. To each tube is added 1.0 mL of a 5% distilled phenol solution (in reagent-grade water) and shaken by vortex (1 min).
5. Tubes are then left at room temperature for 10 min.
6. To each tube is added 5.0 mL of concentrated H_2SO_4 [NOTE 1].
7. Once the acid has been added to the sample, the mixture must be mixed by a vortex for at least 1 min. The solution will assume a color from light yellow to dark brown.
8. Tubes are then closed with a stopper and centrifuged (30 min at 800 × g).
9. The supernatant is analyzed spectrophotometrically in an optical glass cuvette (1 cm optical length) and the absorbance is determined at wavelengths of 485 and 600 nm [NOTE 2] against a blank of reagent-grade water.
10. A blank of reagents (1 mL of reagent-grade water) and at least 3–5 sediment replicates of the sample previously calcinated at 500°C (4 h) are processed in the same way.
11. The sediment is fivefold rinsed with tap water, shaken, and centrifuged for 15 min at 800 × g and, after each centrifugation, the supernatant is removed.
12. The sediment dry weight is assessed as the difference of weights between the empty tube and the tube after sediment desiccation (24 h at 60°C) or, alternatively, by weighing the sediment removed from the tube after accurate scraping.

3.4.4 Calculations

The total carbohydrate concentration in the sediment is therefore calculated by Equations 3.4–3.6:

$$CHO_c\ (mg\ g^{-1}) = \left\{\left\{\left[\left(A_{485c} - A_{485blkr}\right) - \left(1.5 \cdot A_{600} - 0.003\right)\right] \cdot C\right\} + K\right\} \cdot \left(K_{CaCO_3} \cdot P_s\right)^{-1} \cdot 10^{-3} \quad (3.4)$$

$$CHO_{blks}\ (mg\ g^{-1}) = \left\{\left\{\left[\left(A_{485blks} - A_{485blkr}\right) - \left(1.5 \cdot A_{600} - 0.003\right)\right] \cdot C\right\} + K\right\} \cdot \left(K_{CaCO_3} \cdot P_{sc}\right)^{-1} \cdot 10^{-3} \quad (3.5)$$

$$CHO\ (mg\ g^{-1}) = \left(CHO_c - CHO_{blks}\right) \quad (3.6)$$

where

A_{485c} = absorbance of the sample at 485 nm.
$A_{485blkr}$ = absorbance of the reagent blank at 485 nm (mean value of three replicates).
$A_{485blks}$ = absorbance of the sediment blank at 485 nm.
A_{600} = absorbance of the sample at 600 nm.
C = coefficient of the linear interpolation of absorbance (independent variable) and concentration (dependent variable, µg mL^{-1}) of the D(+) glucose standard solutions. C is normally around 100 if the linear interpolation is in the form CONCENTRATION = [ABSORBANCE • C] + K.
K = constant of the linear interpolation of absorbance (independent variable) and concentration (dependent variable, µg mL^{-1}) of the glucose standard solutions.
P_s = weight of the dry sediment after the analysis.
P_{sc} = weight of the calcinated sediment.
K_{CaCO3} = correction factor for carbonates dissolution [NOTE 3].
CHO_c = carbohydrate concentration in the sample.
CHO_{blks} = carbohydrate concentration in the calcinated sediment.

3.5 LABORATORY PROCEDURES: TOTAL LIPIDS

The determination of total lipid concentration in marine sediment samples is generally carried out according to Bligh and Dyer (1959) and Marsh and Weinstein (1966), slightly modified to be applied to the sediment matrix.

3.5.1 Equipment and Instruments

Stainless steel spatula
Graduated glass tubes (15 mL)
Pyrex® (Corning Inc., licensed by Arc International Cookware SAS) 10 mL tubes
Pipette (1–5 mL)
Thermostatic dry bath (up to 200°C)
Optical glass cuvette (1 cm optical length)
Centrifuge (800 × g)
Spectrophotometer (340>λ>750 nm)
Oven (60°C)
Analytical balance (±0.1 mg)

3.5.2 Solutions

Reagent-grade methyl alcohol

Reagent-grade chloroform
Reagent-grade sulfuric acid
Reagent-grade water

3.5.3 Procedure

1. An aliquot of wet or frozen sediment (0.1–2 g) is placed into a glass tube (15 mL), and 1 mL of reagent-grade water is added.
2. The tube is vigorously shaken with a vortex (1 min) and treated with ultrasound (in a sonifier bath, 50–100 W) for 3 min with 30 s of interval between each minute of sonication.
3. To each tube is added 1.25 mL of chloroform and 2.5 mL of methyl alcohol at room temperature.
4. The tubes are vigorously shaken by vortex (1 min), left at 4°C for 10 min, and then centrifuged (800 × g for 10 min).
5. After centrifugation, the supernatant is withdrawn using a Pasteur pipette, placed into Pyrex glass tubes, and 1.25 mL of chloroform and 1.25 mL of reagent-grade water are added.
6. The tubes are vigorously shaken (using a vortex for at least 1 min) in order to allow the formation of a water–methanol–chloroform emulsion.
7. The tubes are centrifuged at 800 × g for 5 min in order to separate the supernatant hydroalcoholic fraction (water and methanol) from the hydrophobic fraction (chloroform).
8. The supernatant hydroalcoholic fraction is pipetted out and the hydrophobic residual (on the bottom of the Pyrex tube) is evaporated in a dry hot bath at 80°C–100°C for 20 min.
9. Once at room temperature, is added 2 mL of concentrated sulfuric acid to each tube.
10. The tubes are then placed in a dry hot bath (at 180°C–200°C) for 15 min. During the reaction, the acid turns to a light yellow or brown color.
11. The tubes are then extracted from the bath and transferred into a wet bath at room temperature (for 15 s), and then into an ice bath (at 0°C for 5 min).
12. To each tube 3 mL of reagent-grade water is added, and the tubes are vigorously shaken by vortex (1 min).
13. The supernatant is analyzed spectrophotometrically in an optical glass cuvette (1 cm optical length) and the absorbance of the sample is determined at a wavelength of 375 nm against a blank of reagent-grade water.
14. A blank of reagents (1 mL of reagent-grade water) and 3–5 sediment replicates of the sample previously calcinated at 500°C (4 h) are processed in the same way.
15. After the analysis, the remaining sediment in the first glass tube after withdrawing the hydroalcoholic fraction is desiccated in the oven at 60°C.
16. The weight of the sediment can be assessed both as the difference of weights of the empty tube and the tube after sediment desiccation (24 h at 60°C) and by directly weighing the sediment removed from the bottom of the tube by accurate scraping.

3.5.4 Calculations

Lipid concentration is calculated from the calibration curves of standard solutions of tripalmitine, processed according to the aforementioned procedure. The detection limit of the method is about 5 μg of tripalmitine per mL^{-1} of reacting solution. For accuracy, variability among replicates of tripalmitine solution must be kept within a variation coefficient (CV) lower than 8%. Total lipid concentration in the sediment is therefore calculated by the following equations (Equations 3.7–3.9):

$$LIP_c\ (mg\ g^{-1}) = \left\{\left[\left(A_{375c} - A_{375blkr}\right)\cdot C\right] + K\right\}\cdot P_s^{-1}\cdot 10^{-3} \quad (3.7)$$

$$LIP_{blks}\ (mg\ g^{-1}) = \{[(A_{375blks} - A_{375blkr}) \cdot C] + K\} \cdot P_s^{-1} \cdot 10^{-3} \tag{3.8}$$

$$LIP\ (mg\ g^{-1}) = LIP_c - LIP_{blks} \tag{3.9}$$

where:

A_{375c} = absorbance of the sample at 375 nm.
$A_{375blkr}$ = absorbance of the reagent blank (mean value of three replicates).
$A_{375blks}$ = absorbance of the calcinated sediment at 375 nm.
C = coefficient of the linear interpolation of absorbance (independent variable) and concentration (dependent variable, μg mL^{-1}) of tripalmitine standard solutions. C normally ranges from 230 to 300 if the linear interpolation is in the form CONCENTRATION = [ABSORBANCE • C] + K.
K = constant of the linear interpolation of absorbance (independent variable) and concentration (dependent variable, μg mL^{-1}) of tripalmitine standard solutions.
P_s = dry sediment weight.
P_{sc} = weight of the calcinated sediment.
LIP_c = lipid concentration in the sample.
LIP_{blks} = lipid concentration in the calcinated sediment.

3.6 LABORATORY PROCEDURES: ENZYMATICALLY HYDROLYZABLE PROTEINS

The protocol was developed from the method proposed by Mayer et al. (1995), which was originally based on the assessment of amino acids released from the sediment after treatment with proteolytic enzymes. The procedure described here allows evaluation of the digestible fraction of proteins in marine sediments. This is estimated as the difference between the total protein pools and the protein remaining after removal of the fraction enzymatically digested.

3.6.1 Equipment and Instruments

Stainless steel spatula
Graduated glass tubes (10 mL)
Pipette (1–5 mL; 20–200 μL)
20 mL glass beakers
Optical glass cuvette (1 cm optical length)
Centrifuge (800 × g)
Spectrophotometer (340>λ>750 nm)
Oven (60°C)
Analytical balance (±0.1 mg)

3.6.2 Solutions

Na-P buffer (0.1 M, pH 7.5)
Proteinase K
Protease
Reagent-grade water
Whatman GF/F membranes
Vacuum filtration apparatus
Vacuum pump

3.6.3 Procedure

1. Aliquots of wet or frozen sediment (0.1–1 g) [NOTE 4] are mixed (using a magnetic stirrer for 5 min at room temperature) in a glass beaker with an adequate volume of Na-P buffer (0.1 M, pH 7.5) in order to have a sediment weight to buffer the volume ratio of 2.5.
2. Six aliquots of equal volume (0.5–1.0 mL) are pipetted out from the beaker and placed in six glass tubes (10 mL), three of which are used as control samples (i.e., not treated with enzymes), whereas the other three (treated) will be added with the enzymes.
3. The volume of these aliquots is adjusted to 2 mL using an adequate volume of Na-P buffer (0.1 M, pH 7.5).
4. The tubes are then treated with ultrasound (in a sonifier bath, 50–100 W) for 3 min with 30 s of interval between each minute of sonication.
5. Aliquots of a proteinase K and a generic protease are added to the three treated tubes at a final concentration sufficient to digest the entire total protein pool in the sample within 6 h [NOTE 5].
6. All tubes (treated and untreated) are placed in an incubator at 37°C for the time needed to obtain the maximum protein release from the sediment after enzyme addition [NOTE 6].
7. At the end of the incubation, sediment samples are concentrated onto Whatman GF/F glass fiber filters (previously calcinated at 450°C, 4 h) and rinsed two times with 5 mL of Na-P buffer (0.1 M pH 7.5) in order to remove hydrolyzed proteins and incubation enzymes.
8. Filters are then analyzed for total protein concentration in the sediment according to the protocol previously described.
9. Three replicates of sediment samples previously calcinated (500°C, 4 h) are processed in the same way as the wet or frozen sediment samples.

3.6.4 Calculations

Enzymatically hydrolyzable protein concentrations are calculated from calibration curves made using BSA standards, processed according to the procedure described for total protein determination. Concentrations of enzymatically hydrolyzable proteins (HPRT) are calculated according to Equation 3.10:

$$HPRT\ (mg\ g^{-1}) = \left(PRT_c - PRT_{cblk}\right) - \left(PRT_h - PRT_{hblk}\right) \tag{3.10}$$

where

PRT_c = average total protein concentration (mg g^{-1}) in the untreated sediment samples.
PRT_{cblk} = average protein concentration (mg g^{-1}) in untreated calcinated sediment samples.
PRT_h = average protein concentration (mg g^{-1}) in treated sediment samples.
PRT_{hblk} = average protein concentration (mg g^{-1}) in treated calcinated sediment samples.

3.7 LABORATORY PROCEDURES: ENZYMATICALLY HYDROLYZABLE CARBOHYDRATES

The procedure described here allows evaluation of the digestible fraction of carbohydrates in marine sediments, estimated as the difference between the total carbohydrate pools and the carbohydrate remaining after removal of the fraction enzymatically digested (Danovaro et al. 2001).

3.7.1 Equipment and Instruments

Stainless steel spatula

Graduated glass tubes (10 mL)
Pipette (1–5 mL; 20–200 μL)
20 mL glass beakers
Optical glass cuvette (1 cm optical length)
Centrifuge (800 × g)
Spectrophotometer (340>λ>750 nm)
Oven (60°C)
Analytical balance (±0.0001 g)

3.7.2 Solutions

Na-P buffer (0.1 M, pH 7.5)
Na-P buffer (0.1 M, pH 6.9)
Proteinase K
Protease
α-Amylase
β-Glucosidase
Lipase
Reagent-grade water

3.7.3 Procedure

1. Aliquots of wet or frozen sediment (0.1–1.0 g) [NOTE 7] are mixed in a glass beaker with an adequate volume of Na-P buffer (0.1 M in EDTA 0.1 M, pH 5) in order to have a sediment weight to buffer volume ratio of 2.5, using a magnetic stirrer for 5 min (at room temperature).
2. Six aliquots of equal volume (0.5–1.0 mL) are pipetted from the beaker and placed in six glass tubes (10 mL), three of which are for control samples (i.e., not treated with enzymes), while the other three (treated) will be supplemented with enzymes.
3. The volume of these aliquots is adjusted to 2 mL using an adequate volume of Na-P buffer (0.1 M, pH 6.9).
4. The tubes are then treated with ultrasound (in a sonifier bath, 50–100 W) for 3 min with 30 s of interval between each minute of sonication.
5. To the three treated tubes, 100 μL of an α-amylase solution, 50 μL of a β-glucosidase solution, 100 μL of a proteinase-K solution, and 100 μL of a generic lipase are added, whereas only 350 μL of Na-P buffer (0.1 M, pH 6.9) are added to the untreated tubes [NOTE 8].
6. After vigorous shaking, the six tubes are incubated at room temperature for a time sufficient to obtain the maximum carbohydrate release.
7. Three replicates of previously calcinated (500°C, 4 h) sediment samples are treated in the same way as the wet or frozen sediment samples and used as blanks.
8. At the end of the incubation, tubes are centrifuged (800 × g, 15 min) and 1 mL of the supernatant is analyzed for total carbohydrate concentrations following the protocol described earlier using D(+) glucose solutions as standards.

3.7.4 Calculations

Carbohydrates released from the sediment into the supernatant after incubation with and without enzymes are determined according to the procedure described earlier, and normalized to the sediment dry weight using Equation 3.11:

$$CHO\ (\mu g\ g^{-1}) = V \cdot CHO_s \cdot P_s^{-1} \tag{3.11}$$

where
V = total volume of the supernatant during incubation.
CHO_s = carbohydrate concentration (μg mL^{-1}) in the aliquot of supernatant utilized for the analysis.
P_s = sediment dry weight of the sample.

The concentration of enzymatically hydrolyzable carbohydrates in the sediment (HCHO) is calculated as the difference between the amount of carbohydrates released from sediments treated with enzymes and the amount released from untreated sediments once the amount of carbohydrates released by treated and untreated calcinated sediment samples, respectively, has been subtracted in both according to the following the formula (Equation 3.12):

$$HCHO_s\ (\mu g\ g^{-1}) = \left[CHO_h - CHO_{hblk}\right] - \left[CHO_c - CHO_{cblk}\right] \tag{3.12}$$

where
CHO_h = mean concentration (μg g^{-1}) of carbohydrates in the sediment treated with enzymes.
CHO_{hblk} = mean concentration (μg g^{-1}) of carbohydrates in the calcinated sediment treated with enzymes.
CHO_c = mean concentration (μg g^{-1}) of carbohydrates in the sediment incubated without enzymes.
CHO_{cblk} = mean concentration (μg g^{-1}) of carbohydrates in the calcinated sediment incubated without enzymes.

3.8 CONVERSION COEFFICIENTS

Concentrations of carbohydrates, proteins, and lipids are transformed into C equivalents using the coefficients 0.40, 0.49 and 0.75 μg C μg^{-1}, respectively, obtained from the relative standards (Fabiano et al. 1995). The sum of the C equivalents of the three main biochemical classes of organic compounds is reported as the biopolymeric fraction of organic carbon (biopolymeric carbon, BPC; Fabiano et al. 1995; Pusceddu et al. 2000). The difference between the total organic carbon (TOC) and BPC is reported as geopolymeric carbon (GPC). The sum of enzymatically hydrolyzable proteins and carbohydrates, plus total lipids, is defined as the bioavailable fraction of organic carbon (bioavailable organic carbon, BAOC, Danovaro et al. 2001; Pusceddu et al. 2003).

3.9 NOTES

1. The resulting reaction is strongly exothermic and catalyzes carbohydrate carbonization (Dubois et al. 1956). The H_2SO_4 must be added very slowly with special caution, possibly using an automatic pipette. The presence of carbonates in the sediment may determine the development of CO_2 bubbles. Subsequently, the carbonate concentration in the sample has to be determined in order to calculate the actual dry sediment weight.
2. The absorbance at 600 nm is an estimate of the turbidity of the supernatant and, therefore, is used as a correction factor.
3. Assuming the carbonate percentage content of each sediment sample remains constant, the correction factor K_{CaCO3} (>1) for carbonate dissolution is calculated as the mean value of the ratios between the weight of dry sediment aliquots (3–5 replicates) before and after treatment with a solution of reagent-grade water, phenol (5%), and concentrated sulfuric acid (1:1:5 in volume, respectively), according to the described protocol.

4. Depending on the sediment grain size.
5. The final concentration of the enzymes in the reacting solution, calculated on the basis of the enzymes' specified selectivity, can vary from 1 to 10 μg of enzyme per milliliter of reacting solution.
6. The time of incubation must be calculated previously, after determination of the degradation kinetics (between 0.5 and 6 h) of treated and untreated samples.
7. Depending on the sediment grain size in the presence of coarser sediments, it is recommended that a larger amount of sediment be used.
8. Concentration of enzyme working solutions is usually of 1 mg mL^{-1}, when total carbohydrate concentrations do not exceed 2 mg g^{-1}. However, concentrations should be arranged according to enzyme kinetics on sediment sample.

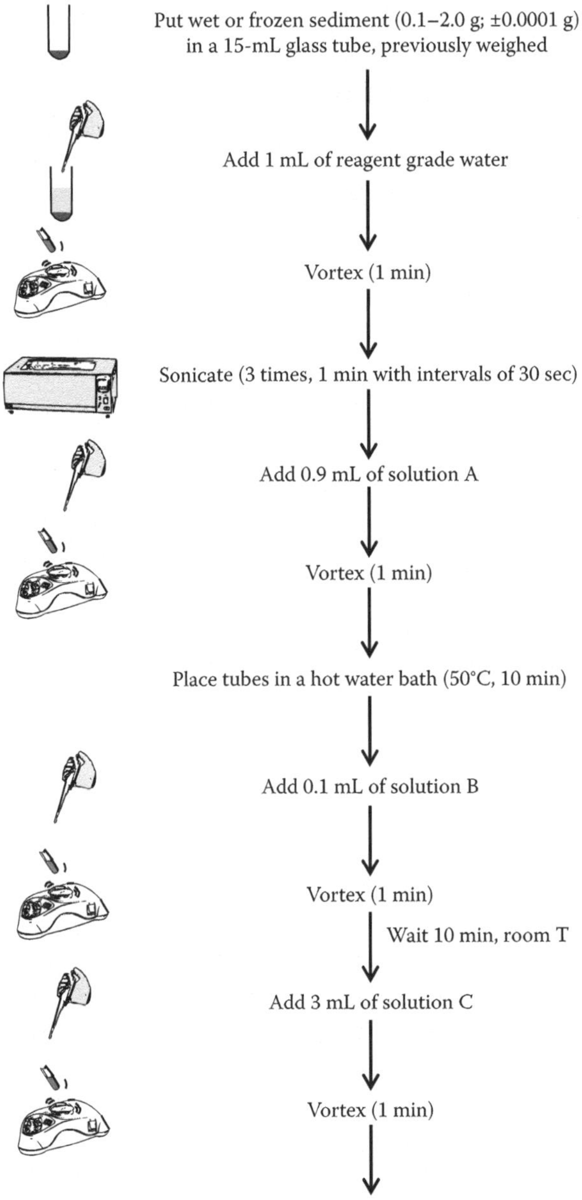

(a)

PROTOCOL 3.1 Quantification of total proteins in marine sediments. *Continued*

Place tubes in a hot water bath (50°C, 10 min)

↓

Centrifuge (800 × g, 15 min)

↓

Transfer the supernatant in
an optical glass cuvette (1 cm optical length)

↓

Measure the absorbance of sample (650 nm wavelength)

↓ Discharge the supernatant

Dry sediment (60°C, 24 h)

↓

Weigh tubes (±0.0001 g)

A blank of reagents (1 mL of reagent grade water)
and 3 sediment replicates of the sample previously muffled at 500°C (4 h)
are processed in the same way

(b)

PROTOCOL 3.1 *Continued.*

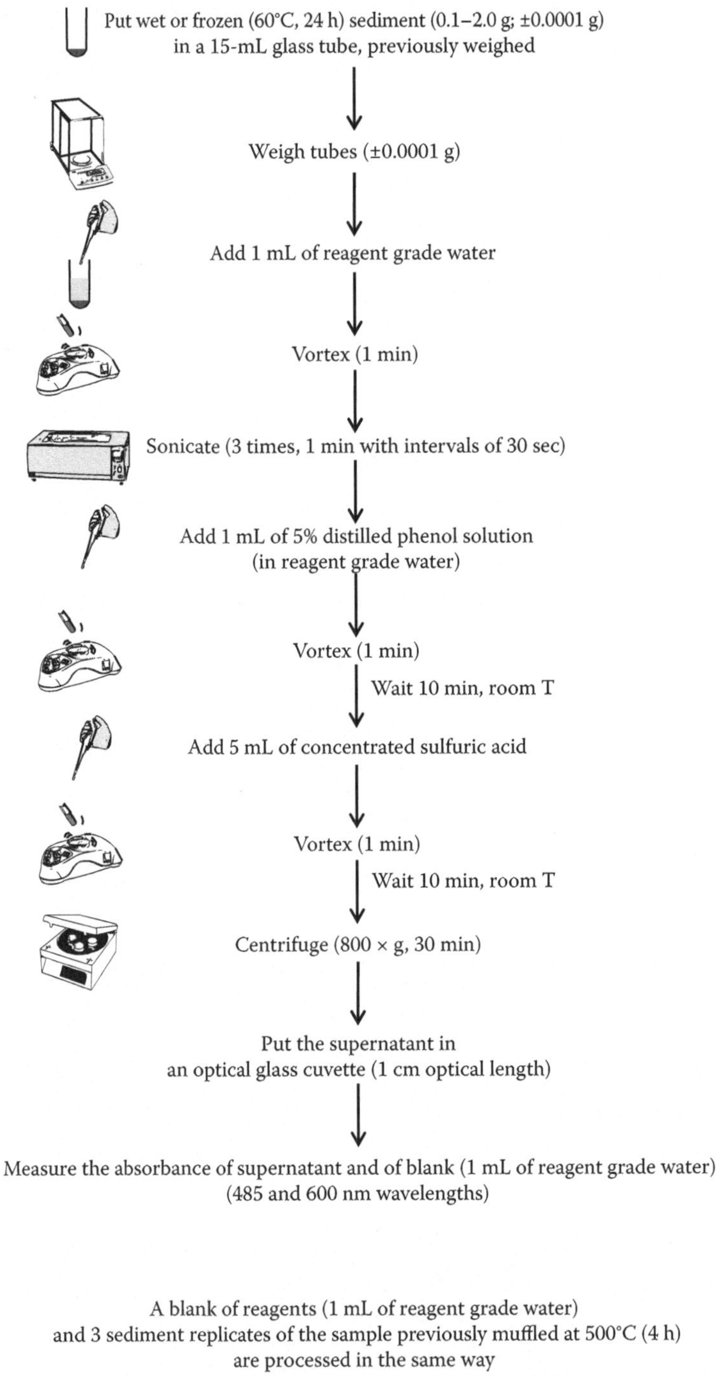

(a)

PROTOCOL 3.2 Quantification of total carbohydrates in marine sediments.

Continued

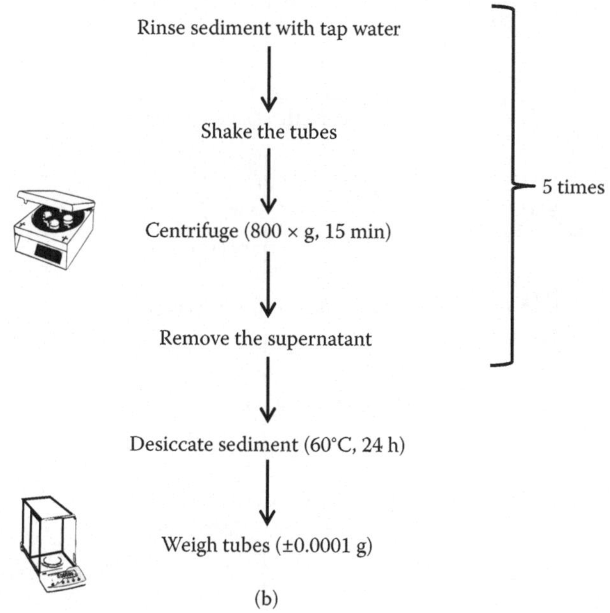

(b)

PROTOCOL 3.2 *Continued.*

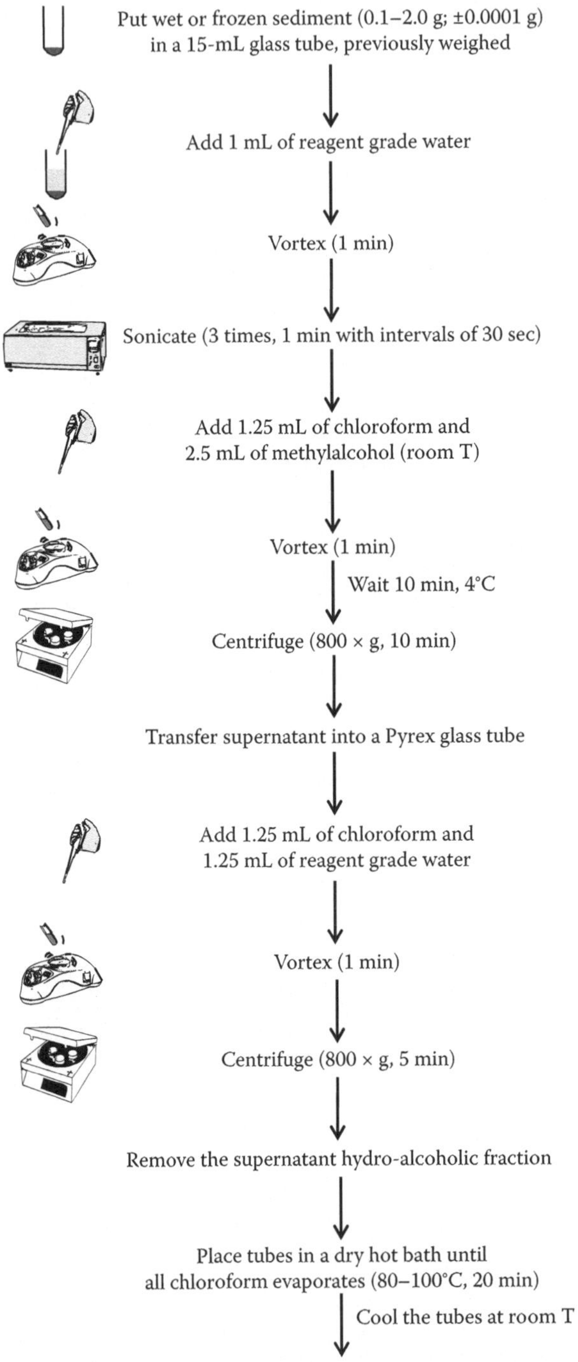

(a)

PROTOCOL 3.3 Quantification of total lipids in marine sediments.

Continued

Add 2 mL of concentrated sulfuric acid

↓

Place tubes in a dry hot bath (200°C, 15 min)

↓

Transfer tubes in a wet bath (room T, 15 sec)

↓

Transfer tubes in an ice bath (0°C, 5 min)

↓

Add 3 mL of reagent grade water

↓

Vortex (1 min)

↓

Put the supernatant in an optical
glass cuvette (1 cm optical length)

↓

Measure the absorbance of supernatant
and of blank (1 mL of reagent grade water)
(375 nm wavelength)

A blank of reagents (1 mL of chloroform)
and 3 sediment replicates of the sample previously
muffled at 500°C (4 h) are processed in the same way

↓ Remove the supernatant

Desiccate sediment in an oven (60°C, 24 h)

↓

Weigh tubes (±0.0001 g)

(b)

PROTOCOL 3.3 *Continued.*

Mix (5 min, room T) a wet or frozen sediment sample (0.1–1.0 g)
+ an equal volume of Na-P buffer (0.1 M, pH 7.5) in a glass beaker

↓

Transfer 6 aliquots of the mixture (0.5–1.0 mL)
to 6 glass tubes (10 mL)

↓

Add 2 mL of Na-P buffer (0.1 M, pH 7.5)

↓

Sonicate (3 times, 1 min with intervals of 30 sec)

↓

Add proteinase K and the generic protease
to 3 of the 6 tubes

↓

Incubate all 6 tubes at 37°C

↓

Concentrate sediment samples onto Whatman GF/F
glass fiber filters (previously calcinated at 450°C, 4 h)

↓

Rinse twofold with 5 mL of Na-P buffer (0.1 M, pH 7.5)

↓

Analyze filters for total protein concentration in the sediment
according to the protocol previously described (1 min)

Three replicates of sediment samples previously muffled (500°C, 4 h) are processed
in the same way as the wet or frozen sediment samples

PROTOCOL 3.4 Determination of enzymatically hydrolyzable proteins.

Mix (5 min, room T) a wet or frozen sediment sample (0.1–1.0 g)
+ an equal volume of Na-P buffer (0.1 M in EDTA 0.1 M, pH 5) in a glass beaker

↓

Transfer 6 aliquots of the mixture (0.5–1.0 mL)
to 6 glass tubes (10 mL)

↓

Add 2 mL of Na-P buffer (0.1 M, pH 6.9)

↓

Sonicate (3 times, 1 min with intervals of 30 sec)

↓

Add 100 μL of the α-amylase solution,
50 μL of the β-glucosidase solution,
100 μL of the proteinase-K solution and
100 μL of the generic lipase to 3 of the 6 tubes

↓

Add only 350 μL of Na-P buffer (0.1 M, pH 6.9)
to the other (control) 3 tubes

↓

Shake vigorously and incubate all 6 tubes (room T, 1 h)

↓

Centrifuge (800 × g; 4 h)

↓

Analyze 1 mL of supernatant for total carbohydrate concentration following the
protocol described above using D(+) glucose solution as standard

Three replicates of previously muffled (500°C, 4 h) sediment samples are treated in
the same way as the wet or frozen sediment samples and used as blanks

PROTOCOL 3.5 Determination of enzymatically hydrolyzable carbohydrates.

REFERENCES

Berges, J. A., A. E. Fisher, and P. J. Harrison. 1993. A comparison of Lowry, Bradford, and Smith protein assays using different protein standards and protein isolated from the marine diatom *Thalassiosira pseudonana*. *Marine Biology* 115: 187–193.

Bligh, E. G. and W. J. Dyer. 1959. A rapid method for total lipid extraction and purification. *Canadian Journal of Biochemistry and Physiology* 37: 911–917.

Bradford, M. 1976. A rapid and sensitive method for the quantification of microgram quantities of protein utilizing the principle of protein-dye binding. *Analytical Biochemistry* 72: 248–254.

Danovaro, R., A. Dell'Anno, and M. Fabiano. 2001. Bioavailability of organic matter in the sediments of the Porcupine Abyssal Plain, northeastern Atlantic. *Marine Ecology Progress Series* 220: 25–32.

Dauwe, B. and J. J. Middelburg. 1998. Amino acids and hexosamines as indicators of organic matter degradation state in North Sea sediments. *Limnology and Oceanography* 43: 782–798.

Dauwe, B., J. J. Middelburg, P. Van-rijswijk, J. Sinke, P. M. J. Herman, and C. H. R. Heip. 1999a. Enzymatically hydrolyzable amino acids in north sea sediments and their possible implication for sediment nutritional values. *Journal of Marine Research* 57: 109–134.

Dauwe, B., J. J. Middelburg, P. M. J. Herman, and C. H. R. Heip. 1999b. Linking diagenetic alteration of amino acids and bulk organic matter reactivity. *Limnology and Oceanography* 44: 1809–1814.

Dell'Anno, A., M. Fabiano, M. L. Mei, and R. Danovaro. 2000. Enzymatically hydrolysed protein and carbohydrate pools in deep-sea sediments: Estimates of the potentially bioavailable fraction and methodological considerations. *Marine Ecology Progress Series* 196: 15–23.

Dubois, M., K. Gilles, J. K. Hamilton, P. A. Rebers, and F. Smith. 1956. Colorimetric method for determination of sugars and related substances. *Analytical Chemistry* 28: 350–356.

Fabiano, M., R. Danovaro, and S. Fraschetti. 1995. A three-year time series of elemental and biochemical composition of organic matter in subtidal sandy sediments of the Ligurian sea (Northwestern Mediterranean). *Continental Shelf Research* 15: 1453–1469.

Gerchacov, S. M. and P. G. Hatcher. 1972. Improved technique for analysis of carbohydrates in sediment. *Limnology and Oceanography* 17: 938–943.

Hartree, E. F. 1972. Determination of proteins: A modification of the Lowry method that gives a linear photometric response. *Analytical Biochemistry* 48: 422–427.

Lowry, O. H., N. J. Rosebrough, A. L. Farr, and R. J. Randall. 1951. Protein measurement with folin phenol reagent. *Journal of Biological Chemistry* 193: 265–275.

Marsh, J. B. and D. B. Wenstein. 1966. A simple charring method for determination of lipids. *Journal of Lipid Research* 7: 574–576.

Mayer, L. M., L. L. Schick, T. Sawyer, C. J. Plante, P. A. Jumars, and R. L. Self 1995. Bioavailable amino acids in sediments: A biomimetic, kinetics-based approach. *Limnology and Oceanography* 40: 511–520.

Mayer, L. M., L. L. Schick, and F. W. Setchell. 1986. Measurement of protein in nearshore marine sediments. *Marine Ecology Progress Series* 30: 159–165.

Middelburg, J. J., J. Nieuwenhuize, and P. Van-Breugel. 1999. Black carbon in marine sediments. *Marine Chemistry* 65: 245–252.

Pusceddu, A., A. Dell'Anno, and M. Fabiano. 2000. Organic matter composition in coastal sediments at Terra Nova bay (Ross Sea) during summer 1995. *Polar Biology* 23: 288–293.

Pusceddu, A., A. Dell'Anno, E. Manini, M. Fabiano, G. Sarà, and R. Danovaro. 2003. Enzymatically hydrolyzable protein and carbohydrate sedimentary pools as indicators of the trophic state of 'detritus sink' systems: A case study in a Mediterranean coastal lagoon. *Estuaries* 26: 641–650.

Rice, D. L. 1982. The detritus nitrogen problem: New observations and perspectives from organic geochemistry. *Marine Ecology Progress Series* 9: 153–162.

Rice, D. L. and D. C. Rhoads. 1989. Early diagenesis of organic matter and the nutritional value of sediment. In *Ecology of Marine Deposit Feeders,* Eds. G. Lopez, G. Taghon, and J. Levinton, 309–317. Berlin, Beidelberg: Springer.

4 Photosynthetic Pigment Concentrations in Marine Sediments

CONTENTS

4.1 INTRODUCTION

In all marine systems, the benthic compartment receives, at varying intensities and durations, discrete inputs of organic matter from the overlying pelagic environment (Graf 1992; Lampitt and Antia 1997; Albertelli et al. 1999; Witte et al. 2003; Danovaro et al. 2000).

Sediment organic matter is composed of an autotrophic fraction (primary organic matter) derived from photosynthesis and a heterotrophic fraction (secondary organic matter). Organic C of autotrophic origin is generally determined through the analysis of photosynthetic pigment concentrations (chlorophyll-a, phaeopigments; Montagna et al. 1983; Plante et al. 1986; MacIntyre et al. 1996). In coastal sediments (i.e., down to 200 m depth), the concentration of phytopigments are the result of vertical inputs and primary production *in situ* by microphytobenthos. Primary organic matter is a key food source for the benthos from small protozoa to meio- and macrofauna (Bianchi and Levinton 1984; Montagna et al. 1983; Plante et al. 1986). Despite several methodological problems in determining the actual concentrations of chlorophyll-a of microphytobenthos (MacIntyre et al. 1996), the determination of phytopigment concentrations in marine sediments is crucial for assessing the amount available for higher trophic levels. In deep-sea sediments (i.e., below 200 m in depth), the photosynthetic pigments are always found as a result of the sedimentation of freshly produced primary organic matter. For this reason, the concentration of phytopigments is a proxy for the material reaching the sea bottom from the upper water layers (Witte et al. 2003).

4.2 FIELD PROCEDURES

4.2.1 SAMPLING

Marine sediment collection for the assessment of the quantity and biochemical composition of organic matter must be carried out by avoiding any manipulation of the sample that could affect the vertical stratification of the sediment or a relevant alteration of the *in situ* physical–chemical

FIGURE 4.1 Particulars of the transparent Plexiglas® liners of the multicorer.

properties of the surface layer (e.g., resuspension, warming, desiccation). For the determination of phytopigment in marine sediments, the most common devices for sediment sampling are box corers and multicorers (Danovaro et al. 1998; Figure 4.1).

4.2.2 Sample Storage

Once the sediment core is extruded from the liner, whenever studies of the vertical distribution of organics are required, the core can be sliced into the layers needed, using a plastic (inert) knife. The thickness of the sediment layers must be chosen depending on the sampling strategy adopted. Generally, the top 5 cm of the corer is sliced into 0.1–1.0 cm thick layers, whereas, deeper in the core, layer thickness can be extended to 2–5 cm. Sediment layers are then immediately homogenized and stored frozen at −20°C until analysis (generally within 3 months from collection).

4.3 LABORATORY PROCEDURES

4.3.1 Equipment and Instruments

Stainless steel spatula
Graduated glass tubes (15 mL)
Pipette (1–5 mL; 20–200 μL)
Optical glass cuvette (1 cm optical length) or quartz cuvette
Centrifuge (800 × g)
Spectrophotometer (340<λ<750 nm) or spectrofluorometer
Oven (60°C)

Analytical balance (±1 mg)
Ultrasound bath
Freezer (–20°C)

4.3.2 Solutions

Reagent-grade $MgCO_3$
Reagent-grade acetone
Reagent-grade water
0.1 N HCl

4.3.3 Procedure

1. An aliquot of wet or frozen sediment (up to 2 g) and 0.1 g of $MgCO_3$ is placed in a glass tube and weighed (precision ±0.1 mg).
2. To tubes containing the sediment, 5–10 mL of 90% acetone is added (5 mL is required for low-chlorophyll sediments), closed with a stopper and vigorously shaken with a vortex (30 s) in the dark.
3. Tubes closed with stoppers are treated in the dark with ultrasound (in an ultrasound bath in ice at 50–100 W for 3 min, with a 30 s interval between each minute of sonication).
4. Store the tubes overnight at 4°C in the dark.
5. Tubes are then centrifuged (10 min at 800 × g).
6. 3 mL of the supernatant is then utilized for photometric (steps 7–8) or fluorometric (step 10) determinations [NOTE 1].
7. The supernatant is pipetted into an optical glass cuvette (1 cm optical length), and the absorbance at the wavelengths of 750 nm and 665 nm are measured (against a blank of 90% acetone).
8. For phaeopigment determination, the acetone extract is acidified directly in the cuvette with 200 µL of 0.1 N HCl and left to react for 20 s. After acidification, the absorbance of samples is read at both 750 and 665 nm [NOTES 2 and 3].
9. The supernatant is removed from the tubes, and the sediment is dried (60°C for 24–36 h, until constant weight is achieved) and then weighed with a tube (precision ±0.1 mg).
10. Next, 3 mL of acetone extract is placed into a quartz cuvette (4 transparent plates) and scanned using the fluorometer, before and after acidification with 200 µL of 0.1 N HCl. Excitation and emission wavelengths are determined from the relative peaks of spectra obtained from scanning the standard solutions. Common values are 430 nm (excitation) and 665 nm (emission) [NOTE 2].
11. The supernatant is removed from the tube, and the tube with its sediment is dried (60°C, 24 h) and then weighed.

4.3.4 Calculations

For the spectrophotometric assay, chlorophyll-a (Chl-a) and phaeopigment (Phaeo) concentrations are then calculated using Equations 4.1 and 4.2

$$Chl-a\ (\mu g\ g^{-1}) = 26.7\left[A_{665O} - A_{665a}\right]\cdot\left[v + \left(PP_u - PP_s\right)\right]\cdot\left(CO\cdot P_s\right)^{-1} \tag{4.1}$$

$$Phaeo\ (\mu g\ g^{-1}) = 26.7\cdot\left[\left(1.7\cdot A_{665a}\right) - A_{665O}\right]\cdot\left[v + \left(PP_u - PP_s\right)\right]\cdot\left(CO\cdot P_S\right)^{-1} \tag{4.2}$$

where

$A_{6650} = [A_{665} - A_{750}]$.
$A_{665a} = [A_{6650a} - A_{750a}]$.
A_{665} = absorbance of the sample at 665 nm before acidification.
A_{750} = absorbance of the sample at 750 nm before acidification.
A_{665a} = absorbance of the sample at 665 nm after acidification.
A_{750a} = absorbance of the sample at 750 nm after acidification.
v = volume of the acetone extract.
CO = optical length (in cm).
PP_u = weight (g) of the tube containing wet sediment and $MgCO_3$.
PP_s = weight (g) of the tube containing dry sediment.
P_s = weight (g) of the dry sediment.

For the fluorometric assay, a calibration curve must be obtained by treating standard solutions of chlorophyll-a. The calibration constant (c) and the acidification factor (x) are calculated using the following formulas Equations 4.3 and 4.4.

$$c = C\ (Chl - a) / F_0 \tag{4.3}$$

$$x = F_0 / F_a \tag{4.4}$$

where

C (Chl-a) = average concentration of the chlorophyll-a standard solutions.
F_0 = average value of the standard fluorescence before acidification.
F_a = average value of the standard fluorescence after acidification.

After this, acetone extracts of the samples are processed, and Chl-a and phaeo concentrations are calculated using Equations 4.5 and 4.6:

$$Chl - a\left(\mu g\ g^{-1}\right) = \left[x/(x-1)\cdot(F_0 - F_a)\cdot c\cdot\left[v + (PP_u - PP_s)\right]\cdot P_s^{-1}\right] \tag{4.5}$$

$$Phaeo\left(\mu g\ g^{-1}\right) = \left\{\left[(F_a\cdot x) - F_0\right]\cdot x/(x-1)\cdot c\cdot(v + PP_u - PP_s)\cdot P_s^{-1}\right\} \tag{4.6}$$

where

v = acetone extract volume (in mL).
PP_u = weight (g) of the tube containing wet sediment and $MgCO_3$.
PP_s = weight (g) of the tube containing the dry sediment.
P_s = weight (g) of the dry sediment.

4.4 NOTES

1. The spectrophotometric assay is generally used for pigment-rich sediments (where concentrations are >0.5 μg Chl-a g^{-1}), whereas the spectrofluorometric assay, having a lower detection limit, is recommended for the analysis of sediments with lower pigment concentrations.
2. Sometimes, at 750 nm, slightly negative values (e.g., −0.001) can be detected.
3. For spectrophotometric determination, the zero is assessed at 750 nm against a blank of 90% acetone. The absorbance of samples at 750 nm indicates centrifugation efficiency (i.e., the turbidity of the supernatant). Values higher than 0.004 and 0.005 indicate that centrifugation has not been efficient and/or that the sediment sample has been resuspended. In this case, the samples need a further centrifugation step.

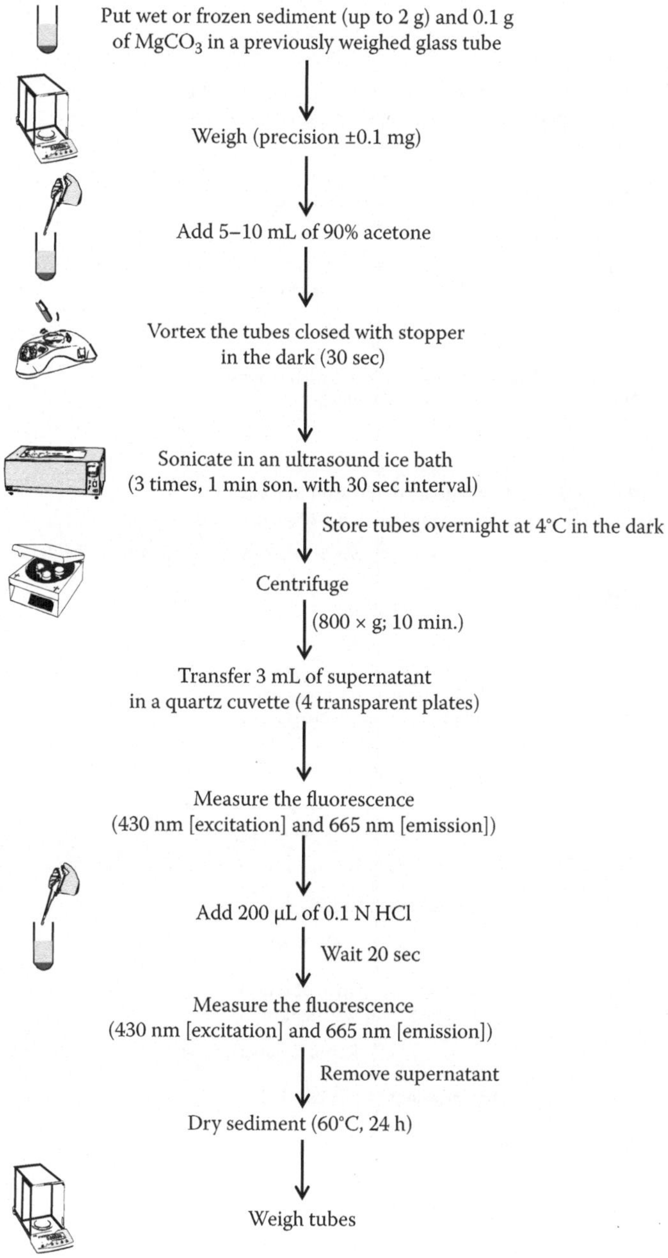

PROTOCOL 4.1 Fluorometric protocol for determination of photosynthetic pigment concentration in marine sediments.

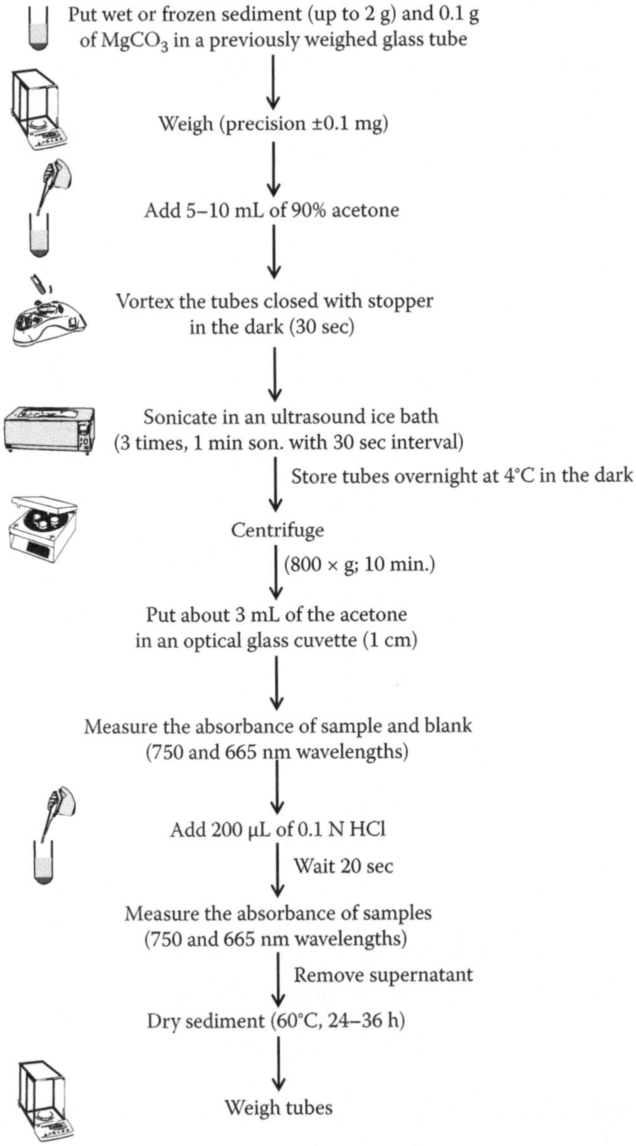

PROTOCOL 4.2 Spectrophotometric protocol for determination of photosynthetic pigment concentration in marine sediments.

REFERENCES

Albertelli, G., A. Covazzi-Harriague, R. Danovaro, M. Fabiano, S. Fraschetti, and A. Pusceddu. 1999. Differential responses of bacteria, meiofauna, and macrofauna in a shelf area (Ligurian Sea, NW Mediterranean): Role of food availability. *Journal of Sea Research* 42: 11–26.

Bianchi, A. J. M. and J. S. Levinton. 1984 The importance of microalgae bacteria and particulate organic matter in the somatic growth of *Hydrobia totteni*. *Journal of Marine Research* 42: 431–443.

Danovaro, R., D. Marrale, N. Della Croce, A. Dell'Anno, and M. Fabiano. 1998. Heterotrophic nanoflagellates, bacteria, and labile organic compounds in continental shelf and deep-sea sediments of the Eastern Mediterranean. *Microbial Ecology* 35: 244–255.

Danovaro, R., D. Marrale, A. Dell'Anno, N. Della Croce, A. Tselepides, and M. Fabiano. 2000. Bacterial response to seasonal changes in labile organic matter composition on the continental shelf and bathyal sediments of the Cretan Sea. *Progress in Oceanography* 46: 345–366.

Graf, G. 1992. Benthic–pelagic coupling: A benthic view. *Oceanography and Marine Biology: An Annual Review* 30:149–190.

Lampitt, R. S. and A. N. Antia. 1997. Particle flux in deep seas: Regional characteristics and temporal variability. *Deep-Sea Research* 44:1377–1403.

MacIntyre, H. L., R. J. Geider, and D. C. Miller. 1996. Microphytobenthos: The ecological role of the "secret garden" of unvegetated, shallow-water marine habitats. I. Distribution, abundance, and primary production. *Estuaries* 19:186–201.

Montagna, P. A., C. B. Coull, T. L. Herring, and B. W. Dudley. 1983. The relationship between abundances of meiofauna and their suspected microbial food (diatoms and bacteria). *Estuarine, Coastal, and Shelf Science* 17: 381–394.

Plante, R., M. R. Plante-Cuny, and J. P. Reys. 1986. Photosynthetic pigments of sandy sediments on the north Mediterranean coast: Their spatial distribution and its effects on sampling strategies. *Marine Ecology Progress Series* 34:133–141.

Witte, U., N. Aberle, M. Sand, and F. Wenzhöfer. 2003. Rapid response of a deep-sea benthic community to POM enrichment: An *in situ* experimental study. *Marine Ecology Progress Series* 251: 27–36.

Section II

Food Supply of Organic Matter to the Deep-Sea Floor

5 Fluxes of Labile Organic Matter to the Sea Floor

CONTENTS

5.1 INTRODUCTION

The downward flux of particulate organic material is an important process in biogeochemical cycles in the marine environment (Jahnke 1996). Downward fluxes of organic particles vary widely in different areas of the world's oceans, they are largely dependent upon the primary production pulses from the upper layers of the water column, and are typically characterized by strong seasonal fluctuations (Honjo 1982; Deuser and Ross 1990). In shallow coastal ecosystems, the export of organic material to the sea floor generally accounts for 30%–50% of the primary productivity of the photic layer, whereas in the open ocean it accounts for 1%–3% (Lampitt and Antia 1997). This "rain" of organic matter supplying the sea floor represents a key process determining the quantity and composition of sedimentary organic matter on the seabed (Danovaro et al. 2003; Dell'Anno et al. 2005) and is one of the major factors controlling the distribution, dynamics, and metabolism of the benthos (Graf 1989; Pfannkuche 1993; Danovaro et al. 1999a; Smith and Kaufmann 1999).

High amounts of material transported by lateral advection processes may deposit on the sea floor (Jahnke et al. 1990; Thomsen and van Weering 1998; Puig et al. 2003). The processes that laterally distribute suspended particles in the near-bed water column can also be as important as the vertical settling in determining the availability of organic matter for benthic metabolism (Monaco et al. 1990; Graf 1992; Jahnke and Jahnke 2000).

As food availability greatly influences standing stock, diversity, and taxonomic composition of the benthos (Tyler et al. 1995; Galeron et al. 2000; Danovaro et al. 2002; Gambi et al. 2003), quantifying organic matter transfer and deposition on the sediment represent key issues for understanding patterns in the structure, functioning, and diversity of the benthic food web.

The biochemical composition and sedimentation rates of the settling particles may change significantly during their sinking as this depends on their diameter, shape, density, and porosity (Banse 1990; Eisma et al. 1996). The properties of organic particles also change as a result of microbial colonization, so that most labile organic compounds associated with the sinking particles (e.g., proteins) may be utilized before they reach the ocean floor (Karl et al. 1988; Turley and Mackie 1995; Wakeham et al. 1997; Vanucci et al. 2001). Therefore, analysis of the biochemical composition of settling particles represents a reliable estimate of the most readily utilizable organic material supplying the benthos (Dell'Anno et al. 1999a; Danovaro et al. 2000; Fabiano et al. 2001; Hedges et al. 2001; Dell'Anno et al. 2005).

5.2 FIELD PROCEDURES

Settling particles can be collected using different typologies of sediment traps (Figure 5.1). The selection of a proper sediment trap and trap configuration is needed to collect representative particle

FIGURE 5.1 Operation on deck prior to the deployment of a mooring line with sediment traps. (Courtesy of Serge Heussner. Used with permission.)

fluxes and sufficient material for the subsequent analyses. PARFLUX and TECHNICAP sediment traps have been extensively utilized in a variety of oceanic systems. The PARFLUX sediment traps have a collecting surface area (surface 0.5 m^2) larger than the one of TECHNICAP (surface 0.125 m^2), so that the former can be reliably utilized in highly oligotrophic oceanic regions in which the fluxes of particles are expected to be low. Sediment traps are generally deployed at different depths (depending on the water column depth) by mounting each device on a single mooring line. Each trap is armed with a number of funnels (typically 12–14) that allow the collection of settling particles at desired time intervals (Figure 5.2). Each trap deployment generally lasts up to 6 months, with a trap open–close period ranging between 7 and 28 d. To better detect short-term variations of particle fluxes, sampling intervals during periods of phytoplankton blooms are typically shorter than under low primary productivity periods.

FIGURE 5.2 Sediment traps. Reported are a sediment trap (Technicap model) deployed along with a mooring line (left half) and a trap associated with a benthic lander (right half). (Courtesy of Serge Heussner. Used with permission.)

For long-term sediment trap deployments, the use of preservatives (e.g., buffered formalin) in the collecting cups is also required to minimize prokaryotic influence (Gardner et al. 1983). Experimental studies demonstrated that the use of formaldehyde as a preservative does not affect the determination of proteins, carbohydrates, and lipids when compared to untreated samples (Wakeham et al. 1997; Danovaro et al. 1999b). Dell'Anno et al. (1999b) demonstrated also that the long-term storage of formalin-preserved trap samples (up to 9 months) does not significantly affect estimates of DNA fluxes, regardless of the quantity and the source of DNA in sinking particles. Conversely, the use of formaldehyde determines an exponential decrease of phytopigment content in trap samples with time (Dell'Anno et al. 1999b). However, the initial amount of chlorophyll-a collected by sediment trap could be estimated after correction by the equations reported in Dell'Anno et al. (1999b).

5.3 LABORATORY PROCEDURES

Once in the laboratory, each of the trap funnels containing settled particles should be treated to remove swimmers, using a two-step procedure (Heussner et al. 1990). The bulk samples are first wet sieved through a mesh nylon net (<1 mm) in order to retain the largest organisms. Swimmers that passed through the sieve are then removed by picking under a dissecting microscope. Subsampling is then carried out by diluting the original sample in a suitable volume of 0.45 μm prefiltered seawater and pouring, while stirring, small equal volumes of the suspension successively in several beakers using a high-precision peristaltic pump. This operation is repeated until the total sample is divided and utilized for further determinations.

REFERENCES

Banse, K. 1990. New view on the degradation and disposition of organic particles as collected by sediment traps in the open sea. *Deep-Sea Research Part I* 37: 1177–1189.

Danovaro, R., A. Dinet, G. Duineveld, and A. Tselepides. 1999a. Benthic response to particulate fluxes in different trophic environments: A comparison between the Gulf of Lions-Catalan Sea (W-Mediterranean) and the Cretan Sea (E-Mediterranean). *Progress in Oceanography* 44: 287–312.

Danovaro, R., D. Martorano, P. Parodi, A. Dell'Anno, D. Marrale, and M. Fabiano. 1999b. Seasonal variation in the biochemical composition of deep-sea nematodes: Bioenergetic and methodological considerations. *Marine Ecology Progress Series* 179: 273–283.

Danovaro, R., N. Della Croce, A. Dell'Anno, M. Fabiano, D. Marrale, and D. Martorano. 2000. Seasonal changes and biochemical composition of the labile organic matter flux in the Cretan Sea. *Progress in Oceanography* 46: 259–278.

Danovaro, R., C. Gambi, and N. Della Croce. 2002. Meiofauna hotspot in the Atacama Trench, eastern South Pacific Ocean. *Deep-Sea Research Part I* 47: 843–857.

Danovaro, R., N. Della Croce, A. Dell'Anno, and A. Pusceddu. 2003. A depocenter of organic matter at 7800 m depth in the SE Pacific Ocean. *Deep-Sea Research Part I* 50: 1411–1420.

Dell'Anno, A., M. Fabiano, M. L. Mei, and R. Danovaro. 1999a. Pelagic–benthic coupling of nucleic acids in an abyssal location of the Northeastern Atlantic Ocean. *Applied and Environmental Microbiology* 65: 4451–4457.

Dell'Anno, A., M. Fabiano, S. Bompadre, M. Armeni, L. Leone, and R. Danovaro. 1999b. Phytopigment and DNA determinations in long-time formalin-preserved trap samples. *Marine Ecology Progress Series* 191: 71–77.

Dell'Anno, A., C. Corinaldesi, S. Stavrakakis, V. Lykousis, and R. Danovaro. 2005. Pelagic–benthic coupling and diagenesis of nucleic acids in a deep-sea continental margin and an open-slope system of the Eastern Mediterranean. *Applied and Environmental Microbiology* 71: 6070–6076.

Deuser, W. G. and E. H. Ross. 1990. Seasonal change in the flux of organic carbon to the Sargasso Sea. *Nature* 283: 364–365.

Eisma, D., A. J. Bale, M. P. Dearnaley et al. 1996. Intercomparison of *in situ* suspended matter (floc) size measurements. *Journal of Sea Research* 36: 3–14.

Fabiano, M., A. Pusceddu, and A. Dell'Anno. 2001. Fluxes of phytopigments and labile organic matter to the deep ocean in the NE Atlantic ocean. *Progress in Oceanography* 50: 89–104.

Galeron, J., M. Sibuet, M. L. Mahaut, and A. Dinet. 2000. Variation in structure and biomass of the benthic communities at three contrasting sites in the tropical Northeast Atlantic. *Marine Ecology Progress Series* 197: 121–137.

Gambi, C., A. Vanreusel, and R. Danovaro. 2003. Biodiversity of nematode assemblages from deep-sea sediments of the Atacama Slope and Trench (South Pacific Ocean). *Deep-Sea Research* 50: 103–117.

Gardner, W. D., K. R. Hinga, and J. Marra. 1983. Observations on the degradation of biogenic material in the deep ocean with implications on accuracy of sediment trap fluxes. *Journal of Marine Research* 41: 195–214.

Graf, G. 1989. Pelagic–benthic coupling in a deep-sea benthic community. *Nature* 341: 437–439.

Graf, G. 1992. Benthic–pelagic coupling: A benthic view. *Oceanography and Marine Biology Annual Review* 30: 149–190.

Hedges, J. I., J. A. Baldock, Y. Gelinas, C. Lee, M. Peterson, and S. G. Wakeham. 2001. Evidence for non-selective preservation of organic matter in sinking marine particles. *Nature* 409: 801–804.

Heussner, S., C. Ratti, and J. Carbonne. 1990. The PPS3 sediment trap and trap sample processing techniques used during the ECOMARGE experiment. *Continental Shelf Research* 10: 943–958.

Honjo, S. 1982. Seasonality and interaction of biogenic and lithogenic particulate fluxes at the Panama Basin. *Science* 218: 883–884.

Jahnke, R. A., C. E. Reimers, and D. B. Craven. 1990. Intensification of recycling of organic matter at the seafloor near ocean margins. *Nature* 348: 50–54.

Jahnke, R. A. 1996. The global ocean flux of particulate organic carbon: Areal distribution and magnitude. *Global Biogeochemical Cycles* 10: 71–88.

Jahnke, R. A. and D. B. Jahnke. 2000. Rates of C, N, P, and Si recycling and denitrification at the U.S. Mid-Atlantic continental slope depocenter. *Deep-Sea Research* 47: 1405–1428.

Karl, D. M., G. A. Knauer, and J. H. Martin. 1988. Downward flux of particulate organic matter in the ocean: A particle decomposition paradox. *Nature* 332: 438–441.

Lampitt, R. S. and A. N. Antia. 1997. Particle flux in deep-seas: Regional characteristics and temporal variability. *Deep-Sea Research* 44: 1377–1403.

Monaco, A., P. Biscaye, J. Soyer, R. Pocklington, and S. Heussner. 1990. Particle fluxes and ecosystem response on a continental margin: The 1985–1988 Mediterranean ECOMARGE experiment. *Continental Shelf Research* 10: 809–839.

Pfannkuche, O. 1993. Benthic response to the sedimentation of particulate organic matter at the BIOTRANS station, 47°N, 20°W. *Deep-Sea Research* 40: 135–149.

Puig, P., A. S. Ogston, B. L. Mullenbach, C. A. Nittrouer, and R. W. Sternberg. 2003. Shelf-to-canyon sediment-transport processes on the Eel continental margin (northern California). *Marine Geology* 193: 129–149.

Smith, K. L. and R. S. Kaufmann. 1999. Long-term discrepancy between food supply and demand in the deep eastern north pacific. *Science* 284: 1174–1177.

Thomsen, L. and T. C. van Weering. 1998. Spatial and temporal variability of particulate matter in the benthic boundary layer at the North East Atlantic Continental Margin (Goban Spur). *Progress in Oceanography* 42: 61–76.

Turley, C. M. and P. J. Mackie. 1995. Bacterial and cyanobacterial flux to the deep NE Atlantic on sedimenting particles. *Deep-Sea Research* 42: 1453–1474.

Tyler, P. A., A. D. Ansell, R. N. Gibson, and M. Barnes. 1995. Conditions for the existence of life at the deep-sea floor: An update. *Oceanography and Marine Biology Annual Review* 33: 221–224.

Vanucci, S., A. Dell'Anno, A. Pusceddu, M. Fabiano, R. Lampitt, and R. Danovaro. 2001. Microbial assemblages associated to sinking particles in the Porcupine Abyssal Plain. *Progress in Oceanography* 50: 105–122.

Wakeham, S. G., J. I. Hedges, C. Lee, M. L. Peterson, and P. J. Hernes. 1997. Compositions and transport of lipid biomarkers through the water column and surficial sediments of the equatorial Pacific Ocean. *Deep-Sea Research* 44: 2131–2162.

6 Total Carbohydrate Flux from Sediment Trap Samples

CONTENTS

6.1 INTRODUCTION

The carbohydrate content in trap material has been determined according to Dubois et al. (1956), and optimized for sediments by Gerchacov and Hatcher (1972), after minor modifications. The colorimetric assay is based on the reaction between sugars and phenol in presence of concentrated sulfuric acid. The method is not specific and allows determining concentrations of total carbohydrates, cellulose included. The method here described is generally preferred because molar absorbance for most sugars lies within a limited range and because anomalous colors are avoided. Moreover, the procedure allows higher reproducibility than the protocol described by Dubois et al. (1956).

6.2 LABORATORY PROCEDURES

6.2.1 Equipment and Instruments

Filtration apparatus
Potter or glass sticks [NOTE 1]
Pipettes
Vortex
Centrifuge (must be suitable for 10–15 mL test tubes)
Spectrophotometer
Chemical air flow

6.2.2 Solutions

5% Phenol aqueous solution
Concentrated sulfuric acid (96%)

6.2.3 Procedure

1. Resuspend trap material.
2. Filter an aliquot of trap samples onto Whatman GF/F membrane (0.45 μm nominal pore size).
3. Wash the filter two times with reagent-grade water.
4. Transfer the filter into a 15 mL glass tube.

5. Add 1 mL of reagent-grade water.
6. Homogenize the filter [NOTE 1].
7. Add 1 mL of 5% phenol solution (in reagent-grade water).
8. Vortex for 1 min.
9. Add 5 mL of concentrated sulfuric acid (96%).
10. Vortex vigorously.
11. Wait for 10–15 min for cooling.
12. Centrifuge at 3000 × g for 10 min.
13. Read the absorbance of the supernatant at 485 and 600 nm wavelengths [NOTE 2].

Carbohydrate concentration in the analytical subsample is calculated by the following equation:

$$C_{CHO}\ (\mu g\ mL^{-1}) = \{\{[(A_{485c} - A_{485blkr}) - (1.5 \cdot A_{600} - 0.003)] \cdot C\} + K\} \cdot V_f^{-1} \qquad (6.1)$$

where

C_{CHO} = carbohydrate concentration in the analytical subsample.
A_{485c} = absorbance of the sample at 485 nm.
$A_{485blkr}$ = absorbance of the reagent blank at 485 nm (mean value of three replicates).
A_{600} = absorbance of the sample at 600 nm.
C = coefficient of the linear interpolation of absorbance (independent variable) and concentration (dependent variable, μg mL^{-1}) of the D(+) glucose standard solutions. C is normally around 100 if the linear interpolation is in the form CONCENTRATION = [ABSORBANCE • C] + K.
K = constant of the linear interpolation of absorbance (independent variable) and concentration (dependent variable, μg mL^{-1}) of the glucose standard solutions.
V_f = the volume of the filtered trap material.

Carbohydrate fluxes are estimated using the following formula:

$$F_{CHO} = C_{CHO}\ (\mu g\ mL^{-1}) \cdot V_s \cdot F_s^{-1} \cdot T_d^{-1} \cdot S_c^{-1} \qquad (6.2)$$

where

F_{CHO} = the flux of carbohydrates.
C_{CHO} = the concentration (μg mL^{-1}) of carbohydrates in the analytical subsample.
V_s = the total volume of the aliquot (mL).
F_s = the fraction of the total volume collected by the trap.
T_d = the time of trap deployment (days).
S_c = the collecting surface of the trap (m^2).

6.3 NOTES

1. Filters can be homogenized by a potter (for small volume samples) at 2500 rpm for 3 min (on ice) or alternatively by a glass stick (till the filters are completely destroyed).
2. The absorbance at 600 nm is an estimate of the turbidity of the supernatant and, therefore, is used as the correction factor. Three blanks (1 mL reagent-grade water) are processed in the same way as samples. Using a glass Pasteur pipette, 2.5 mL of the supernatant are placed in an optical glass cuvette (1 cm optical length). The absorbance of the sample is determined at the wavelengths of 485 and 600 nm against a blank of reagent-grade water. Sometimes, carbohydrate concentration in the sample is so high that the absorbance falls

out of the instrumental range. In this case, it is necessary to repeat the analysis using lower trap amounts and enhancing the number of replicates (from 3 to 5). In the case this procedure would give the same problem, it is recommended to dilute the sample using the reacting solution to form the reagent blank. However, the resulting absorbance is not linear, so that a correction factor must be calculated experimentally.

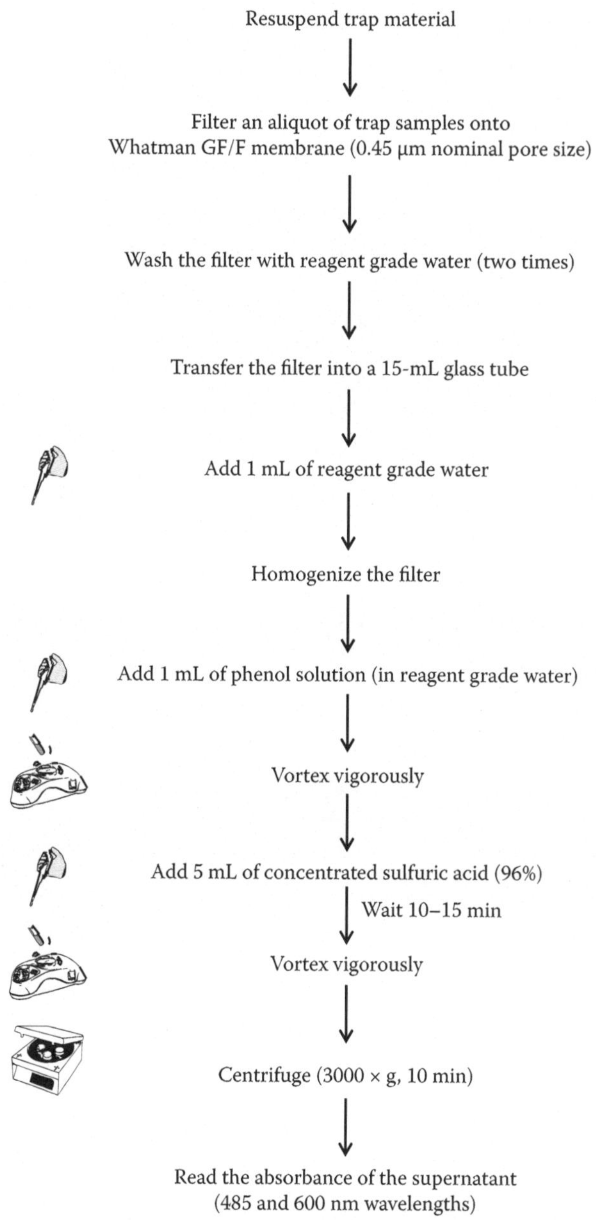

PROTOCOL 6.1 Determination of total carbohydrates in trap samples.

REFERENCES

Dubois, M., K. A. Gilles, J. K. Hamilton, P. A. Rebers, and F. Smith. 1956. Colorimetric method for determination of sugars and related substances. *Analytical Chemistry* 28: 350–356.

Gerchacov, S. M. and Hatcher P. G. 1972. Improved technique for analysis of carbohydrates in sediments. *Limnology and Oceanography* 17: 938–943.

7 Total Protein Flux from Sediment Trap Samples

CONTENTS

7.1 INTRODUCTION

Protein content in environmental samples can be determined using different analytical protocols (Bradford 1976; Rice 1982; Mayer et al. 1986). Among these, the most frequently selected are the Bradford (1976) and the Hartree (1972) protocols, the latter being a modification of the widely used procedure by Lowry et al. (1951), rearranged by Rice (1982) for sediment samples. Laboratory experiments carried out on algal cultures have demonstrated that the Bradford (1979) and the Hartree (1972) protocols provide similar results in terms of bovine serum albumin equivalents (Berges et al. 1993). In the present handbook, details are provided about the Hartree (1972) protocol after modifications by Rice (1982), slightly modified. This colorimetric method allows the reaction of proteins with rameic tartrate and the Folin–Ciocalteau reactive in basic environment (pH 10). The reaction provides a stable blue coloration, whose intensity is proportional to the protein concentrations in the reaction solution.

7.2 LABORATORY PROCEDURES

7.2.1 Equipment and Instruments

Filtration apparatus
Potter or glass sticks [NOTE 1]
Pipettes
Vortex
Centrifuge (must be suitable for 10–15 mL test tubes)
Water bath
Spectrophotometer
Chemical air flow

7.2.2 Solutions

Solution A—Dissolve 2 g of NaK tartrate and 100 g of Na_2CO_3 anhydrous in 500 mL of 1 N NaOH (20 g in 500 mL of reagent-grade water); the solution, uncolored and transparent, is diluted to 1000 mL with reagent-grade water. Solution A should be stored at 4°C, and it will remain stable for 2–3 months, after which the Na_2CO_3 tends to precipitate. In this case, the solution cannot be further utilized.

Solution B—Dissolve 2 g of NaK tartrate and 1 g of $CuSO_4$ in 90 mL of reagent-grade water; the slightly blue solution is diluted with 10 mL of 1 N NaOH. The final solution of a transparent electric blue can be stored at 4°C; it will remain stable for several weeks.

Solution C—1 mL of the Folin–Ciocalteau reagent is diluted with 15 mL of reagent-grade water. The solution, of a transparent yellow, must be prepared a few minutes before being used.

BSA standard stock solution—Dissolve 1 mg of bovine serum albumin (BSA) in 1 mL of reagent-grade water. Dilute the stock solution according to the expected range of concentrations in the sample.

7.2.3 Procedure

1. Resuspend trap material.
2. Filter an aliquot of trap samples onto Whatman GF/F membrane (0.45 μm nominal pore size).
3. Wash the filter two times with reagent-grade water.
4. Transfer the filter into a 15 mL glass tube.
5. Add 1 mL of reagent-grade water.
6. Homogenize the filter [NOTE 1].
7. Add 0.9 mL of solution A.
8. Vortex for 10 s.
9. Incubate the sample at 50°C for 10 min.
10. Wait for 5 min for cooling.
11. Add 0.1 mL of solution B.
12. Vortex for 10 s.
13. Incubate the sample at room temperature for 10 min.
14. Add 3.0 mL of solution C.
15. Vortex for 10 s.
16. Incubate the sample at 50°C for 10 min.
17. Wait for 5 min for cooling.
18. Centrifuge at 3000 × g for 10 min.
19. Read the absorbance of the supernatant at 650 nm wavelength [NOTE 2].

Protein concentration in the analytical subsample is calculated by the following equation:

$$PRT_c\ (\mu g\ mL^{-1}) = \{[(A_{650c} - A_{650blkr}) \cdot C] + K] \cdot V_f^{-1} \quad (7.1)$$

where

PRT_c = protein concentration in the analytical subsample.
A_{650c} = absorbance of the sample at 650 nm.
$A_{650blkr}$ = absorbance of the reagents blank (average of three replicates).
C = coefficient of the linear interpolation of absorbance (independent variable) and concentration (dependent variable, μg mL^{-1}) of the BSA standard solutions. C is normally around 300, if the linear interpolation is in the form CONCENTRATION = [ABSORBANCE C] + K.
K = constant of the linear interpolation of absorbance (independent variable) and concentration (dependent variable, μg mL^{-1}) of the BSA standard solutions.
V_f = the volume of the filtered trap material.

Protein fluxes are estimated using the following formula:

$$F_{PRT} = C_{PRT}\ (\mu g\ mL^{-1}) \cdot V_s \cdot F_s^{-1} \cdot T_d^{-1} \cdot S_c^{-1} \tag{7.2}$$

where

F_{PRT} = the flux of proteins
C = the concentration (μg mL^{-1}) of proteins in the analytical subsample
V_s = the total volume of the aliquot (mL)
F_s = the fraction of the total volume collected by the trap
T_d = the time of trap deployment (days)
S_c = the collecting surface of the trap (m^2)

7.3 NOTES

1. Filters can be homogenized by a potter (for small volume samples) at 2500 rpm for 3 min (the tube is in a ice bath) or alternatively by a glass stick (until the filters are completely destroyed).
2. The absorbance of the sample is determined against a blank of reagent-grade water.

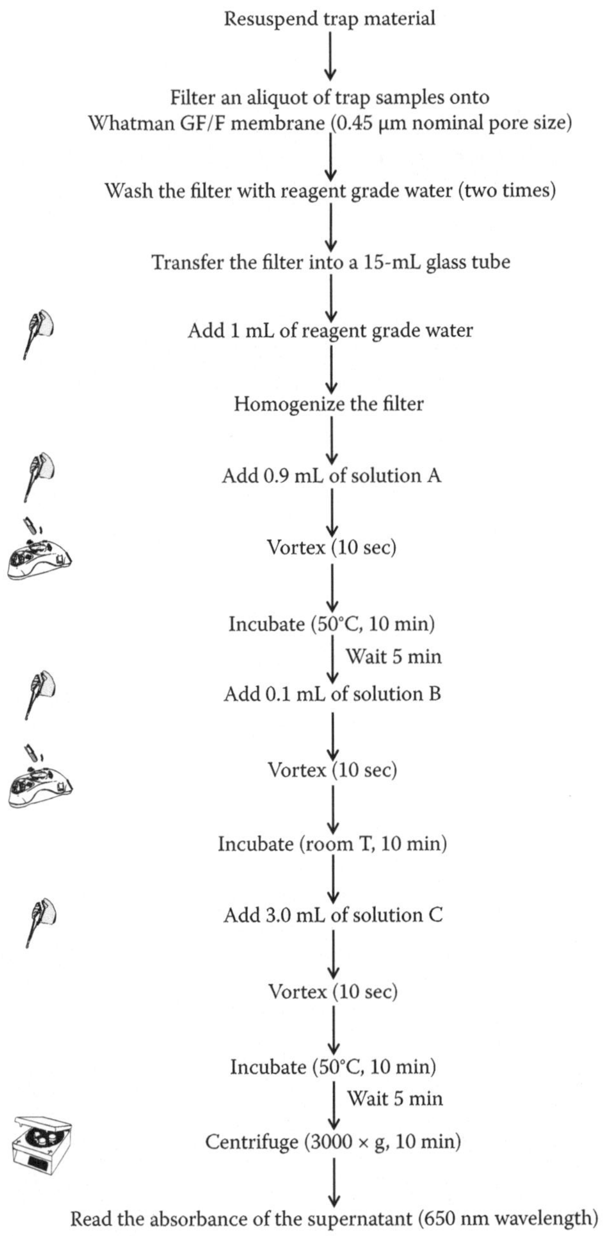

PROTOCOL 7.1 Determination of total proteins in trap samples.

REFERENCES

Berges, J. A., A. E. Fisher, and P. J. Harrison. 1993. A comparison of Lowry, Bradford, and Smith protein assays using different protein standards and protein isolated from the marine diatom *Thalassiosira pseudonana*. *Marine Biology* 115: 187–193.

Bradford, M. 1976. A rapid and sensitive method for the quantification of microgram quantities of protein utilizing the principle of protein-dye binding. *Analytical Biochemistry* 72: 248–254.

Hartree, E. F. 1972. Determination of proteins: A modification of the Lowry method that give a linear photometric response. *Analytical Biochemistry* 48: 422–427.

Lowry, O. H., N. S. Rosenbrough, A. L. Farr, and R. J. Randall. 1951. Protein measurement with the Folin-phenol reagent. *Journal of Biological Chemistry* 193: 265–275.

Mayer, L. M., L. Schick, and W. Setchell. 1986. Measurement of protein in nearshore marine sediment. *Marine Ecology Progress Series* 30: 159–165.

Rice, D. L. 1982. The detritus nitrogen problem: New observations and perspectives from organic geochemistry. *Marine Ecology Progress Series* 9: 153–162.

8 Total Lipid Flux from Sediment Trap Samples

CONTENTS

8.1 INTRODUCTION

The determination of total lipid concentrations in trap samples is carried out according to Bligh and Dyer (1959) and Marsh and Weinstein (1966), slightly modified to be applied to the sediment trap material.

8.2 LABORATORY PROCEDURES

8.2.1 Equipment and Instruments

Filtration apparatus
Potter or glass sticks [NOTE 1]
Pipettes
Vortex
Centrifuge (must be suitable for 10–15 mL test tubes)
Thermal block
Spectrophotometer
Chemical air flow

8.2.2 Solutions

Methanol (i.e., methyl alcohol)
Chloroform
Concentrated sulfuric acid (96%)
Reagent-grade water

8.2.3 Procedure

1. Resuspend trap material.
2. Filter an aliquot of trap samples onto Whatman GF/F membrane (0.45 μm nominal pore size).
3. Wash the filter two times with reagent-grade water.
4. Transfer the filter into a 15 mL glass tube.
5. Add 1 mL of reagent-grade water.
6. Homogenize the filter [NOTE 1].

7. Add 1.25 mL of chloroform and 2.5 mL of methanol.
8. Vortex vigorously for 1 min.
9. Incubate at 4°C for 10 min.
10. Centrifuge at 3000 × g for 10 min.
11. Transfer the supernatant using a Pasteur pipette into Pyrex® glass tubes.
12. Add 1.25 mL of chloroform and 1.25 mL of reagent-grade water.
13. Vortex vigorously for 1 min [NOTE 2].
14. Centrifuge the sample at 800 × g for 5 min.
15. Pipette out the supernatant hydroalcoholic fraction.
16. Evaporate the hydrophobic residual fraction (on the bottom of the Pyrex) in a thermal block at 80°C–100°C for 20–30 min.
17. Cool the tubes on ice for 5 min.
18. Add 2 mL of concentrated sulfuric acid (96%).
19. Vortex.
20. Incubate the tubes at 180°C–200°C for 15 min in a thermal block [NOTE 3].
21. Cool the samples in an ice bath for 5 min.
22. Add 3.0 mL of reagent-grade water.
23. Read the absorbance of the supernatant at 375 nm wavelength [NOTE 4].

Lipid concentration in the analytical subsample is calculated by the following equation:

$$LIP_c\ (\mu g\ mL^{-1}) = \{[(A_{375c} - A375_{blkr}) \cdot C] + K\} \cdot V_f^{-1} \tag{8.1}$$

where

LIP_c = lipid concentration in the analytical subsample.
A_{375c} = absorbance of the sample at 375 nm.
$A_{375blkr}$ = absorbance of the reagents blank (average of three replicates).
C = coefficient of the linear interpolation of absorbance (independent variable) and concentration (dependent variable, μg mL^{-1}) of tripalmitine standard solutions. C normally ranges from 230 and 300, if the linear interpolation is in the form CONCENTRATION = [ABSORBANCE • C] + K.
K = constant of the linear interpolation of absorbance (independent variable) and concentration (dependent variable, μg mL^{-1}) of tripalmitine standard solutions.
V_f = the volume of the filtered trap material.

Lipid fluxes are estimated using the following formula:

$$F_{LIP} = C_{LIP}\ (\mu g\ mL^{-1}) \cdot V_s \cdot F_s^{-1} \cdot T_s^{-1} \cdot S_c^{-1} \tag{8.2}$$

where

F_{LIP} = the flux of lipids.
C = the concentration (μg mL^{-1}) of lipids in the analytical subsample.
V_s = the total volume of the aliquot (mL).
F_s = the fraction of the total volume collected by the trap.
T_d = the time of trap deployment (days).
S_c = the collecting surface of the trap (m^2).

8.3 NOTES

1. The filters can be homogenized by a potter (for small volume samples) at 2500 rpm for 3 min (on ice) or alternatively by a glass stick (till the filters are completely destroyed).
2. The tubes are vigorously shaken (using a vortex for at least 1 min) in order to allow the formation of a water–methanol–chloroform emulsion, and centrifuged at 800 × g at least for 5 min in order to separate the supernatant hydroalcoholic fraction (water and methanol) from the hydrophobic fraction (chloroform).
3. During the reaction, the color of the sulfuric acid turns to a light yellow or up to a brown color. Tubes are then extracted from the bath and transferred into a wet bath at room temperature (for 15 s) and, then, in an ice bath (at 0°C for 5 min). Finally, 3 mL of reagent-grade water are added to the sample, and the tube is vigorously shaken by vortex.
4. The absorbance of the sample is determined at the wavelength of 375 nm against a blank of reagent-grade water.

Resuspend trap material

↓

Filter an aliquot of trap samples onto
Whatman GF/F membrane (0.45 μm nominal pore size)

↓

Wash the filter with reagent grade water (two times)

↓

Transfer the filter into a 15-mL glass tube

↓

Add 1 mL of reagent grade water

↓

Homogenize the filter

↓

Add 1.25 mL of chloroform and 2.5 mL of methanol

↓

Vortex (1 min)

↓

Incubate (4°C, 10 min)

↓

Centrifuge (3000 × g, 10 min)

↓

(a)

PROTOCOL 8.1 Determination of total lipids in trap samples. *Continued*

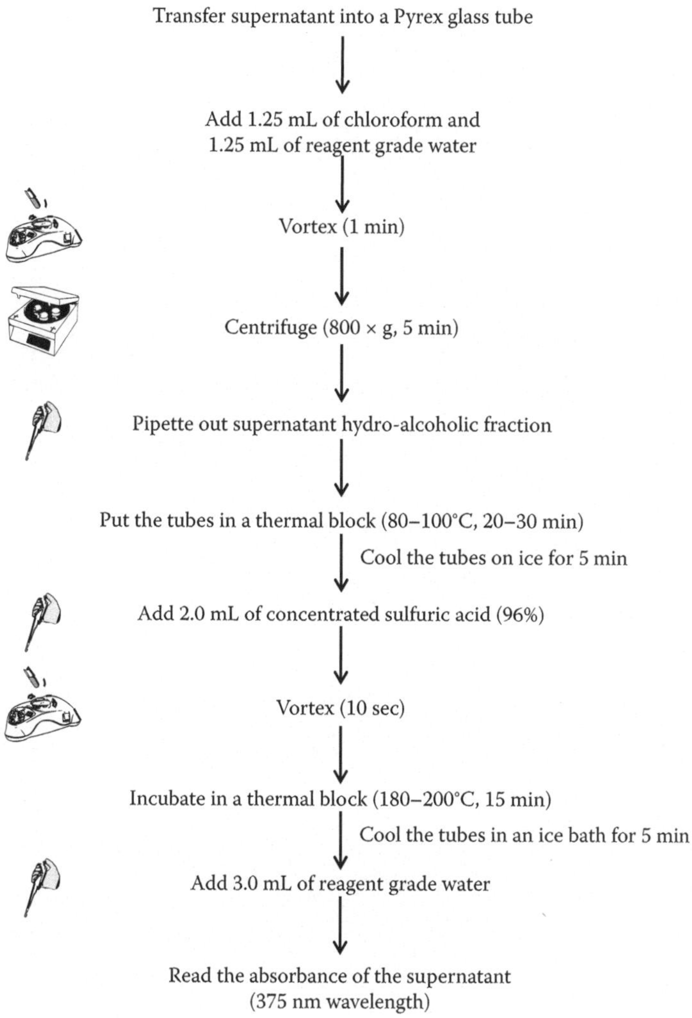

(b)

PROTOCOL 8.1 *Continued.*

REFERENCES

Bligh, E. G. and W. J. Dyer. 1959. A rapid method for total lipid extraction and purification. *Canadian Journal of Biochemistry and Physiology* 37: 911–917.

Marsh, J. B. and D. B. Weinstein. 1966. A simple charring method for determination of lipids. *Journal of Lipid Research* 7: 574–576.

9 Total DNA from Sediment Trap Samples

CONTENTS

9.1 INTRODUCTION

The determination of total DNA concentrations in trap samples is carried out according to Bailiff and Karl (1991). This procedure, based on the use of diaminobenzoic acid, allows the determination of DNA concentrations in long-term formalin-preserved trap samples (up to 9 months; Dell'Anno et al. 1999), without any significant interference of the preservative or loss of DNA associated with sinking particles.

9.2 LABORATORY PROCEDURES

9.2.1 Equipment and Instruments

Filtration apparatus
Centrifuge (must be suitable for 2 mL test tubes, i.e., Eppendorf)
Vortex
Pipettes
Speed vacuum
Oven
Spectrofluorometer
Chemical and laminar air flow

9.2.2 Solutions

Absolute acetone
90% Acetone
95% Ethanol
10% Trichloroacetic acid
1.0 M Ammonium hydroxide
40% Solution (W:V) of 3,5 diaminobenzoic di-hydrochloride acid (DABA 2HCl)
1.0 N HCl

9.2.3 Procedure

1. Resuspend the trap material.
2. Filter an aliquot of trap samples through a 0.2 μm pore size cellulose acetate filter (e.g., GS membrane 0.2 μm pore size).
3. Transfer the filter into a 2-mL DNase- and RNase-free tube.
4. Add 1.6 mL of 100% acetone.
5. Incubate the samples for 1 h at −20°C.
6. Centrifuge at 10,000 × g for 10 min.
7. Discharge the supernatant.
8. Resuspend the pellet with 1.6 mL of 100% acetone.
9. Incubate the samples for 30 min at −20°C.
10. Centrifuge at 10,000 × g for 10 min.
11. Discharge the supernatant [NOTE 1].
12. Resuspend the pellet with 1.6 mL of 90% acetone.
13. Centrifuge at 10,000 × g for 10 min.
14. Discharge the supernatant.
15. Resuspend the pellet with 1.5 mL of cold 10% trichloroacetic acid.
16. Sonicate the samples three times for 1 min (with 30 s of interval between each minute of sonication).
17. Centrifuge at 10,000 × g for 10 min.
18. Discharge the supernatant.
19. Add 1.5 mL of 95% ethanol to the pellet.
20. Centrifuge at 10,000 × g for 5 min.
21. Remove the supernatant [NOTE 2].
22. Dry the supernatant under vacuum (by using a speed vacuum) or at room temperature in a laminar air-flow chamber.
23. Resuspend the pellet with 200 μL of 1.0 M ammonium hydroxide.
24. Transfer the samples into 10–15 mL sterile tubes.
25. Dry the sample at 60°C.
26. Add 100 μL of DABA 2HCl and incubate for 1 h at 60°C [NOTE 3].
27. Add 4.9 mL of 1 N HCl to the sample.
28. Read the fluorescence at 405 nm excitation and 520 nm emission wavelengths.

DNA concentration in the analytical subsample is calculated by the following equation:

$$DNA_c\ (\mu g\ mL^{-1}) = \{[(Fluo_c - Fluo_{blkr}) \cdot C] + K\} \cdot V_f^{-1} \tag{9.1}$$

where

DNA_c = DNA concentration in the analytical subsample.

$Fluo_c$ = fluorescence of the sample.

$Fluo_{blkr}$ = fluorescence of the reagents blank (average of three replicates).

C = coefficient of the linear interpolation of fluorescence (independent variable) and concentration (dependent variable, μg mL^{-1}) of DNA standard solutions.

K = constant of the linear interpolation of fluorescence (independent variable) and concentration (dependent variable, μg mL^{-1}) of DNA standard solutions.

V_f = the volume of the filtered trap material.

DNA fluxes are estimated using the following formula:

$$F_{DNA} = C_{DNA}\ (\mu g\ mL^{-1}) \cdot V_s \cdot F_s^{-1} \cdot T_d^{-1} \cdot S_c^{-1} \quad (9.2)$$

where

F_{DNA} = the flux of DNA.
C = the concentration (μg mL^{-1}) of DNA in the analytical subsample.
V_s = the total volume of the aliquot (mL).
F_s = the fraction of the total volume collected by the trap.
T_d = the time of trap deployment (days).
S_c = the collecting surface of the trap (m^2).

9.3 NOTES

1. Repeat steps 8–11 twice. These washing steps allow the complete dissolution of cellulose acetate filters.
2. Repeat steps 19–21 one more time.
3. DABA must be freshly prepared at a concentration of 40% (W:V). To reduce the background fluorescence of DABA, the solution can be purified with activated charcoal.

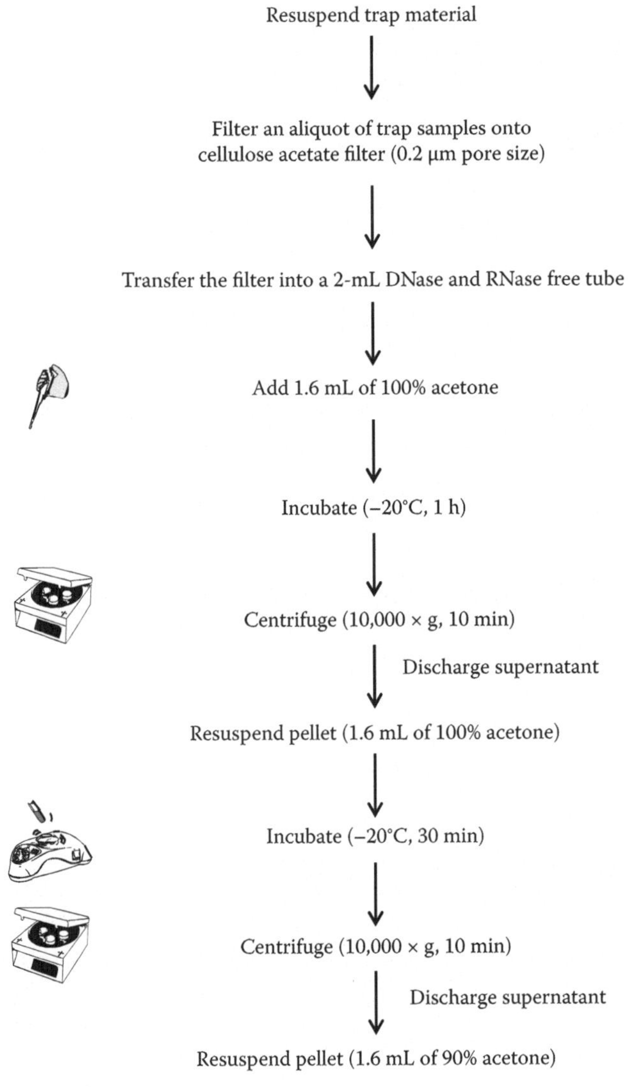

(a)

PROTOCOL 9.1 Determination of total DNA in trap samples. *Continued*

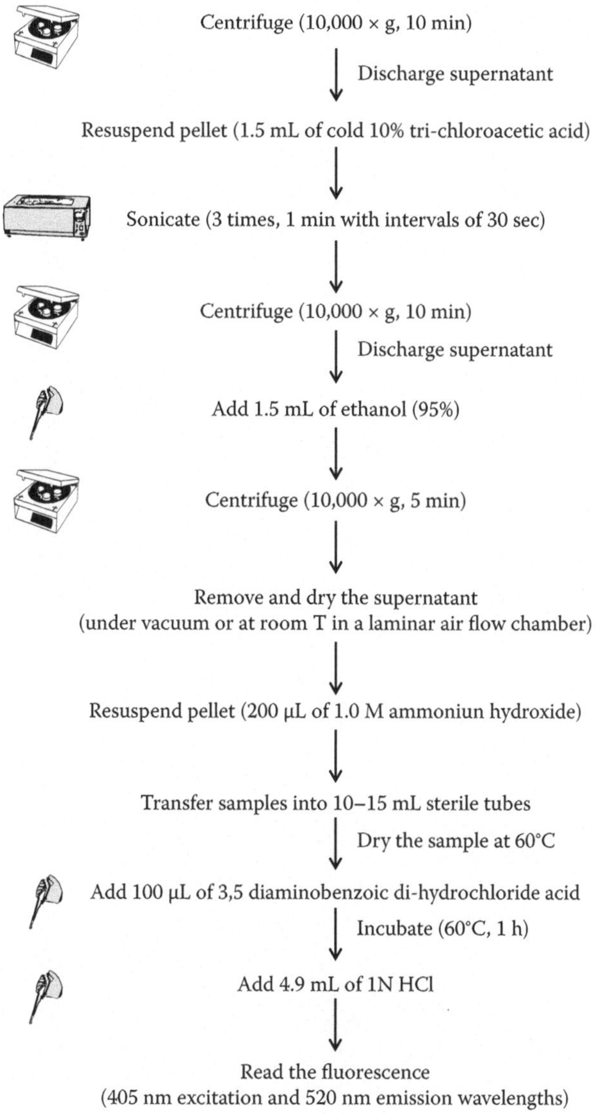

(b)

PROTOCOL 9.1 *Continued.*

REFERENCES

Bailiff, D. M. and D. M. Karl. 1991. Dissolved and particulate DNA dynamics during a spring bloom in the Antarctic Peninsula region, 1986–1987. *Deep-Sea Research* 38: 1077–1095.

Dell'Anno, A., M. Fabiano, S. Bompadre, M. Armeni, L. Leone, and R. Danovaro. 1999. Phytopigment and DNA determinations in long-time formalin-preserved trap samples. *Marine Ecology Progress Series* 191: 71–77.

10 Phytopigment Flux from Sediment Trap Samples

CONTENTS

10.1 INTRODUCTION

The use of preservatives can cause interference in phytopigment quantification. Chloropigments in trap samples can be identified and quantified by means of reverse-phase high performance liquid chromatography (HPLC) (Mantoura and Llewellyn 1983), following the procedure described by Dell'Anno et al. (1999).

10.2 LABORATORY PROCEDURES

10.2.1 Equipment and Instruments

Filtration apparatus
Potter or glass sticks [NOTE 1]
Centrifuge (must be suitable for 10–15 mL test tubes)
Vortex
Fridge
Pipettes
HPLC system [NOTE 2]
Chemical air flow

10.2.2 Solutions

90% Acetone
Solutions for the HPLC system should be prepared according to the specific chromatographic separation procedure

10.2.3 Procedure

1. Resuspend the trap material.
2. Filter an aliquot of trap samples through a Whatman GF/F membrane (0.45 μm nominal pore size).
3. Transfer the filter into a 10–15 mL glass tube.
4. Add 2–3 mL of 90% acetone.

5. Homogenize the filter [NOTE 1].
6. Incubate the sample in the dark at 4°C for 12 h.
7. Centrifuge the samples at 3000 × g for 10 min.
8. Filter the supernatant through a 0.5 μm Millex-SR filter.
9. Inject an aliquot of the supernatant into the HPLC system [NOTE 2].

The flux of any identified pigment in the trap sample is estimated using the following formula:

$$F_i = C_f\ (\mu g\ mL^{-1}) \cdot V_s \cdot F_s^{-1} \cdot T_s^{-1} \cdot S_c^{-1} \tag{10.1}$$

where

F_i = the flux of the *i*-compound.
C_f = the concentration ($\mu g\ mL^{-1}$) of the *i*-compound in the analytical subsample.
Vs = the total volume of the aliquot (mL).
Fs = the fraction of the total volume collected by the trap.
T_d = the time of trap deployment (days).
S_c = the collecting surface of the trap (m^2).

10.3 NOTES

1. Filters can be homogenized by a potter (for small volume samples) at 2500 rpm for 3 min (on ice) or alternatively by a glass stick (till the filters are completely destroyed).
2. Different HPLC systems can be used for the determination of phytopigments in trap samples. In the basic configuration, a reversed-phase chromatographic C18 column (25 cm × 4.6 mm ID, 5 μm particle size), a UV-Vis detector, pumps, injector, and software for the system management, data acquisition, and elaboration are required. Previous studies carried out on trap samples have utilized mobile-phase acetone:methanol [10:90 (V/V, solvent A)] and acetone:methanol [40:60 (V/V, solvent B)], pumped at a flow rate of 1 mL min^{-1} (Dell'Anno et al. 1999). In this case, pigments are eluted using solvent A for 10 min, followed by a linear gradient from A to B over 5 min, and an isocratic hold of B at 100% for 10 min. Pigments can be detected at 436 nm for chlorophylls and 405 nm for phaeophytins (Wright et al. 1991) and can be identified by comparing their retention times with those of standard solutions of commercial chlorophyll-a and -b and phaeophytin-a and -b obtained by acidification (1 M HCl) of the corresponding chlorophylls (Mantoura and Llewellyn 1983). The identified pigments in the samples are then converted into concentrations from calibration curves obtained from the relationships between peak surface of standards and their relative concentrations.

Resuspend trap material

↓

Filter an aliquot of trap samples onto
Whatman GF/F membrane (0.45 μm nominal pore size)

↓

Transfer the filter into a 10–15-mL glass tube

↓

Add 2–3 mL of acetone (90%)

↓

Homogenize the filter

↓

Incubate in the dark (4°C, 12 h)

↓

Centrifuge (3000 × g, 10 min)

↓

Filter the supernatant (0.5 μm Millex-SR filters)

↓

Inject an aliquot of the supernatant into the HPLC system

PROTOCOL 10.1 Determination of phytopigments in trap samples.

REFERENCES

Dell'Anno, A., M. Fabiano, S. Bompadre, M. Armeni, L. Leone, and R. Danovaro. 1999. Phytopigment and DNA determinations in long-time formalin-preserved trap samples. *Marine Ecology Progress Series* 191: 71–77.

Mantoura, R. F. C. and C. A. Llewellyn. 1983. The rapid determination of algal chlorophyll and carotenoid pigments and their breakdown products in natural waters by reverse-phase high performance liquid chromatography. *Analytica Chimica Acta* 151: 297–314.

Wright, S. W., S. W. Jeffrey, R. F. C. Mantoura et al. 1991. Improved method for the analysis of chlorophylls and carotenoids from marine phytoplankton. *Marine Ecology Progress Series* 77: 183–196.

Section III

Deep-Sea Benthic Life

11 Viral Abundance

CONTENTS

11.1 INTRODUCTION

Viruses are by far the most abundant biological entities on the planet. The estimated overall abundance in the world's oceans is ca. 4×10^{30} (Suttle 2005), a value that exceeds bacterial abundance by one order of magnitude (Whitman et al. 1998; Suttle 2005). Every second, approximately 10^{23} viral infections occur in the ocean, being a major source of mortality of a wide range of organisms from prokaryotes to whales (Suttle 2005). As a result, viruses play critical roles in shaping aquatic communities and determining ecosystem dynamics (Fuhrman 1999; Suttle 2005).

Compared to virioplankton (i.e., viruses in aquatic systems), viruses in the benthos have received little attention (Danovaro and Serresi 2000; Mei and Danovaro 2004). Viral abundances in surface and subsurface sediments exceed those in the water column (10^5 and 10^8 particles mL^{-1}) by 10–1000 times (Danovaro and Serresi 2000; Danovaro et al. 2002; Corinaldesi and Danovaro 2003; Maranger and Bird 1996; Paul et al. 1993). The highest viral abundances have been reported for nutrient-rich surface coastal sediments (ca. fivefold higher than in deep-sea sediments) although high numbers (12.1 to 23.8×10^8 viruses g^{-1} dry sediment) have also been reported in deep-sea trenches (Danovaro et al. 2001) and even 100 m below the sediment surface (Bird et al. 2001).

Knowledge of virus abundances in marine sediments is important for understanding the viral global distribution but provides only the most basic information to assess their importance in marine ecosystems (Danovaro et al. 2008). In fact, accurate estimates of viral production and infection are also needed to assess viral dynamics and understand their role in benthic food webs and biogeochemical cycles.

Generally, approaches and procedures used for determining viriobenthos abundances are derived from those applied to water samples. However, these adaptations have often been difficult because the physical and chemical matrix that characterizes the sediments often interferes with analyses.

The traditional method for viral counting in environmental samples is transmission electron microscopy (TEM). However, viral counting by TEM is extremely laborious, even with water samples, and it is particularly difficult when applied to sediment samples (Bettarel et al. 2006). At present, epifluorescence microscopy (EFM) is the preferred method for counting viruses in marine samples because of its higher accuracy and precision (Hennes and Suttle 1995; Noble and Fuhrman 1998) than the TEM approach. EFM counting requires that viruses to be dislodged from sediments and stained with fluorochromes. Different stains (Yo-Pro-1, DAPI, SYBR Green I, SYBR Green

II, and SYBR Gold) have been used to enumerate viruses in sediment samples (Hennes and Suttle 1995; Weinbauer and Suttle 1997; Noble and Fuhrman 1998; Chen et al. 2001). At present, SYBR Green I–II and SYBR Gold are currently the most utilized stains (Noble and Fuhrman 1998; Chen et al. 2001; Middelboe et al. 2006). Here, we report the procedure for benthic viral counting based on EFM and the use of SYBR Gold as the staining fluorochrome.

11.2 FIELD PROCEDURES

11.2.1 Sampling

Surface sediment samples may be collected using multicorers or box corers, depending on the sampling depth. Since the vertical distribution of viruses in sediments could be strongly heterogeneous (Mei and Danovaro 2004; Corinaldesi et al. 2007), to provide a more realistic estimate of viral abundance and its vertical distribution, it is strongly recommended that the sediment layers are not mixed during sampling and recovery of samples. This implies that the sampling must be performed limiting sediment perturbation and maintaining an effective isolation of sediment samples from the surrounding waters. Slicing of sediment horizons must be performed using sterile tools (i.e., tips and spatulas). Finally, all of the materials and solutions used for the subsequent processing of sediment samples must be sterilized by autoclave in order to avoid any contamination.

11.2.2 Sample Storage

The standard method to preserve samples for viral counting is the addition of virus-free formaldehyde or glutaraldehyde (2% final concentration in seawater; Wommack and Colwell 2000). However, both preservatives have been repeatedly reported to affect viral abundance in water and sediment samples (Xenopoulos and Bird 1997; Danovaro et al. 2001; Wen et al. 2004). Significant reductions in virus abundance can already occur after 24 h from the sample fixation (Wen et al. 2004). After an initial decline for up to 1 week, counts remain relatively constant for up to 3 months (Danovaro et al. 2001), suggesting that the sample fixation is possible only if viruses are counted for comparative purposes. Conversely, accurate counts in absolute terms require either the application of proper correction factors for viral losses during sample storage or, preferably, immediate analysis of fresh samples. If benthic viruses cannot be counted immediately after sampling, to avoid significant losses of viruses, fresh samples can be filtered (after viral extraction from the sediment matrix) and the filters stored at −20°C. In this case, counts should be carried out as far as possible (within 30 d of storage). Alternatively, whole sediment cores may also be frozen until analysis (Helton et al. 2006).

11.3 LABORATORY PROCEDURES

11.3.1 Equipment and Instruments

Autoclave
Laminar air flow hood
Filtration apparatus: Sterile filter holder for 25 mm diameter filters with 15 mL funnel (Millipore), vacuum pump, and vacuum flask (for discharge)
Sonicator
Centrifuge
Epifluorescence microscopy

11.3.2 Materials

1. Sterile 15 mL tubes
2. Sterile 10 mL syringes

3. Virus-free seawater [NOTE 1]
4. Tetrasodium pyrophosphate solution (SIGMA)
5. 0.02 µm Pore size, 25 mm ø, Anodisc, Al_2O_3 filters (Whatman)
6. 0.02 µm Pore size, 25 mm ø disposable syringe filters (Anotop, Whatman)
7. Sterile Petri dishes
8. 2–20 µL, 100–1000 µL, and 1–5mL pipettes and relative sterile tips
9. Sterilized forceps
10. Glass slides and coverslips
11. DNase I from bovine pancreas
12. SYBR Green or Gold stain (Invitrogen) [NOTE 2]
13. Antifade solution

11.3.3 Solutions

Tetrasodium pyrophosphate solution: Prepare 100 mM tetrasodium pyrophosphate (SIGMA) in reagent-grade water prefiltered through 0.02 µm filters (Anotop, Whatman).

DNase I working solution: Dilute the stock solution to 100 U mL^{-1} final concentration. For example, if the initial concentration of the stock solution is 10 U μL^{-1}, the working solution can be prepared diluting 100 µL of the stock solution in 10 mL virus-free reagent-grade water (working solution) [NOTE 3].

TE buffer: 10 mM Tris-HCl, 1 mM EDTA in reagent-grade water prefiltered onto 0.02 µm filters (Anotop, Whatman), pH 7.4–7.6.

SYBR Gold working solution: The manufacturer's stock concentration of the SYBR Gold (Invitrogen) is in concentrated form (10,000×). An ideal working solution of SYBR Gold has to be diluted till to a 2× final concentration. Generally, the stock solution (10,000×) is diluted in two further steps: first, it is diluted 100-fold (100×) with 0.02 µm filtered TE buffer, then the solution obtained is again diluted to another 50-fold (2×). The working solution (2×) must be prepared fresh on the day of the sample filtration and slide preparation.

Antifade solution: 50% Glycerol, 50% phosphate-buffered saline (FLUKA, pH 7.4–7.6), and 0.5% ascorbic acid (weight/volume) [NOTE 4].

11.3.4 Procedure

1. Collect ca. 0.5 g of sediment in a sterile tube.
2. Add virus-free seawater for a total volume of 5 mL.
3. Add tetrasodium pyrophosphate at a final concentration of 5 mM (i.e., 250 µL of the 100 mM solution).
4. Incubate the samples in the dark and at room temperature for 15 min [NOTE 5].
5. Sonicate the samples for 3 min with 30 s intervals between each minute. During time intervals, samples are manually shaken [NOTE 6].
6. Centrifuge at 800 × g speed for 1 min [NOTE 7].
7. Dilute the supernatant with virus-free seawater from 10× to 1000× depending on the expected concentration of viruses. Generally, we make the last dilution (250×) in a final volume of 5 mL [NOTE 8].
8. Using sterile forceps, place 0.02 µm anodisc filters on the filter holders. Place the 15 mL funnel on the filter holder, taking care that filter holders and funnels are perfectly overlapped.
9. Transfer the diluted sample (250×) into the funnel, add DNase I to 2–5 UmL^{-1} final concentration (i.e., 100/250 µL of the 100 UmL^{-1} working solution in a 5 mL volume) and incubate for 15 min in the dark at room temperature.

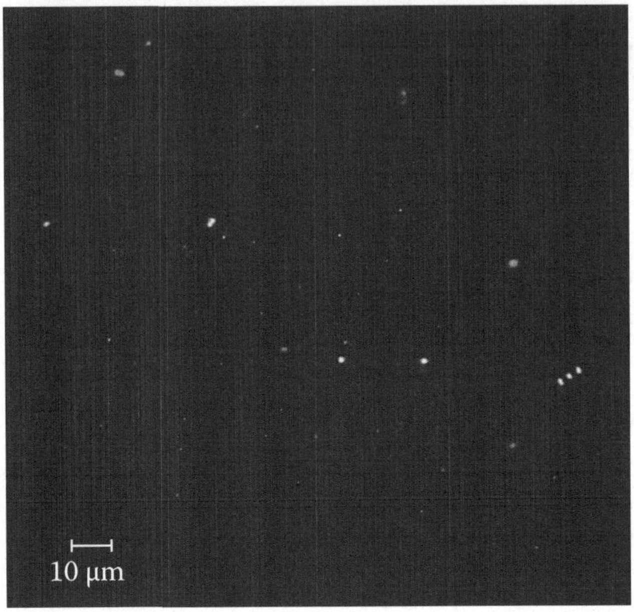

FIGURE 11.1 *(A color version of this figure follows page 234.)* A typical view of marine microbial community stained with SYBR.

10. Filter with vacuum pressure no higher than 20 kPa (i.e., 150 mmHg) to avoid destroying of viral particles.
11. When the sample has passed through, remove the filter and place it on a sterile Petri dish.
12. Stain with 100 μL of SYBR Gold (2×), distributing the solution equally on the filter's surface.
13. Incubate in the dark for 20 min.
14. Wash the filter 3 times with 3 mL of virus-free seawater.
15. Place a drop (20 μL) of antifade solution onto a clean slide, and place the filter on the drop. Spot 20 μL antifade solution onto a clean coverslip, flip it over to cover the sample filter. Avoid the formation of bubbles.
16. Add one drop of immersion oil to the top of the coverslip and examine the slide under blue-green light excitation (Fitch filter, excitation at 480–495 nm) using an epifluorescence microscope (1000× magnification).
17. Count at least 10 optical fields or 200 viruses (Figure 11.1).
18. Slices can be preserved up to 3 months at –20°C without sensible viral loss.
19. After using, filter holders and funnels should be washed with 1 N HCl.

Viral abundance calculations:

$$\text{Viral abundance (viruses g}^{-1}) = [\text{Average virus number for each optical field}] \cdot \frac{\text{Optical field coefficient} \cdot \text{extraction coefficient} \cdot \text{dilution factor}}{\text{sediment dry weight}} \tag{11.1}$$

where

Optical field coefficient = (filtration area/counting area)

Extraction coefficiency = 1.66 [NOTE 9]

Sediment dry weight: sediment weight after desiccation (60°C at least for 24 h)

11.4 NOTES

1. Virus-free seawater (i.e., autoclaved and filtered through Anotop 0.02 μm filters) should be prepared well before the beginning of the sample processing.
2. SYBR Gold is a sensitive fluorescence stain to detect both double- and single-stranded DNA and RNA, whereas SYBR Green I yields greatest absorbance with double-stranded DNA (Molecular Probes). The latter can also be used to detect single-stranded DNA and RNA, but the sensitivity is lower than for double-stranded DNA (Fischer et al. 2005). Conversely, SYBR Green II gives brightest fluorescence with RNA and single-stranded DNA (Molecular Probes). SYBR Gold is becoming the most common fluorochrome to enumerate viruses in sediments. It typically yields a bright and stable yellow-green fluorescence under EFM although the brightness can vary due to different genome sizes among viruses (Chen et al. 2001). In comparison with SYBR Green I, the fluorescence of SYBR Gold-stained viruses is typically brighter, whereas background noise is sufficiently reduced, thus facilitating virus counting (Fischer et al. 2005). It is preferable to use freshly prepared working solutions because these fluorochromes are unstable at low concentration.
3. Extracellular DNA could interfere with viral counts. This effect can be circumvented by treating sediment samples with DNases (Danovaro et al. 2001). However, we recommend to add low concentrations of DNases (2–5 UmL^{-1}) as high concentration could damage virus particles.
4. We recommend to add glycerol after PBS in order to avoid difficulties in the mixing of the reagents.
5. Incubations in the dark are performed in order to limit viral decay due to UV light.
6. Do not increase the sonication time because this could determine the bursting of prokaryotic cells and interfere with virus counting.
7. This step could be omitted when sediments characterized by coarse grain size are analyzed.
8. Generally, the sediment dilution ranges from 500× to 1000× for coastal and nutrient-rich samples, whereas from 100× to 250× for deep-sea samples.
9. Empirically determined by testing different extraction protocols (Danovaro et al. 2001).

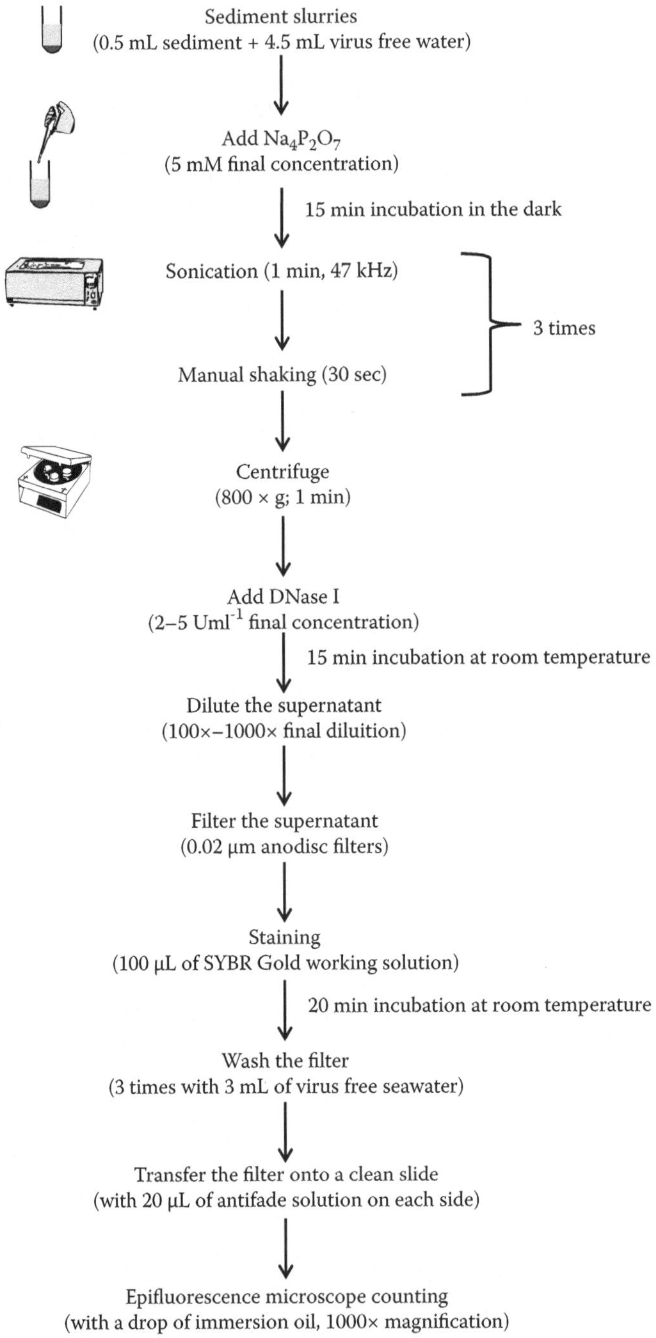

PROTOCOL 11.1 Viral counting in marine sediments.

REFERENCES

Bettarel, Y., M. Bouvy, C. Dumont, and T. Sime-Ngando. 2006. Virus–bacterium interactions in water and sediment of West African inland aquatic systems. *Applied and Environmental Microbiology* 72: 5274–5282.

Bird, D. F., S. K. Juniper, M. Ricciardi-Rigault, P. Martineu, Y. T. Praire, and S. E. Calvert. 2001. Subsurface viruses and bacteria in Holocene/Late Pleistocene sediments of Saanich Inlet, BC: ODP holes 1033b and 1034b, Leg 169s. *Marine Geology* 174: 227–239.

Chen, F., J. R. Lu, B. Binder, and R. E. Hodson. 2001. Enumeration of viruses in aquatic environments using SYBR Glod stain: Application of digital image analysis and flow cytometer. *Applied and Environmental Microbiology* 67: 539–545.

Corinaldesi, C. and R. Danovaro. 2003. Sunscreen increase virus production through prophage induction in marine bacterioplankton. *Microbial Ecology* 45: 109–118.

Corinaldesi, C., A. Dell'Anno, and R. Danovaro. 2007. Viral infection plays a key role in extracellular DNA dynamics in marine anoxic systems. *Limnology and Oceanography* 52: 508–516.

Danovaro, R., A. Dell'Anno, A. Trucco, and S. Vannucci. 2001. Determination of virus abundance in marine sediments. *Applied and Environmental Microbiology* 67: 1384–1387.

Danovaro, R. and M. Serresi. 2000. Viral abundance and virus-to-bacterium ratio in deep-sea sediments of the Eastern Mediterranean. *Applied and Environmental Microbiology* 66: 1857–1861.

Danovaro, R., E. Manini, and A. Dell'Anno. 2002. Higher abundance of bacteria than viruses in deep Mediterranean sediments. *Applied and Environmental Microbiology* 68: 1468–1472.

Danovaro, R., C. Corinaldesi, M. Filippini et al. 2008. Viriobenthos in freshwater and marine sediments: A review. *Freshwater Biology* 53 (6): 1186–1213

Fischer, U. R., A. K. T. Kirschner, and B. Velimirov. 2005. Optimization of extraction and estimation of viruses in silty freshwater sediments. *Aquatic Microbial Ecology* 40: 207–216.

Fuhrman, J. A. 1999. Marine viruses and their biogeochemical and ecological effects. *Nature* 399: 541–548.

Helton, R. R., L. Liu, and K. E. Wommack. 2006. Assessment of factors influencing direct enumeration of viruses within estuarine sediments. *Applied and Environmental Microbiology* 72: 4767–4774.

Hennes, J. E. and C. A. Suttle. 1995. Direct counts of viruses in natural waters and laboratory cultures by epifluorescence microscopy. *Limnology and Oceanography* 40: 1050–1055.

Maranger, R. and D. F. Bird. 1996. High concentrations of viruses in the sediments of Lac Gilbert, Quebec. *Microbial Ecology* 31: 141–151.

Mei, M. L. and R. Danovaro. 2004. Virus production and life strategies in aquatic sediments. *Limnology and Oceanography* 49: 459–470.

Middelboe, M., R. N. Glud, F. Wenzhöfer, K. Ogurib, and H. Kitazato. 2006. Spatial distribution and activity of viruses in the deep-sea sediments of Sagami Bay, Japan. *Deep-Sea Research Part I* 53: 1–13.

Noble, R. T. and J. A. Fuhrman. 1998. Use of SYBR Green I for rapid epifluorescence counts of marine viruses and bacteria. *Aquatic Microbial Ecology* 14: 113–118.

Paul, J. H., J. B. Rose, S. C. Jiang, C. A. Kellogg, and L. Dickson. 1993. Distribution of viral abundance in the reef environment of Key Largo, Florida. *Applied and Environmental Microbiology* 59: 718–724.

Suttle, C. A. 2005. Viruses in the sea. *Nature* 437: 356–361.

Wen, K., A. Ortmann, and C. A. Suttle. 2004. Accurate estimation of viral abundance by epifluorescence microscopy. *Applied and Environmental Microbiology* 70: 3862–3867.

Weinbauer, M. G. and C. A. Suttle. 1997. Comparison of epifluorescence and transmission electron microscopy for counting viruses in natural marine waters. *Aquatic Microbial Ecology* 13: 225–232.

Weinbauer, M. G. 2004. Ecology of prokaryotic viruses. *FEMS Microbiology Reviews* 28: 127–81

Whitman, W. B., D. C. Coleman, and W. J. Wiebe. 1998. Prokaryotics: The unseen majority. *Proceedings of the National Academy of Sciences of United States of America* 95: 6578–6583.

Wommack, K. E. and R. R. Colwell. 2000. Virioplankton: Viruses in aquatic ecosystems. *Microbiology and Molecular Biology Reviews* 64: 69–114.

Xenopoulos, M. A. and D. F. Bird. 1997. Virus à la sauce Yo-Pro: Microwave enhanced staining for counting viruses by epifluorescence microscopy. *Limnology and Oceanography* 42: 1648–1650.

12 Prokaryotic Abundance

CONTENTS

12.1 INTRODUCTION

Enumeration of prokaryotic cells is the most basic task in aquatic microbial ecology (Karner and Fuhrman 1997). Heterotrophic and autotrophic prokaryotes are known to play a major role in channeling the flux of energy in aquatic environments (Azam et al. 1983; Zimmermann et al. 1978). It is therefore a major aim of microbial ecologists to develop and improve methods to estimate their abundance (Fischer and Velimirov 2000). For three decades, the standard method used for total prokaryote enumeration is direct counting, using epifluorescence microscopy (EFM), after staining cells with fluorochromes. EFM counting allows more accurate and precise estimates compared to the CFU (colony-forming units) counts used earlier. Plate counts were supposed to show the numbers of culturable and, by extrapolation, active bacteria, whereas total EFM counts, showing all bacterium-shaped particles, were usually higher by 2 to 3 orders of magnitude than plate counts (Lee and Fuhrman 1991). This, as well as theoretical considerations (Pomeroy 1974), led to the view that most marine bacteria *in situ* must be viable but nonculturable under the conditions used. Total-count methods (Hobbie et al. 1977; Porter and Feig 1980) became the standard for bacterial enumeration. Different stains (Yo-Pro-1, DAPI, Acridine Orange, SYBR Green I, SYBR Green II, and SYBR Gold) have been used to enumerate prokaryotes in sediment samples. To date, Acridine Orange and SYBR Green I are the most-utilized fluorochromes in deep-sea sediments (Danovaro et al. 1994, 2002; Kepner and Pratt 1994; Luna et al. 2002; Manini et al. 2003). Generally, approaches and procedures used for determining prokaryotic abundances derived from those applied to water samples. However, this adaptation has often been difficult because the physical and chemical matrix that characterizes sediments and other benthic environments often interferes with analyses. Therefore, EFM counting requires that prokaryotes are first dislodged from sediments, and then stained and counted.

Here, we report the procedure for benthic prokaryotic enumeration based on Acridine Orange and SYBR Green I stains and counts under epifluorescence microscope.

12.2 FIELD PROCEDURES

12.2.1 Sampling

Surface sediment samples are actually collected using multiple corers or box corers. Immediately after retrieval, sediment subsamples are gently removed from the core by inserting a cutoff 5 mL sterile plastic syringe along the axis of the core. The syringe plunger is held fixed at the sediment surface, whereas the barrel is pushed into the sediment for 1 cm, a procedure that is analogous to piston coring. Sediment subsamples (ca. 0.5–1 mL) are then transferred into sterile tubes (15 mL) and fixed with filter-sterilized (0.2 μm) 2% formalin in seawater solution.

Since the vertical distribution of prokaryotes in sediments could be strongly heterogeneous, in order to provide a more realistic estimate of prokaryotic abundance, it is strongly recommended that the sediment layers are not mixed during sampling and recovery of samples. This implies that sampling must be performed for limiting sediment perturbation and maintaining an effective isolation of sediment samples from the surrounding waters. Prior to analyses, all material and solutions used for the subsequent processing of sediment samples should be sterilized by autoclave in order to avoid any contamination.

12.2.2 Sample Storage

Sediment sample are fixed in formalin (2% final concentration; Luna et al. 2002) and frozen or stored at 4°C.

12.3 LABORATORY PROCEDURES: SYBR GREEN I STAINING

12.3.1 Equipment and Instruments

Autoclave
Laminar air flow
Filtration apparatus: Sterile filter holder for 25 mm diameter filters with 15 mL funnel (Millipore), vacuum pump, and vacuum flask (for discharge)
Sonicator
Centrifuge
Epifluorescence microscopy

12.3.2 Materials

Sterile 15 mL tubes
Sterile 5 mL syringes
Sterile seawater [NOTE 1]
Tetrasodium pyrophosphate solution (SIGMA)
Anodisc filters, 0.2 μm pore size, 25 mm ø (Whatman)
Syringe filters, 0.2 μm pore size (Whatman Puradisc AS 25 mm ø)
Sterile Petri dishes
2–20 μL, 100–1000 μL, and 1–5 mL pipettes and sterile tips
Sterilized forceps
Glass slides and coverslips
SYBR Green I stain (Invitrogen) [NOTE 2]

Formalin
Antifade solution

12.3.3 Solutions

Buffered formalin solution is prepared by mixing pure formalin in artificial seawater, sterilized and prefiltered onto 0.2 μm filters (Whatman Puradisc) 2% [v/v] final concentration, then borax (sodium tetraborate, SIGMA; 20 gL^{-1}, pH 8.8 to 9.0) is added. The borax concentration is calculated on pure formalin volume.

Tetrasodium pyrophosphate solution: Stock solution is prepared by dissolving 500 mM tetrasodium pyrophosphate (SIGMA) in reagent-grade water (prefiltered onto 0.2 μm filters, Whatman Puradisc). The working solution is 5 mM.

SYBR Green I solution: SYBR Green I (Invitrogen) in concentrated form (10,000×) is dissolved in dimethyl sulfide (DMS) and should be preserved in the dark at −20°C. An ideal working solution of SYBR Green I has to be diluted 20-fold, in order to obtain 500× final concentration. The stock solution (10,000×) must be defrozen and diluted with sterile reagent-grade water in order to avoid contaminations. Since the fluorochrome and the DMS have different cryoscopic points, be sure that both are in the liquid state. The working solution (500×) can be stored in the dark at 4°C, and it will remain stable for a few weeks.

Antifade solution: It is prepared by mixing 50% glycerol, 50% phosphate-buffered saline (FLUKA, pH 7.4–7.6), and 0.5% ascorbic acid (weight/volume) [NOTE 3].

Immersion oil for microscopy.

12.3.4 Procedure

1. Collect ca. 0.5–1 mL of sediment in a sterile tube.
2. Add sterile formalin solution (2% final concentration) for a final volume of 5 mL [NOTE 4].
3. Add tetrasodium pyrophosphate at a final concentration of 5 mM (i.e., 50 μL of the 500 mM solution).
4. Incubate the samples in the dark and at room temperature for 15 min.
5. Sonicate the samples for 3 min with 30 s intervals between each minute. During intervals, samples are manually shaken [NOTE 5].
6. Wait until the suspended sediment is settled at the bottom of the tube. At the end of this step, a clear supernatant should be recognized [NOTE 6].
7. Dilute the supernatant with sterile seawater first at 10×. Then, depending on the expected abundance of prokaryotes, dilute the samples from 100× to 1000× [NOTE 7].
8. Using forceps, place 0.2 μm anodisc filters on the filter holders. Place the funnel on the filter holder taking care that filter holders and funnels are perfectly overlapped.
9. Transfer the supernatant into the funnel and filter with vacuum pressure no greater than 20 kPa (or 150 mmHg) to avoid destruction of cells.
10. When the sample has passed through, remove the filter and place it on a sterile Petri dish.
11. Stain with 20 μL of SYBR Green I working solution [NOTE 8].
12. Incubate in the dark for 20 min.
13. Wash the filter three times with 3 mL of sterile seawater.
14. Place a drop (20 μL) of antifade solution onto a clean slide and mount the filter on the drop. Spot 20 μL antifade solution onto a clean coverslip, flip it over to cover the sample filter. Try to avoid the formation of bubbles.
15. Add one drop of immersion oil to the top of the coverslip and examine the slide under blue-green light excitation (Fitch filter, excitation at 450–490 nm) using an epifluorescence microscope (1000× magnification).
16. Count at least 10–20 optical fields or 200–400 cells (Figure 12.1).

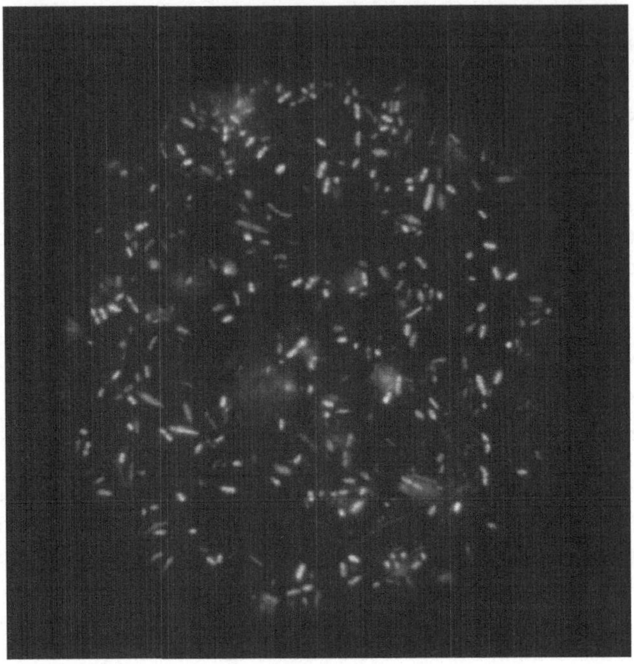

FIGURE 12.1 *(A color version of this figure follows page 234.)* Visualization of prokaryote cells with SYBR Green I staining. Prokaryotic cells appear green/yellow.

17. Slides can be preserved up to 3 months at −20°C.
18. After using, filter holders and funnels should be washed with 1 N HCl.
19. Discard the liquid from the sample tube by centrifuging at 800 × g for 1 min.
20. Place the sediment in an oven at 60°C for 24–48 h, until it is dry.
21. Weigh the dry sediment using an analytical balance.

12.4 LABORATORY PROCEDURES: ACRIDINE ORANGE (AO) STAINING

12.4.1 Equipment and Instruments

Autoclave
Laminar air flow
Filtration apparatus: Sterile filter holder for 25 mm diameter filters with 15 mL funnel (Millipore), vacuum pump, and vacuum flask (for discharge)
Sonicator
Centrifuge
Epifluorescence microscopy

12.4.2 Materials

Sterile 15 mL tubes
Sterile 5 mL syringes
Sterile seawater [NOTE 1]
Tetrasodium pyrophosphate solution (SIGMA)
Formalin (SIGMA)
3,6-bis[dimethylamino]acridinium chloride (Acridine Orange [AO])
Sterilized forceps

Glass slides and coverslips
0.2 μm filters for syringe Whatman Puradisc AS 25 mm ø
Polycarbonate filters black (Whatman) 0.2 μm pore size, 25 mm ø [NOTE 9]
2–20 μL, 100–1000 μL, and 1–5mL pipettes and sterile tips
10 mL sterile syringe

12.4.3 Solutions

Buffered formalin solution is prepared by mixing pure formaldehyde in artificial seawater, sterilized and prefiltered onto 0.2 μm filters (Whatman Puradisc) 2% v/v final concentration, then borax (sodium tetraborate, SIGMA; 20 gL^{-1} pH 8.8 to 9.0) is added. The borax concentration is calculated on pure formalin [NOTE 4].

Tetrasodium pyrophosphate solution: stock solution is prepared by dissolving 500 mM tetrasodium pyrophosphate (SIGMA) in reagent-grade water (prefiltered onto 0.2 μm filters, Whatman Puradisc). The working solution is 5 mM.

Acridine Orange (AO) 3,6-bis[dimethylamino]acridinium chloride working solution 0.025% [w/v] is prepared by dissolving AO stock in reagent-grade water. Working solution should be stored in the dark at 4°C till 3 weeks.

Immersion oil for microscopy.

12.4.4 Procedure

1. Collect ca. 0.5–1 g of sediment in a sterile tube.
2. Add sterilized formalin 2% [v/v], for a final volume of 5 mL. Three replicates are prepared from each sample to minimize count variance (Kirchman et al. 1982).
3. Add tetrasodium pyrophosphate at a final concentration of 5 mM (i.e., 50 μL of the 500 mM solution).
4. Incubate the samples in the dark and at room temperature for 15 min.
5. Sonicate the samples for 3 min with 30 s interval between each minute. During intervals, samples are manually shaken [NOTE 5].
6. Wait until the suspended sediment has settled at the bottom of the tube. At the end of this step, a clear supernatant should be recognized [NOTE 6].
7. Dilute the supernatant with sterile seawater first at 10×. Then, depending on the concentration of prokaryote cells, dilute the samples from 100× to 1000× [NOTE 7].
8. Stain the sample diluted with 0.5 mL of Acridine Orange working solution.
9. Samples are manually shaken for 30 s, and then incubate in the dark at room temperature for a total time of 5 min [NOTE 10].
10. Using forceps, place 0.2 μm black polycarbonate membrane filters (Whatman) on the filter holders. Place the funnel on the filter holder taking care that filter holders and funnels are perfectly overlapped.
11. Transfer the sample stained into the funnel and filter with vacuum pressure no greater than 20 kPa (or 150 mmHg) to avoid destroying of prokaryotic cells.
12. When the sample has passed through, wash the filter three times with 3 mL of reagent-grade water.
13. Remove the filter and place it on glass slides and under coverslips. Block the coverslip with small pieces of clear cellophane tape.
14. Add a reagent-grade water drop before examining the slide. [NOTE 11].
15. Add one drop of immersion oil to the top of the coverslip, and examine the slide under blue-green light excitation (excitation at 450–490 nm) using an epifluorescence microscope (1000× magnification).
16. Count at least 10–20 optical fields or 200–400 cells (Figure 12.2 A,B).

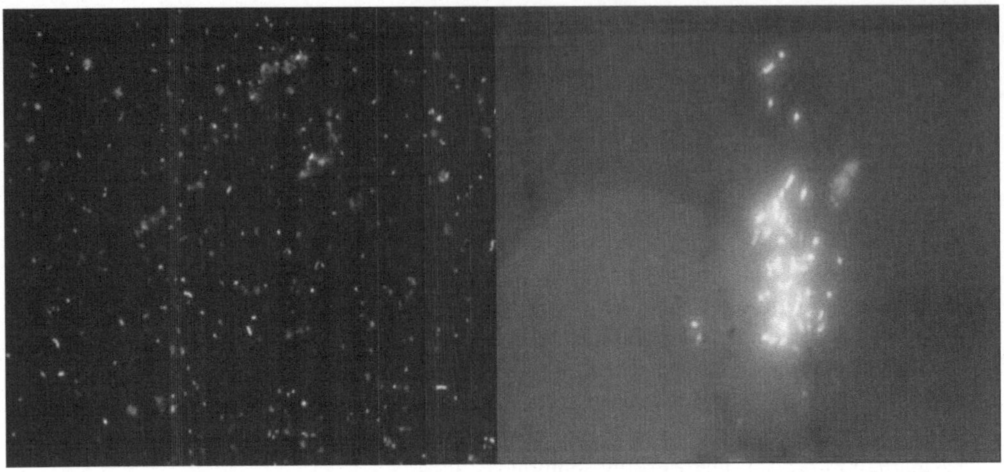

FIGURE 12.2 *(A color version of this figure follows page 234.)* Visualization of the prokaryote cells with AO staining. (Left) Sediment particles appear orange/red and prokaryotic cells appear green/yellow. Sometimes cells could appear orange/red, these should be counted anyway identified on the base of their prokaryotic shape and dimension. (Right) Visualization of the prokaryotic cells with a strong background. The background is red, and the cells show a strong color.

17. Slides can be preserved up to 3 months at −20°C.
18. After using, filter holders and funnels should be washed with 1 N HCl.
19. Discard the liquid from the sample tube by centrifuging at 800 × g for 1 min.
20. Place the tubes in an oven at 60°C for 24–48 h, until the content sediment weight is obtained.
21. Weigh the dry sediment using an analytical balance.

12.4.5 Calculations

Total prokaryotic numbers are calculated by using the following formula:

(Average cell number for each optical field) × [(Optical field coefficient × extraction coefficient × dilution factor of the sediment)/sediment dry weight]

where

Optical field coefficient = (filtration area/counting area).
Correction for the extraction coefficiency = 1.44.
Sediment dry weight-sediment weight after desiccation (60°C at least for 24 h).

12.5 NOTES

1. Sterile seawater (i.e., autoclaved and filtered through 0.2 μm pore size filters) should be prepared before the beginning of sample processing.
2. SYBR Green I, a sensitive fluorescence, yields greatest absorbance with double-stranded DNA. It typically yields a bright and stable yellow-green fluorescence under EFM, although the brightness can vary due to different genome sizes and sediment interference. It is recommended to use correctly preserved working solutions because, at room temperature and/or in the daylight, the fluorochromes rapidly degrade.
3. We recommend adding glycerol after PBS in order to avoid difficulties in the mixing of the reagents.

4. The formalin solution should be prepared with seawater, in order to avoid the osmotic cell burst.
5. Do not increase the sonication time because the consequent increase of temperature could alter the sample by determining the bursting of prokaryotic cells.
6. Deposition of sediment requires from 30 s up to a few minutes, depending on the grain size. The deposition of small-size sediments could require more time but no longer than 15 min, in order to avoid prokaryotic losses.
7. Generally, the sediment dilution ranges from 500× to 1000× for mud sediments of coastal and nutrient-rich samples, whereas it is from 100× to 250× for deep-sea samples and sand sediments of coastal samples.
8. With the dilution of the SYBR Green I stock, it is possible that a fraction of DMS will precipitate. Be sure not to transfer this precipitate on the filter during the staining step: It might interfere with EFM counting.
9. In order to have a better background between cells and sediment/organic matter, it is preferable to use black polycarbonate filters. Sediment and detritus particles will appear orange/red and prokaryotic cells will appear green/yellow. Orange/red particles with a well-defined border should be counted as well as prokaryotes.
10. Incubations in the dark are performed in order to limit prokaryotic cell decay due to light.
11. Attention must be given during the addition of the reagent-grade water to the slide microscope examination. Avoid excess of reagent-grade water that could dislodge cells and particles and alter the background.

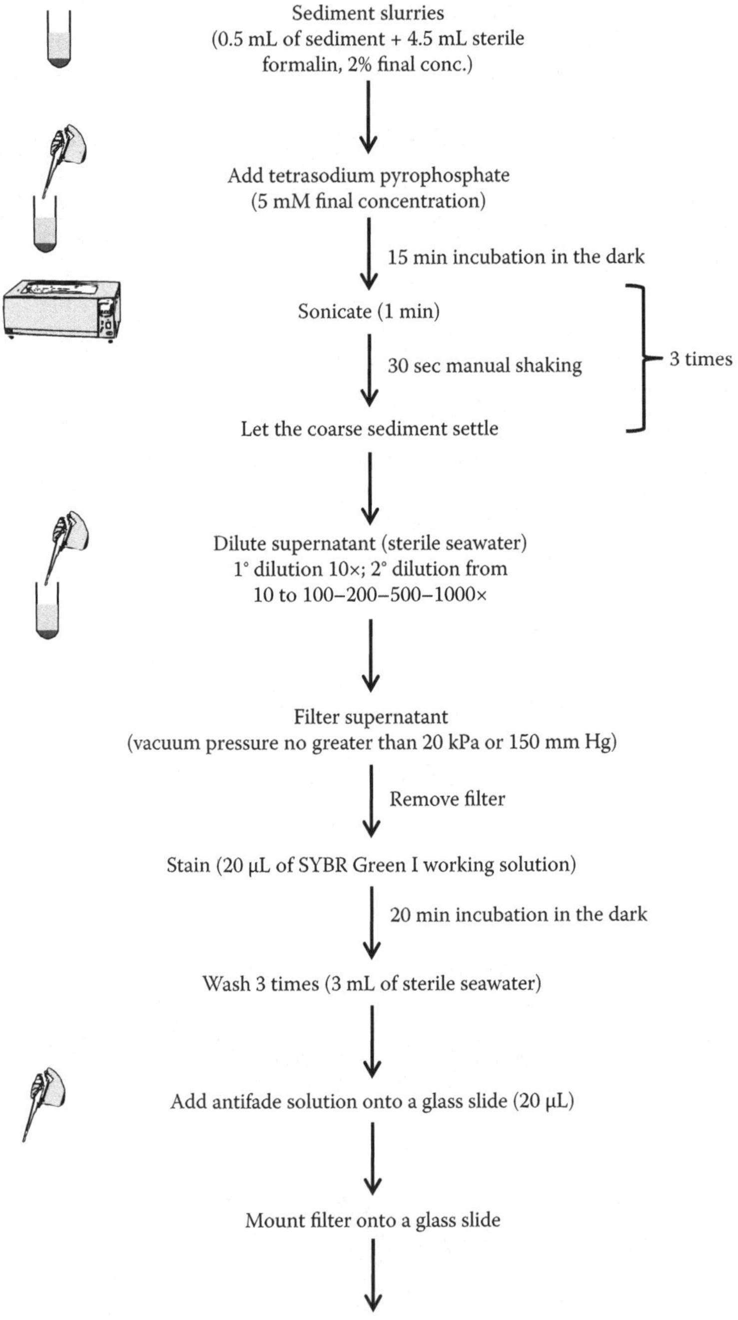

(a)

PROTOCOL 12.1 Prokaryote counting based on SYBR Green. *Continued*

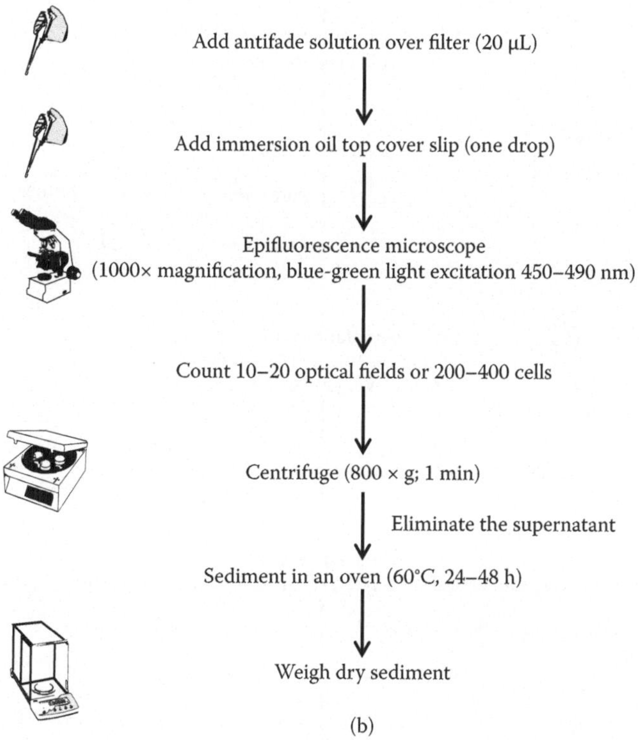

(b)

PROTOCOL 12.1 *Continued.*

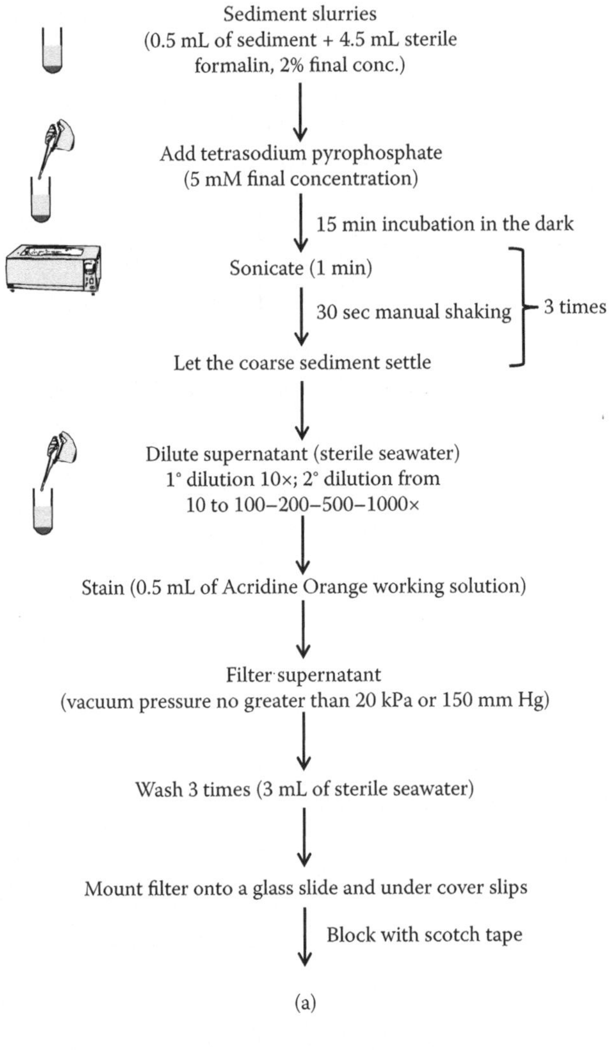

(a)

PROTOCOL 12.2 Prokaryote counting based on Acridine Orange. *Continued*

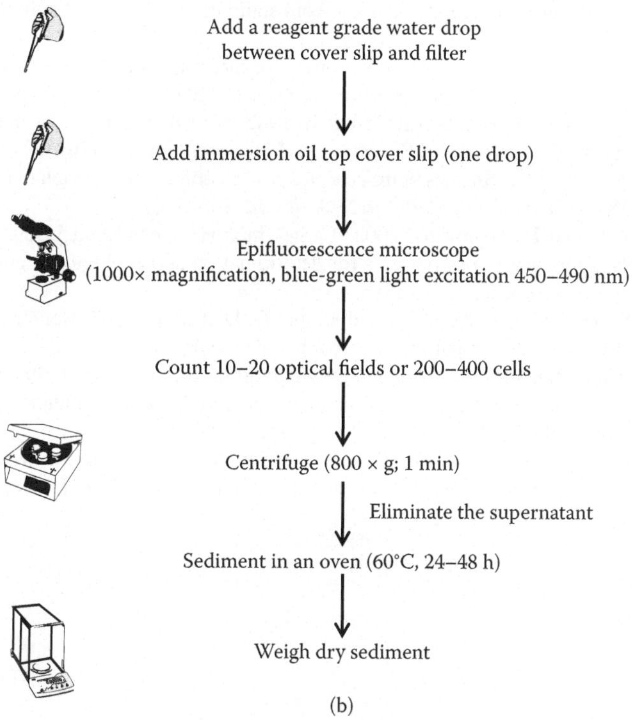

(b)

PROTOCOL 12.2 *Continued*

REFERENCES

Azam, F., T. Fenchel, J. G. Field, L. A. Meyer-Reil, and F. Thingstad. 1983. The ecological role of water-column microbes in the sea. *Marine Ecology Progress Series* 10: 257–263.

Danovaro, R., M. Fabiano, and M. Boyer. 1994. Seasonal changes of benthic bacteria in a seagrass bed (*Posidonia oceanica*) of the Ligurian Sea in relation to origin, composition, and fate of the sediment organic matter. *Marine Biology* 119: 489–500.

Danovaro, R., E. Manini, and A. Dell'Anno. 2002. Higher abundance of bacteria than of viruses in deep Mediterranean sediments. *Applied and Environmental Microbiology* 68: 1468–1472.

Fischer, U. R. and B. Velimirov. 2000. Comparative study of the abundance of various bacterial morphotypes in an eutrophic freshwater environment determined by AODC and TEM. *Journal of Microbiological Methods* 39: 213–224.

Hobbie, J. E., R. J. Daley, and S. Jasper. 1977. Use of Nuclepore filters for counting bacteria by epifluorescence microscopy. *Applied and Environmental Microbiology* 33: 1225–1228.

Karner, M. and J. A. Fuhrman. 1997. Determination of active marine bacterioplankton: A comparison of universal 16S rRNA probes, autoradiography, and nucleoid staining. *Applied and Environmental Microbiology* 63: 1208–1213.

Kepner, R. L. and J. R. Pratt. 1994. Use of fluorochromes for direct enumeration of total bacteria in environmental samples: Past and present. *Microbiological Reviews* 58: 603–615.

Kirchman, D., J. Sigda, R. Kapuscinski, and R. Mitchell. 1982. Statistical analysis of the direct count method for enumerating bacteria. *Applied and Environmental Microbiology* 44: 376–382.

Lee, S. H. and J. A. Fuhrman. 1991. Species composition shift of confined bacterioplankton studied at the level of community DNA. *Marine Ecology Progress Series* 79: 195–201.

Luna, G. M., E. Manini, and R. Danovaro. 2002. Large fraction of dead and inactive bacteria in coastal marine sediments: Comparison of protocols for determination and ecological significance. *Applied and Environmental Microbiology* 68: 3509–3513.

Manini, E., C. Fiordelmondo, C. Gambi, A. Pusceddu, and R. Danovaro. 2003. Benthic microbial loop functioning in coastal lagoons: A comparative approach. *Oceanologica Acta* 26: 27–38.

Pomeroy, L. R. 1974. The ocean's food web, a changing paradigm. *Bio-Science* 24: 499–504.

Porter, K. and Y. S. Feig. 1980. The use of DAPI for identifying and counting aquatic microflora. *Limnology and Oceanography* 25: 943–948.

Zimmermann, R., R. Iturriaga, and J. Becker-Birck. 1978. Simultaneous determination of the total number of aquatic bacteria and the number thereof involved in respiration. *Applied and Environmental Microbiology* 36: 926–935.

13 Fluorescence *In Situ* Hybridization and Catalyzed Reporter Deposition for Benthic Prokaryote Assemblage Structure

CONTENTS

13.1 INTRODUCTION

Fluorescence *in situ* hybridization (FISH) of bacteria with rRNA-targeted probes allows the phylogenetic identification of prokaryotes in mixed assemblages without prior cultivation of Bacteria and Archaea (Giovannoni et al. 1988; DeLong et al. 1989; Amann et al. 1990a,b; Amann et al. 1995; Llobet-Brossa et al. 1998; Murray et al. 1998). Most bacteria in aquatic environments are small, slowly growing, or starving, and the signal intensities of hybridized bacterioplankton cells were frequently below the detection limit. During the last few years, numerous efforts have been made to increase the sensitivity of FISH (Glökner et al. 1996). Recently, a novel FISH protocol has been developed for the identification of small prokaryotic cells in both marine plankton and benthos (Pernthaler et al. 2002, 2004; Teira et al. 2004; Ishii et al. 2004; Danovaro et al. 2009). This approach is based on the use of horseradish peroxidase (HRP)-labeled oligonucleotide probes and tyramide signal amplification, also termed catalyzed reporter deposition (CARD). CARD is based on the deposition of a large number of labeled tyramide molecules by peroxidase activity. Each HRP conferred by a probe catalyzes the deposition of many labeled tyramides. This technique allows obtaining an enhanced FISH sensitivity when compared to probes carrying a single fluorochrome (Schönhuber et al. 1997; Amann and Fuchs 2008). CARD–FISH is a multiple-step procedure that includes (1) preparation of the biological material, (2) selection of the nucleic acid probe, (3) hybridization with a labeled probe, (4) detection of a cytochemical probe, and (5) microscopical analysis. The preparation of the specimen for CARD–FISH involves fixation to obtain the maximum retention of target nucleic acid sequences and morphology. Usually, the cells are concentrated and/or immobilized. Seawater and sediment samples can be filtered onto polycarbonate membrane filters, then immobilized with low-gelling-point agarose (Teira et al. 2004). Depending on the cell walls of the target organisms and the fixation, different permeabilization strategies can be applied, such as treatment with enzymes, diluted HCl, or detergents, thereby facilitating access of the labeled nucleic acid probes and cytochemical detection molecules. Specificity of probe binding to the target site depends on the hybridization and washing conditions. Hybridization probes are added to a defined, stringency-determining buffer at saturation concentrations (50 ng μL^{-1}) to maximize probe binding.

13.2 FIELD PROCEDURES

Sediment sampling can be carried out as described previously for deep-sea prokaryotes. Three independent replicates must be collected according to a randomized, stratified sampling strategy. Immediately after sediment collection, tubes can be carried onboard and processed under sterile conditions to avoid contamination. Sediment cores are sliced into different sections and stored in sterile Falcon test tubes (50 mL), and immediately frozen at −20°C (without any fixation) until processing.

13.3 LABORATORY PROCEDURES

13.3.1 Equipment and Instruments

13.3.1.1 General

Laminar flow hood
Dry-type incubator or hybridization oven
Water bath
Freezer
Fridge
Vortex
High-quality epifluorescence microscope equipped with filter set for the fluorochrome utilized

13.3.1.2 Fixation and Preparation of Sediment Samples

Centrifuge for 15 mL tubes
Vacuum pump
Microwave oven
Sonicator
15 mL screw-top microfuge tubes
15 mL microfuge tubes
Plastic Petri dishes (diameter 5 cm)
White polycarbonate membrane filters (diameter 25 mm; pore size 0.2 μm)
Cellulose nitrate support filters (diameter 25 mm; pore size 0.45 μm)
Filter tower for 25 mm membrane filters
Reagent-grade water
PBS (145 mM NaCl; 1.4 mM NaH_2PO_4; 8 mM Na_2HPO_4; pH 7.4)
Ethanol (96% [v/v])
2% (v/v) formalin (40% formaldehyde solution)
Low-gelling-point agarose (0.1% to 0.2% [w/v] in reagent-grade water)

13.3.1.3 Permeabilization and Inactivation of Peroxidase

Plastic Petri dishes
Microscopic slides + coverslips
Reagent-grade water
Lysozyme (Sigma)
Proteinase K (Fluka)
Stock solution: 1 M Tris-HCl, pH 7.5
Stock solution: 0.5 M EDTA, pH 8.0
Stock solution: 0.1 M HCl

13.3.1.4 Hybridization with Horseradish Peroxidase (HRP)-Labeled Probes and Washing

Razor blades
Blotting paper
Plastic Petri dishes
50 mL polyethylene tubes and rack
Glass beaker (50 mL)
HRP-labeled oligonucleotide probe [50 ng μL^{-1}] (www.biomers.net) (see Table 13.1)
PBS containing 0.05% Triton X-100

13.3.2 Solutions

13.3.2.1 Stock Solutions: Hybridization Buffer

20% [w/v] Sodium dodecyl sulfate (SDS)
1 M Tris-HCl, pH 7.5
5 M NaCl
20% [w/v] SDS
Dextran sulfate (Fluka)
10% [w/v] blocking reagent (PerkinElmer)
Formamide (ultra for molecular biology (Fluka) (see Tables 13.1 and 13.2)

13.3.2.2 Stock Solutions: Washing Buffer

0.5 M EDTA, pH 8.0
1 M Tris-HCl, pH 7.5

TABLE 13.1
List and Sequence of Probes Utilized for the Identification of Different Prokaryotic Groups

Probe	Sequence 5′—>3′	FAa (%)	Thb (°C)	Target Group	Reference
EUB338-I	GCTGCCTCCCGTAGGAGT	55	35	Most bacteria	Amann et al. 1990a
EUB338-II	GCAGCCACCCGTAGGTGT	55	35	Planctomycetales	Daims et al. 1999
EUB338-III	GCTGCCACCCGTAGGTGT	55	35	Verrucomicrobiales	Daims et al. 1999
NON338	ACTCCTACGGGAGGCAGC	55	35	Control probe	Amann et al. 1995
CF319a	TGGTCCGTGTCTCAGTAC	55	35	Cytophaga/Flavobacterium/ Bacteroidetes	Manz et al. 1996
BET42a	GCCTTCCCACTTCGTTT	55	35	Betaproteobacteria	Manz et al. 1992
GAM42a	GCCTTCCCACATCGTTT	55	35	Gammaproteobacteria	Manz et al. 1992
ALF968	GGTAAGGTTCTGCGCGTT	60	35	Alphaproteobacteria	Manz et al. 1992
SRB385	CGGCGTCGCTGCGTCAGG	55	35	Sulfate reducers	Manini et al. 2008
DSS658	TCCACTTCCCTCTCCCAT	70	35	Sulfate reducers	Manz et al. 1998
PLA886	GCCTTGCGACCATACTCCC	55	35	Planctomycetales	Neef et al. 1998
ARCH915	GTGCTCCCCCGCCAATTCCT	55	35	Archaea	Manini et al. 2008
EURY806	CACAGCGTTTACACCTAG	20	35	Euryarchaea	Teira et al. 2004
CREN537	TGACCACTTGAGGTGGTGCTG	20	35	Crenarchaea	Teira et al. 2004

TABLE 13.2
Volumes of Formamide and Water (reagent grade) for 20 mL Hybridization Buffer

% Formamide in Hybridization Buffer	Water (mL)	Formamide (mL)
20	10	4
25	9	5
30	8	6
35	7	7
40	6	8
45	5	9
50	4	10
55	3	11
60	2	12
65	1	13
70	0	14

Source: Data from A. Pernthaler and J. Pernthaler (2005). Simultaneous fluorescence *in situ* hybridization of mRNA and rRNA for the detection of gene expression in environmental microbes, *Methods in Enzymology*, 397: 352–371.

5 M NaCl
20% [w/v] SDS

Tyramide signal amplification (TSA) and microscopic evaluation

TSA™ Fluorescence system (containing tyramide-Cy3 and amplification diluent; PerkinElmer)
H_2O_2, 30 wt.% sol. in water (Sigma)
Glycerol 99.5+% (Sigma)
L-ascorbic acid (Sigma)
Immersion oil for microscopic evaluation

13.3.3 Preparation of Solutions

13.3.3.1 Preparation of Hybridization Buffer (Final Volume 20 mL)—Working Solution

3.6 mL 5 M NaCl (final concentration 0.9 M NaCl)
0.4 mL 1 M Tris-HCl (final concentration 20 mM Tris-HCl)
20 μL SDS (final concentration 0.02% SDS)
x mL reagent-grade water (see Table 13.1)
x mL formamide (see Table 13.2)
2.0 mL blocking reagent (final concentration 1% blocking reagent)
2.0 g dextran sulfate (final concentration 10% dextran sulfate)

Add and mix by stirring in a glass beaker (50 mL) dextran sulfate, NaCl, SDS, and Tris-HCl at 60°C. Stir until the dextran sulfate has dissolved completely. Cool in ice. Small portions of the buffer can then be stored at −20°C for 3 months.

13.3.3.2 Preparation of Washing Buffer (Final Volume 50 mL)

0.5 mL EDTA (final concentration 5 mM EDTA)
1.0 mL Tris-HCl (final concentration 20 mM Tris-HCl)
See Table 13.3 for (NaCl amount to be added)
Reagent-grade water (add to obtain a final volume of 50 mL)
25 μL SDS (final concentration 0.01 SDS)

The NaCl concentration in the washing buffer as well as the formamide concentration of the hybridization buffer determine the stringency of the hybridization at a certain temperature.

13.3.4 Procedure

13.3.4.1 Fixation and Preparation of Sediment Samples

1. Suspend 0.5 g of freshly collected sediment in 4.5 mL of buffered 2% formalin solution in a 15 mL screw-top microfuge tube. Wait for 1–2 h at room temperature or alternatively overnight at 4°C.
2. Centrifuge at 16,000 × g for 5 min, pour off supernatant.
3. Add 4.5 mL of PBS and resuspend the sample.
4. Centrifuge at 16,000 × g for 5 min, pour off supernatant.
5. Add 4.5 mL of a 1:1 mixture of PBS/ethanol 96% and store sample at −20°C until further processing.
6. Sediment sample is treated with ultrasound 3 times for 1 min with 30 s of interval after each treatment. Samples are also manually gently shaken. Perform ultrasound treatment in an ice bath (4°C).
7. After sonication, dilute the sediment samples by 100–500 times using PBS/ethanol 96%.

TABLE 13.3
Volumes of 5 M NaCl in 50 mL of Washing Buffer with Corresponding Formamide Concentration in the Hybridization Buffer. The Na^+ concentration is calculated for stringent washing at 37°C after hybridization at 35°C

% Formamide in Hybridization Buffer	5 M NaCl (µm)
20	1350
25	950
30	640
35	420
40	270
45	160
50	90
55	30
60	0
65	0
70	0

Source: Data from A. Pernthaler and J. Pernthaler. (2005) Simultaneous fluorescence *in situ* hybridization of mRNA and rRNA for the detection of gene expression in environmental microbes, *Methods in Enzymology*, 397: 352–371.

8. Place cellulose in nitrate support filters, beneath the membrane filters to improve the distribution of cells. Filter the diluted sediment sample onto 0.2 µm pore size Nucleopore® polycarbonate filters under low vacuum (<100 mmHg).
9. Air-dry filtered preparations and store in Petri dishes at –20°C until hybridization.

13.3.4.2 Attachment

1. Prepare 0.1% [w/v] low-gelling-point agarose in reagent-grade water.
2. Boil the agarose before each use in a microwave oven.
3. Fill the agarose in a Petri dish and let it cool down to 35°C.
4. Dip filters in 0.1% low-gelling-point agarose, place filters face up onto glass plates, and air-dry at 35°C for approximately 10 to 30 min.
5. Remove the filters from the glass plates and dehydrate them in ethanol 96% (1 min at room temperature).
6. Let the filters air dry on a paper tissue.

13.3.4.3 Inactivation of Endogenous Peroxidase

1. Incubate the filter section in 50 mL of 0.01 M HCl for 20 min at room temperature.
2. Wash filters in 1 × PBS, then in reagent-grade water, and in ethanol 96% (1 min at room temperature).
3. Filters can then be further processed or air-dried and stored in a refrigerator (–20°C).

Filters are cut into several sections (optional) according to the probes.

13.3.4.3.1 Permeabilization with Lysozyme for Bacteria

1. Prepare a fresh lysozyme solution (10 mg mL^{-1} [Fluka] in 0.05 M EDTA, pH 8.0; 0.1 M Tris-HCl, pH 7.5).
2. Incubate the filters in the lysozyme solution for 60 min at 37°C.
3. Wash the filters three times in reagent-grade water, then in ethanol 96% (1 min at room temperature). Let filters air dry.

13.3.4.3.2 Permeabilization with Proteinase K for Archaea

1. Prepare a fresh proteinase K solution (0.4 mU/mL [Sigma] in 0.05 M EDTA, pH 8.0; 0.1 M Tris-HCl, pH 7.5).
2. Incubate the filters in the proteinase K solution for 60 min at 37°C.
3. Wash the filters three times in reagent-grade water, then in ethanol 96% (1 min at room temperature). Let filters air dry.

After permeabilization, filters can be stored at −20°C up to several months.

13.3.4.4 Hybridization

1. Place filter sections in several Petri dishes according to the probes. Mix hybridization buffer and HRP-PROBE working solution (200:2) and add to filter sections.
2. Hybridize on a rotation shaker (10 rpm) in humidity chamber for at least 2 h (e.g., EUB338 and ARCH915 probes 2 h; CREN537 and EURY806 probes 8–12 h) at 35°C in accordance with probes utilized.

13.3.4.5 Washing

1. For stringent washing, prepare washing buffer (3 mM NaCl, 5 mM EDTA, pH 8; 20 mM Tris-HCl, pH 7.5; 0.01% [p/v] SDS) and preheat at 37°C.
2. Wash sections after hybridization for 10 min at 37°C.
3. Remove excess liquid with blotting paper but do not let filters run dry (this will reduce the activity of the HRP).
4. Equilibrate the probe-HRP. Incubate filters in PBS amended with 0.05% Triton X-100 for 15 min at room temperature.
5. Remove excess liquid with blotting paper but do not let filters dry.

13.3.4.6 Catalyzed Reporter Deposition

1. Place filter sections in a Petri dish and incubate filters in substrate mix (1 part Cy3-tyramide and 40 parts of amplification diluent, and 100 × H_2O_2 solution in a ratio of 100:1 [v/v]) for 10–30 min in the dark at 37°C).
2. Remove excess liquid by dabbing filters onto blotting paper. Wash sections in PBS amended with 0.05% Triton X-100 for 15 min at room temperature in the dark.
3. Wash sections in reagent-grade water (1 min at room temperature), then in 96% ethanol (1 min at room temperature). Let sections air dry.

13.3.4.7 Microscopic Analyses

Microscopy is performed after embedding in low-fluorescence glycerol mountant containing an antibleaching agent (50% PBS [v/v]; 50% glycerol, [v/v]; 0.5% l-ascorbic acid [w/v]). Sections can also be stored at −20°C until further processing.

Microscopic evaluation is obtained by epifluorescence with an appropriate filter set for Cy3 fluorescence (green excitation). The cells are visible in red (see Figure 13.1).

FIGURE 13.1 *(A color version of this figure follows page 234.)* Microscopic evaluation of the bacteria (probe: EUB338).

13.3.4.8 Total Number Cells

1. After microscopic evaluation of the hybridized cells, a filter section for each replicate is removed from a slide, washed with reagent-grade water for three times, and stained with SYBR GREEN I (add 20 μL of SYBR [diluted 1:20] to the filter section and incubate for 20 min at the dark and at room temperature) in a Petri dish.
2. Wash the filter section with reagent-grade water for three times and embed in antibleaching agent (50% PBS, [v/v]; 50% glycerol, [v/v]; 0.5% l-ascorbic acid, [w/v]).
3. Microscopic evaluation is obtained by epifluorescence with an appropriate filter set for SYBR GREEN I fluorescence (blue excitation). The cells are visible in yellow (see Figure 13.1).

13.4 POSSIBLE PROBLEMS FOR CARD–FISH PROCEDURE

Sometimes there can be high or uneven background during epifluorescence excitation, which lead to possible "false positive" signals of the cells (see Figure 13.2).

Prepare only small aliquots of probe working solutions. Use of a repeatedly frozen and thawed probe working solutions may cause the appearance of numerous brightly fluorescent particles at green excitation (Cy3). Hybridization signals become dim, and background is high. For HRP-labeled probes, do not freeze working solutions but store them at 4°C. Stock solution aliquots can be stored in a freezer.

Utilize only white polycarbonate filters. In our experience, black filters show high levels of background fluorescence at green excitation.

Possible causes for low cell-staining intensity: (1) low ribosome content of the target cells. Increase the tyramide concentration or the temperature during tyramide signal amplification (up to 46°C). A prolonged hybridization time (up to 15 h) may also help. (2) Check the probe. The probe should be thawed only once and should not be stored in the fridge for

FIGURE 13.2 ***(A color version of this figure follows page 234.)*** Example of a high background sample.

more than 6 months. (3) Check the pH of the PBS; it should be around 7.6. (4) The HRP-labeled probe cannot penetrate the cell wall. Modify the permeabilization protocol.

For any questions, we suggest a visit to the SILVA Web site: www.arb-slva.de; Section: FISH and Probes.

(1) Fixation and preparation of sediment samples

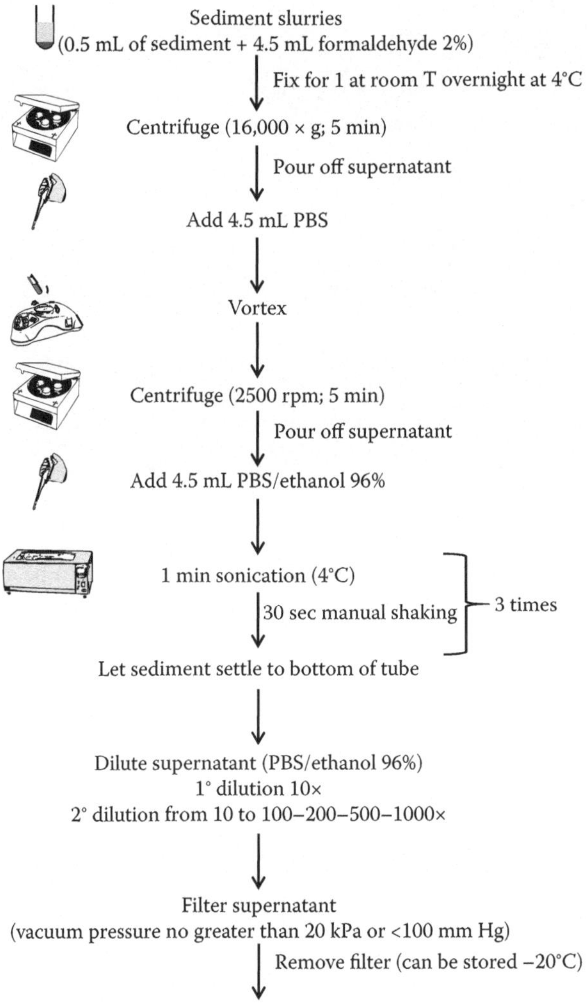

(a)

PROTOCOL 13.1 Determination of prokaryotic abundance by CARD–FISH. *Continued*

(2) Attachment

Boil filters face up onto gel
(0.1% w/v) in microwave oven

↓ Fill a Petri dish with agarose

Place filters face up on glass plates
and air dry (35°C; 10–30 min)

↓ Remove filter

Dehydrate filters (ethanol 96%)

↓ 1 min, room T, air dry on a paper tissue

(3) Inactivation of endogenous peroxidase

Incubate filter (0.01 M HCl)

↓ 20 min, room T

Wash filter
(1× PBS, reagent grade water and ethanol 96%)

↓ 1 min, room T, air dry on a paper tissue

Cut filters in sections

(4) Permeabilization for 60 minutes at 37°C

For bacteria
Incubate filter sections
(lysozyme 10 mg/mL)

For Archaea
Incubate filter sections
(proteinase k 0.4 mV/mL)

↓

Wash filters
(three times reagent grade water and ethanol 96%)

↓ 1 min, room T
Let filters air dry
Can be stored −20°C for several months

(5) Hybridization

Add to filter sections mix hybridization buffer and HRP-PROBE
(5ng/μL; dilution 200:2)

↓

Hybridize for at least 2 h at 35°C

↓

(b)

PROTOCOL 13.1 *Continued.*

(6) Washing

Remove excess liquid

↓

Wash in buffer (10 min, 37°C)

↓

Remove excess liquid

↓

Equilibrate probe-HRP
(PBS+0.05% Triton X-100; 15 min, in situ T)

↓

Remove excess liquid with blotting paper

(7) Catalyzed reporter deposition

Incubate filters in substrate mix (1:40)
(10 min in the dark, 37°C)

↓

Remove excess liquid

↓

Wash PBS and Triton (15 min)

↓

Wash in reagent grade water (1 min)

↓

Embed in antibleaching agent

↓

Wash in ethanol (1 min)

↓

Microscopy counting

(c)

PROTOCOL 13.1 *Continued.*

(8) Total number of cells

Filter section is removed from a slide

↓

Wash filter section with reagent grade water (3 times)

↓

Stain with SYBR GREEN I (20 μL)

↓ 20 min incubation in the dark, room T

Wash filter section with reagent grade water (3 times)

↓

Embed in antibleaching agent

↓

Microscopy counting

(d)

PROTOCOL 13.1 *Continued.*

REFERENCES

Amann, R. I., B. J. Binder, R. J. Olson, S. W. Chisholm, R. Devereux, and D. A. Stahl. 1990a. Combination of 16S rRNA-targeted oligonucleotide probes with flow cytometry for analyzing mixed microbial populations. *Applied and Environmental Microbiology* 56: 1919–1925.

Amann, R. I., L. Krumholz, and D. A. Stahl. 1990b. Fluorescent-oligonucleotide probing of whole cells for determinative, phylogenetic, and environmental studies in microbiology. *Journal of Bacteriology* 172: 762–770.

Amann, R. I., W. Ludwig, and K. H. Schleifer. 1995. Phylogenetic identification and *in situ* detection of individual microbial cells without cultivation. *Microbiology and Molecular Review* 59: 143–169.

Amann, R. and B. M. Fuchs. 2008. Single-cell identification in microbial communities by improved fluorescence *in situ* hybridization techniques. *Nature* 6: 339–348.

Daims, H., A. Bruhl, R. Amann, K. H. Schleifer, and M. Wagner. 1999. The domain-specific probe EUB338 is insufficient for the detection of all bacteria: Development and evaluation of a more comprehensive probe set. *Systematic and Applied Microbiology* 22: 434–444.

Danovaro, R., C. Corinaldesi, G. M. Luna, M. Magagnini, E. Manini, and A. Pusceddu. 2009. Prokaryote diversity and viral production in deep-sea sediments and seamounts. *Deep-Sea Research Part II* 56: 738–747.

DeLong, E. F., G. S. Wickham, and N. R. Pace. 1989. Phylogenetic stains: Ribosomal RNA-based probes for the identification of single cells. *Science* 243: 1360–1363.

Giovannoni, S. J., E. F. DeLong, G. J. Olsen, and N. R. Pace. 1988. Phylogenetic group-specific oligodeoxynucleotide probes for identification of single microbial cells. *Journal of Bacteriology* 170: 720–726.

Glökner, F. O., R. Amann, A. Alfreider et al. 1996. An *in situ* hybridization protocol for detection and identification of planktonic bacteria. *Systematic and Applied Microbiology* 19: 403–406.

Ishii, K., M. Mussmann, B. J. MacGregor, and R. Amann. 2004. An improved fluorescence *in situ* hybridization protocol for the identification of bacteria and archaea in marine sediment. *FEMS Microbiology Ecology* 50: 203–212.

Llobet-Brossa, E., R. Rosselló-Mora, and R. Amann. 1998. Microbial community composition of Wadden sea sediments as revealed by fluorescence *in situ* hybridization. *Applied and Environmental Microbiology* 64: 2691–2696.

Manini, E., G. M. Luna, C. Corinaldesi et al. 2008. Prokaryote diversity and virus abundance in shallow hydrothermal vents of the Mediterranean sea (Panarea Island) and the Pacific Ocean (North Sulawesi-Indonesia). *Microbial Ecology* 55: 626–639.

Manz, W., R. Amann, W. Ludwig, M. Wagner, and K. H. Schleifer. 1992. Phylogenetic oligodeoxynucleotide probes for the major subclasses of proteobacteria—problems and solutions. *Systematic and Applied Microbiology* 15: 593–600.

Manz, W., R. Amann, W. Ludwig, M. Vancanneyt, and K. H. Schleifer. 1996. Application of a suite of 16S rRNA-specific oligonucleotide probes designed to investigate bacteria of the phylum cytophaga-flavobacter-bacterioides in the natural environment. *Microbiology—UK* 142: 1097–1106.

Manz, W., M. Eisembrecher, T. R. Neu, and U. Szewzyk. 1998. Abundance and spatial organization of Gram-negative sulphate-reducing bacteria in activated sludge investigated by *in situ* probing with specific 16S rRNA targeted oligonucleotides. *FEMS Microbiology Ecology* 25: 43–61.

Murray, A. E., C. M. Preston, R. Massana et al. 1998. Seasonal and spatial variability of bacterial and archaeal assemblages in the coastal waters near Anvers island, Antarctica. *Applied and Environmental Microbiology* 64: 2585–2595.

Neef, A., R. Amann, H. Schlesner, and K. H. Schleifer. 1998. Monitoring a widespread bacterial group: *In situ* detection of planctomycetes with 16S rRNA-targeted probes. *Microbiology (UK)* 144: 3257–3266.

Pernthaler, A., J. Pernthaler, and R. I. Amann. 2002. Fluorescence *in situ* hybridization and catalysed reporter deposition (CARD) for the identification of marine bacteria. *Applied and Environmental Microbiology* 68: 3094–3101.

Pernthaler, A., J. Pernthaler, and R. I. Amann. 2004. Sensitive multi-color fluorescence *in situ* hybridization for the identification of environmental miroorganisms. In *Molecular microbial ecology manual*, ed. Kowalchuk, F., J. De Bruin, I. M. Head, A. D. L. Akkermans, and J. D. Van Elsas, 711–726. Dordrecht, The Netherlands: Kluwer Academic Publishers.

Schönhuber, W., B. Fuchs, S. Juretschko, and R. Amann. 1997. Improved sensitivity of whole-cell hybridization by the combination of horseradish peroxidase-labeled oligonucleotides and tyramide signal amplification. *Applied and Environmental Microbiology* 63: 3268–3273.

Teira, E., T. Reinthaler, A. Pernthaler, J. Pernthaler, and G. J. Herndl. 2004. Combining catalyzed reporter deposition-fluorescence *in situ* hybridization and microautoradiography to detect substrate utilization by bacteria and archaea in the deep ocean. *Applied and Environmental Microbiology* 70: 4411–4414.

14 Prokaryotic Abundance by Real-Time PCR

CONTENTS

14.1 INTRODUCTION

Most of the molecular techniques currently used in microbial ecology is qualitative or semiquantitative (e.g., traditional PCR, molecular fingerprinting techniques). However, quantification of prokaryotic gene sequences is of great relevance in understanding the microbial community structure and dynamics in natural environments.

Real-time quantitative PCR (RTQ-PCR) is a highly sensitive and specific technique for detecting and quantifying prokaryotic and eukaryotic genetic sequences, down to the genus and species levels, from a large number of environmental samples (Takai and Horikoshi 2000; Skovhus et al. 2004). Therefore, RTQ-PCR can also provide an indirect estimate of the absolute and relative prokaryote abundance of Bacteria and Archaea to the total prokaryotic community.

Studies concerning the quantification of 16S rDNA prokaryotic genes by real-time PCR carried out in marine sediments are rather limited (Schippers et al. 2004; Inagaki et al. 2006; Schippers and Neretin 2006). Recent investigations reported indirect estimations of prokaryote abundance in surface sediments of the Pacific Ocean, ranging from 10^6 to 10^8 cells cm^{-3}. These values decreased along the vertical profile from 10^6 cells cm^{-3} at ca. 40 m depth up to 10^5 cells cm^{-3} below this depth (Schippers et al. 2004; Inagaki et al. 2006). In the same investigations, archaeal abundance ranged from 0.01% to 30% of the total prokaryote abundance (Schippers et al. 2004; Inagaki et al. 2006). High prokaryote abundances (10^{10} cells cm^{-3}) were obtained in deeply buried marine sediments on the Peru continental margin with Archaea abundance from 1 to 3 orders lower (Schippers and Neretin 2006).

RTQ-PCR measures in real time the increase in fluorescence resulting from the amount of DNA amplified (Vaerman et al. 2004). The fluorescence is generated from double-stranded DNA-binding dyes (e.g., SYBR Green I) or fluorogenic probes (e.g., TaqMan probes). In this protocol, we describe the TaqMan fluorogenic system because it is more specific and sensitive than SYBR Green I [NOTE 1] and, therefore, more suitable for detection and quantification of small amounts of the targeted genetic sequences as prokaryotic 16S rDNAs in marine sediments samples. TaqMan probes are oligonucleotides that contain a fluorescent dye (i.e.,

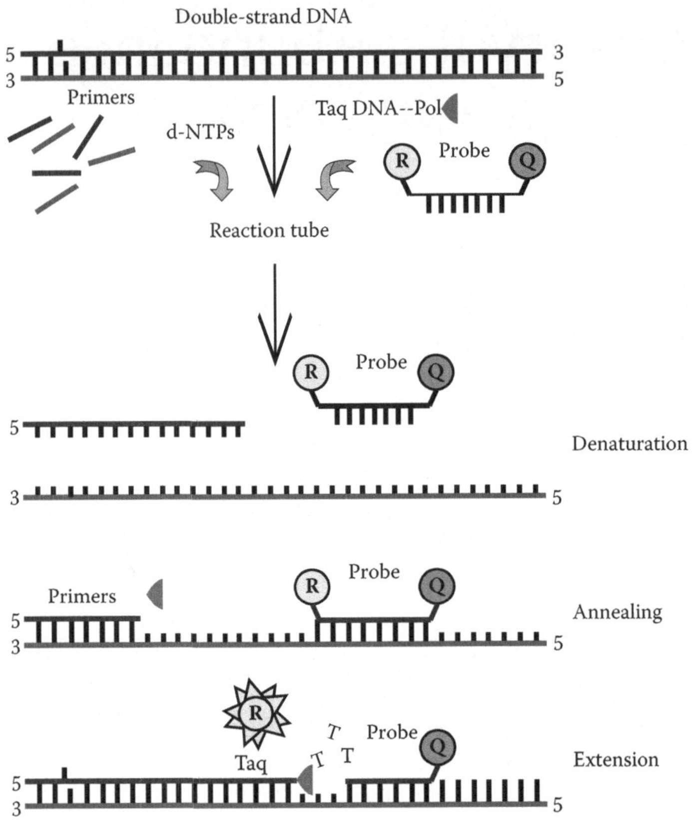

FIGURE 14.1 Schematic description of the molecular steps during real-time PCR processing based on the use of the TaqMan fluorogenic system.

6-carboxyfluorescein, FAM), typically on the 5′ base (defined Reporter, R), and a quenching dye (i.e., tetramethylrhodamine, TAMRA), typically located on the 3′ base (defined Quencher, Q). TaqMan probes (generally 15–30 bases long) are designed to hybridize to an internal region of a specific genetic sequence. Therefore, only specific PCR products can generate fluorescent signals in TaqMan PCR. After DNA denaturation, both primers and TaqMan probe anneal to the DNA target (Figure 14.1). Before replication of a new DNA strand, the fluorescence emitted by the reporter is absorbed by the quencher dye, and no fluorescence is detected. When the replication begins, the 5′–3′ exonuclease activity of the Taq polymerase degrades the probe. This separates the fluorescent and quenching dyes, and the fluorescence increases in each cycle proportional to the rate of probe cleavage.

A basic RTQ-PCR run can be divided into three sequential phases (Figure 14.2):

1. Exponential phase: Exact doubling of the PCR product that accumulates at every cycle (when the reaction efficiency is 100%). All reagents (e.g., primers, probe, Taq polymerase) are at the optimum level. This reaction is very specific and precise.
2. Linear phase: The components of the reaction begin to be consumed and products begin to degrade. The reaction starts to slow down, and the PCR product is no longer doubled at each cycle. The three replicates of the sample begin to diverge in their quantities during this phase, which is, therefore, characterized by high variability.
3. Plateau (end point, where traditional PCR takes its measurement): No more products are made and, if left long enough, the PCR products will begin to degrade.

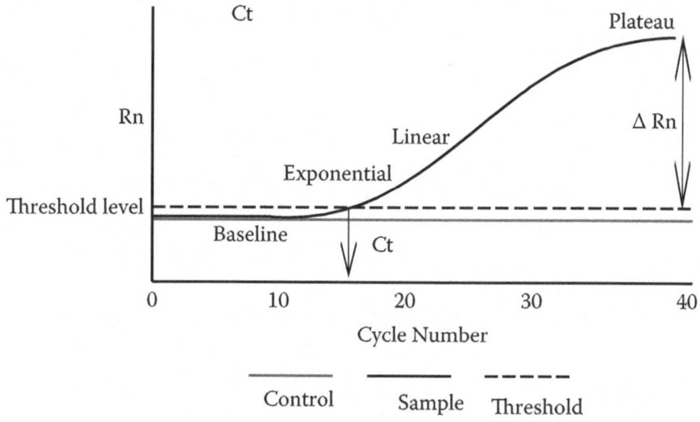

FIGURE 14.2 Exponential, linear, and plateau phases of real-time PCR in linear view. The number of cycles and the fluorescence due to the reporter (R_n) are reported in the x and y axis, respectively. The figure also shows the threshold line, baseline, and Ct value. ΔR_n is the difference between R_n and the baseline.

Amplification curves are constructed by plotting the increase in fluorescence versus cycle number (Figure 14.2). In real-time PCR, the quantitative information that is present during the exponential phase is extracted from the amplification curve [NOTE 2].

To perform the quantification of prokaryotic genes (i.e., absolute quantification), a series of dilutions of an external standard, which is amplified together with unknown samples, has to be carried out. Once PCR is run, each amplification curve is defined by its threshold cycle number (Ct). Ct is the number of cycles at which the system begins to detect the increase in fluorescence and corresponds to the crossing between threshold line and fluorescence curve (Figure 14.2). Fluorescence threshold levels are selected and set by the operator so that the threshold line crosses (over the background and below the plateau values) all reaction curves of the samples compared (Figure 14.2). Standard curves are then constructed by plotting the log of the initial standard copy numbers on the x-axis and the corresponding Ct values on y-axis. A linear regression is obtained. Ct values of analyzed samples are converted to concentration using the calibration curve equation [NOTE 3].

Theoretically, during the exponential growth phase, there is a quantitative relationship between the amount of starting target sequence and the amount of PCR product at any given cycle:

$$X_n = X_0 \cdot \left(1+E\right)_n \tag{14.1}$$

where

X_n is the amount of target sequence at cycle n.
X_0 is the initial amount of target sequence.
E is the amplification efficiency.

The efficiency of each reaction can be calculated from the slope of the line during the linear phase of the reaction (Vaerman et al. 2004). Efficiency can be calculated by the formula

$$E = 10^{(-1/slope)} - 1$$

The E values range from 0 (no amplification) to 1 (every amplicon is doubled every cycle). The efficiency of the RTQ-PCR should be measured between 90% and 100% (slope coefficient ranging from −3.6 to −3.1). For 100% efficiency, there will be a doubling of the amount of DNA at each cycle;

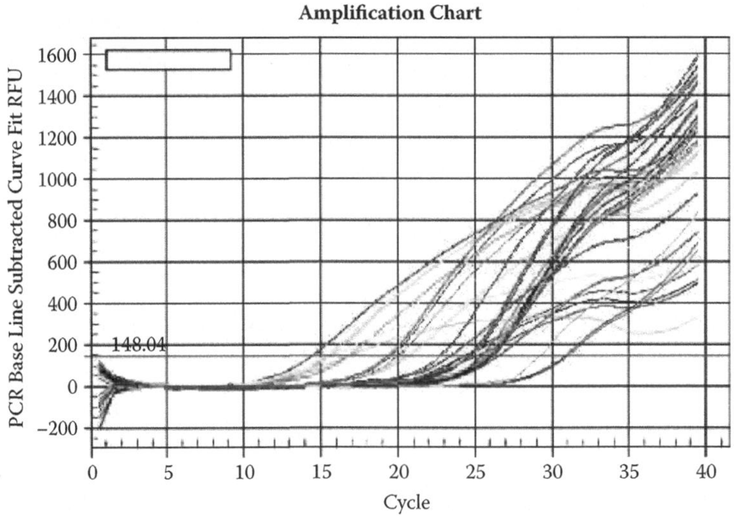

FIGURE 14.3 *(A color version of this figure follows page 234.)* Real-time amplification chart: Pattern of amplification with increasing number of cycles.

for an efficiency of 90%, the amount of DNA will increase from 1 to 1.9 at each cycle, and similarly for efficiencies of 80% and 70%.

Figure 14.3 shows the amplification chart obtained after a real-time PCR run.

14.2 FIELD PROCEDURE

14.2.1 Sampling

Surface sediment samples are collected as described for other microbiological variables. Since the vertical distribution of prokaryotes in sediments could be strongly heterogeneous, in order to provide a more realistic estimate of prokaryotic abundance (i.e., bacteria and Archaea), perturbation and mixing of sediment layers should be avoided during sampling. Slicing of sediment horizons must be performed using sterile tools (i.e., tips, spatulas, etc.). Finally, all materials and solutions used for the subsequent processing of sediment samples must be sterilized by autoclave in order to avoid any contamination.

14.2.2 Sample Storage

If DNA cannot be extracted directly after sampling, sediment samples or cores may be frozen at −80°C until analysis. After extraction (see Chapter 20 about the extraction and purification of DNA), DNA has to be stored at −80°C until analyses.

14.3 LABORATORY PROCEDURE

We report here a model protocol for quantifying 16S rDNA genes belonging to total prokaryotes and estimating their abundance in the sediments. Therefore, this must be only intended as an example since procedures for this type of analysis can vary depending on the real-time detection system and reagents (e.g., Taq polymerase, primers, and probe) utilized.

14.3.1 Equipment and Instruments

Autoclave

Laminar airflow hood
Microcentrifuge
Real-time PCR detection system (i.e., iQ5 real-time PCR)

14.3.2 Materials

Latex gloves
Sterile 1.5 mL tubes
2–20 μL, 100–1000 μL, and 1–5 mL pipettes and sterile tips
Sterile 5 mL syringes
0.02 μm pore size, 25 mm ø disposable syringe filters (Anotop, Whatman)
Sterile PCR test tubes (or strip or 96-well plate)
Adhesive film for well plate, or cups for strips
Rack for Eppendorf
Ice

14.3.3 Solutions

DNA extracted from marine sediment with a concentration ranging between 100 fg/μL and 20 pg/μL
DNA standard (e.g., *Escherichia coli*) at known concentration
0.02 μm prefiltered and autoclaved reagent-grade water [NOTE 4]
Real-time mix [e.g., iQ™ Supermix (2x, Bio-Rad) containing 100 mM KCl; 40 mM Tris-HCl, pH 8.4; 0.4 mM each dNTP (dATP, dCTP, dGTP, dTTP); 50 units per mL hot-start iTaq™ DNA polymerase; 6 mM $MgCl_2$] [NOTE 5]
Oligonucleotide primers (e.g., Uni340F 5′-CCTACGGGRBGCASCAG-3′ and Uni806R 5′-GGACTACNNGGGTATCTAAT-3′ at the final concentration of 0.8 μM)
Probe (e.g., Uni516F 5′-TGYCAGCMGCCGCGGTAAHACVNRS-FAM at the final concentration of 0.2 μM) [NOTE 6]

14.3.4 Procedure

1. Carry out serial 1:10 dilutions of DNA standard from 20 fg to 200 pg μL^{-1} concentrations in sterile PCR test tubes as follows (Figure 14.4):
   1° Standard concentration = 200 pg/μL
   2° Standard concentration (1:10) = 20 pg/μL

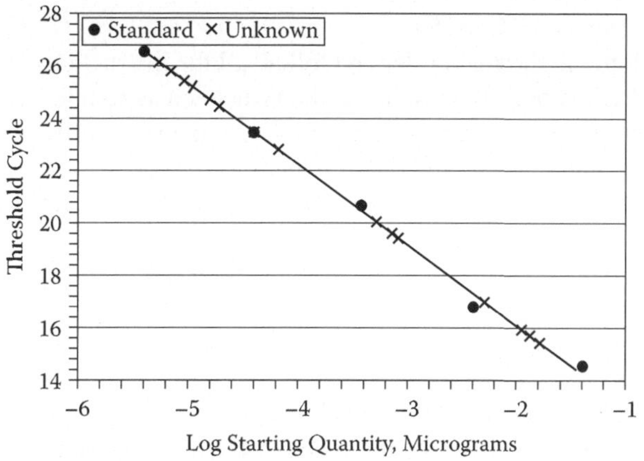

FIGURE 14.4 Standard curve chart.

TABLE 14.1
Example of a 15 μL Final Volume of Reaction Mix

	X 1 Reaction (μL)	X 45 Reactions (μL)
Supermix	7.5	337.5
Primer forward	0.12	5.4
Primer reverse	0.12	5.4
Probe	0.03	1.35
Reagent-grade water	6.23	280.35

3° Standard concentration (1:100) = 2 pg/μL
4° Standard concentration (1:1000) = 0.2 pg/μL
5° Standard concentration (1:10000) = 20 fg/μL

2. Calculate the actual number of samples and total replicates to analyze in order to prepare the reaction mix with the correct volumes of each ingredient:
   Example
   6 Standard samples × 3 replicates = 18 +
   1 Negative control × 3 replicates = 3 +
   6 DNA samples × 3 replicates = 18 =
   Wells/tubes to fill = 39

Consider a number of total samples in excess for preparing a larger volume of the reaction mix (Table 14.1). Generally, besides the actual number of samples, we count other two additional samples (e.g., 39 + (2 × 3 replicates) = 45 total samples).

3. Collect the reaction mix in a 2 mL sterile tube.
4. Dispense 14 μL of reaction mix in each well of the plate (or sterile PCR tube), changing the tip every time and avoiding creating bubbles in the well.
5. Add 1 μL of DNA standard and then DNA sample to the wells.
6. Add 1 μL of reagent-grade water to negative control.
7. Cover the plate with adhesive film (or the tubes with stoppers), keeping the film for the two ends. Do not touch the central part of the film.
8. Program the thermal protocol of the real-time PCR for quantifying 16S rDNA in Archaea (Figure 14.5):
   Taq activation: 95°C for 3 min
   40 cycles of denaturation at 95°C for 25 s
   Annealing/extension at 59°C for 5 min
9. Insert the plate into the thermal cycler and follow all the instructions for running PCR.
10. Once real-time PCR is over, the results can be visualized as weight (e.g., pg, ng, mg) or in number of copies of 16S rDNA, depending on the setup of the software.
11. In order to convert the weight of 16S rDNA in number of copies, we assume that 1 pg of 16S rDNA contains 1.29×10^6 copies and that each cell contains 3.8 ± 2.9 copies of 16S rDNA.

14.4 NOTES

1. SYBR-Green can bind any double-stranded DNA, with the production of false-positive results.
2. Real-time PCR allows the detection of PCR amplification during the early phases of the reaction. Measuring the kinetics of the reaction in the early phases of PCR provides a distinct advantage over traditional PCR detection. Traditional methods use agarose gels for the detection of PCR amplification at the final phase or end point of the PCR reaction.

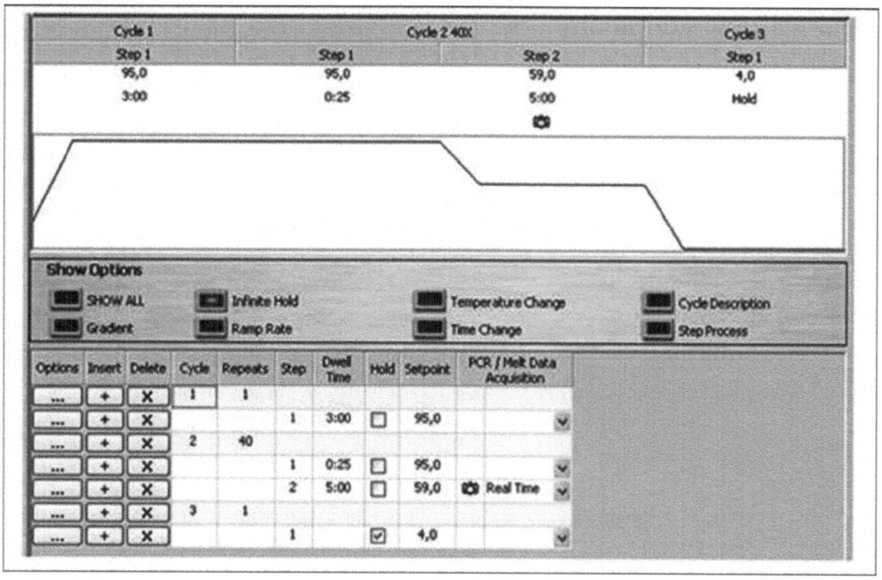

FIGURE 14.5 Example of thermal protocol.

However, end-point detection is very time consuming, and the results are based on size discrimination on electrophoresis gel, which may be inaccurate.

3. The most important parameter for gene quantification is Ct. The higher the initial amount of genomic DNA, the sooner the accumulated product is detected in the PCR process, and the lower the Ct value. The threshold line should be placed above the baseline and within the exponential increase phase (which looks linear in the log transformation). This provides better run-to-run reproducibility. A Ct value of 40 or higher means no amplification, and this value cannot be included in the calculations.
4. Reagent-grade water should be prepared before the beginning of the sample processing.
5. This mix solution for real time already contains a hot-start Taq polymerase, which is active only at 95°C for 3 min.
6. Generally, primers and probes are delivered as lyophilized powders, desalted and purified. These powders can be resuspended with 0.02 μm prefiltered reagent-grade water till a known concentration (e.g., 100 μM, that is, 100 pmol/μL) is obtained. The correct volume of these solutions in the reaction mix will depend upon the final concentration of each primer and probe (e.g., if the final concentration of each primer is 0.8 μM, for a volume of reaction mix of 15 μL, we have to add to the reaction mix 0.12 μL of each primer for each reaction). The stock solutions should be divided into aliquots, frozen, and the probes must be kept in the dark.

Make DNA standard dilutions
(1:10, DNA: sterile 0.02 μm prefiltered water)
in sterile PCR tubes

↓

Calculate the volume of each ingredient (Supermix, primers, probe and water)
of the reaction mix for a total number of samples equal to
(the actual number plus other 2 samples) considering 3 replicates for each sample.
Consider a final volume for each replicate from 15 to 50 μL

↓

Prepare the reaction mix in a 2 mL sterile tube

↓

Dispense the reaction mix in each well (e.g., when a final volume of 15 μL is considered,
add 14 μL reaction mix plus 1 μL DNA) of the plate (or PCR tubes)

↓

Add 1 μL standard DNA and DNA sample to wells

↓

Add 1 μL sterile water to negative controls

↓

Cover the plate with adhesive film (or the tubes with stoppers)
and program the thermal protocol of the real time PCR

↓

Insert the plate into the thermal cycler and
follow all of the instructions for the PCR run

↓

Once the real time PCR is over, check the calibration curve,
the efficiency and final results

↓

Convert the copy number of 16S rDNA in cell numbers

PROTOCOL 14.1 Quantification of specific prokaryotes by RT-PCR.

REFERENCES

Inagaki, F., T. Nunoura, S. Nakagawa et al. 2006. Biogeographical distribution and diversity of microbes in methane hydrate-bearing deep marine sediments on the Pacific Ocean margin. *Proceedings of the National Academy of Sciences USA* 103: 2815–2820.

Schippers, A. and L. N. Neretin. 2006. Quantification of microbial communities in near-surface and deeply buried marine sediments on the Peru continental margin using real-time PCR. *Environmental Microbiology* 8: 1251–1260.

Schippers, A., L. N. Neretin, J. Kallmeyer et al. 2004. Prokaryotic cells of the deep sub-seafloor biosphere identified as living bacteria. *Nature* 433: 861–864.

Skovhus, T. L., N. B. Ramsing, C. Holmstro, S. Kjelleberg, and I. Dahllo. 2004. Real-time quantitative PCR for assessment of abundance of *Pseudoalteromonas* species in marine samples. *Applied and Environmental Microbiology* 70: 2373–2382.

Takai, K. and K. Horikoshi. 2000. Rapid detection and quantification of members of Archaeal community by quantitative PCR using fluorogenic probes. *Applied and Environmental Microbiology* 66(11): 5066–5072.

Vaerman, J. L., P. Saussoy, and I. Ingargiola. 2004. Evaluation of real-time PCR data. *Journal of Biological Regulators and Homeostatic Agents* 18: 212–214.

15 Abundance of Heterotrophic Benthic Protists

CONTENTS

15.1 INTRODUCTION

Protists play an important role in the organic matter fluxes in both pelagic and benthic environments of the oceans (Azam et al. 1983; Fenchel 1982). Benthic protozoans consume bacteria and microalgae, thus enhancing primary and secondary productions and degradation of organic material (Sherr et al. 1982; Fenchel 1986; Hondeveld et al. 1994; Boenigk and Arndt 2002), and they are consumed by meio and macrofauna (Patterson et al. 1989; Epstein 1997a,b). In addition, Thraustochytrids (marine fungoid protists) are adapted to solid substrate pervasion and are involved in the breakdown, scavenging, and mineralization of highly refractory organic matter through extracellular digestion (Santangelo et al. 2000; Bongiorni and Dini 2002; Raghukumar 2002; Bongiorni et al. 2005a,b). Knowledge of the abundance and distribution of these organisms is required to improve our understanding of the role of protistis in the benthic microbial food-web dynamics. However, studies of the *in situ* abundance of protists have been hampered by inadequate methodologies. Ciliates are typically extracted from sediment following Uhlig's ice-water filtration method (Uhlig 1964), and live-sample counts are performed just after collection (Fenchel 1986; Lucchesi and Santangelo 1995; Santangelo and Lucchesi 1995), while the numbers of benthic thraustochytrids have been estimated by baiting fresh sediment samples with pine pollen according to the Most Probable Number (MPN) technique modified by Gaertner (1968). Direct counts have resulted in underestimates (e.g., seawater ice, pipetting, decantation) or overestimates (laboratory cultivation) of protozoan densities (Alongi 1986). In addition, the several limitations imposed by these techniques (labor intensive analysis of fresh material and cultures) have allowed examination of only a small number of replicated samples (Santangelo et al. 2000). In the last decades, count methodologies have been improved by staining protist cells with fluorochromes and counting them by epifluorescence microscopy (Sherr et al. 1993; Raghukumar and Schaumann 1993; Danovaro et al. 1998). However, in order to avoid background fluorescence due to sediment, different kinds of extraction, mainly by centrifugation in a nonlinear silica-gel gradient, have been described (Alongi 1993; Starink et al. 1994; Epstein, 1995; 1997a,b; Tao and Taghon 1997). Recently, in an attempt to simultaneously describe different components of the benthic protists community, a combined protocol that is based on an improvement and a compromise between those expounded by Raghukumar and Schaumann (1993) for thraustochytrids and by Epstein (1995) for nano-, micro-, and meiobenthos has been developed and tested (Santangelo et al. 2000; Bongiorni et al. 2005a). This method, described here in detail, is based on

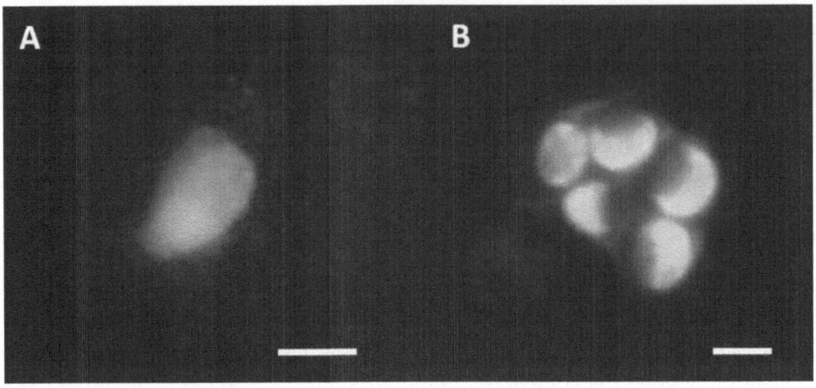

FIGURE 15.1 *(A color version of this figure follows page 234.)* Ciliate (B) stained with DAPI and FITC (A) and a cluster of thraustochytrid cells stained with Acriflavin; scale bars are 10 and 5 µm in A and B, respectively.

glutaraldehyde sample fixation, extraction in a nonlinear silica Percoll gradient, and cell staining with three different fluorochromes (Acriflavin, DAPI, and FITC) (Figure 15.1A and B) [NOTES 1 and 2]. This method enables extraction of nano- and microbentos from sediment and also allows collection and examination of a high number of samples.

15.2 FIELD PROCEDURES

Sediment samples are collected as previously described for other microbial variables. Aliquots of ca. 1 cm^3 from each sediment cores are collected with sterile spatulas and fixed in 5 mL of 1.5% glutaraldehyde–seawater solution for counts. Samples are stored at 4°C until analysis. Glutaraldehyde fixation (Epstein 1995) is preferred to formalin fixation (Raghukumar and Schaumann 1993) as it is compatible with Percoll and works well with thraustochytrids, flagellates, and ciliates.

15.3 LABORATORY PROCEDURES

15.3.1 Equipment and Instruments

Autoclave
Laminar air flow
Filtration apparatus: Sterile filter holder for 25 mm diameter filters with 15 mL funnel Millipore, vacuum pump, vacuum flask (for discharge)
Centrifuge
Epifluorescence microscopy

15.3.2 Materials

Sterile 15 mL tubes
Sterile 15 mL falcon tubes
Sterile 50 mL falcon tubes
0.2 µm pore size Anotop filters
Nucleopore polycarbonate filters, 2.0 µm pore size, 25 mm Ø (Whatman)
2–20 µL, 100–1000 µL, 1-5 mL pipettes and sterile tips
Glass slides and coverslips
Percoll (SIGMA P1644)
Sodium chloride [NaCl]

Sodium bicarbonate [$NaHCO_3$]
Sodium carbonate [Na_2CO_3]
Potassium phosphate [KH_2PO_4]
Dipotassium hydrogenphosphate [K_2HPO_4]
DAPI stain ([4′,6′-diamino-2-phenylindole hydrochloride]; Sigma D-9542)
FITC stain ([fluorescein isothiocyanate]; Sigma F-1522)
Acriflavin hydrochloride (Sigma A-8251)
Calcofluor White (Sigma F6259)
Immersion oil

15.3.3 Solutions and Preparation of Solutions

1.5% glutaraldehyde solution is prepared by diluting 1.5 mL of pure glutaraldehyde (Sigma) in 98.5 mL of natural or artificial seawater (kept at the same salinity as that measured in the field, 35 psu sea salt dissolved in1 L distilled water, dW), prefiltered onto 0.2 μm filters (particle-free seawater, PFSW).

50% Percoll solution is prepared by mixing equal volume of Percoll solution and 70 psu seawater (prepared dissolving 70 g sea salts in 1 L dW).

Carbonate buffer (pH 9.5) is prepared by mixing 100 mL of $NaHCO_3$ solution (4.2 g $NaHCO_3$ in 100 mL dW) with 60 mL of Na_2CO_3 solution (5.3 g Na_2CO_3 in 100 mL distilled water) and adjusting pH to 9.5. The buffer is then sterilized by 0.2 μm filtration and can be stored at 4°C.

Phosphate buffer (pH 7.2) is prepared by mixing 72 mL of KH_2PO_4 solution (0.87 g KH_2PO_4 in 500 mL dW) with 6.2 mL of K_2HPO_4 solution (0.87 g K_2HPO_4 in 500 mL dW) and adjusting pH to 7.2. The buffer is then sterilized by 0.2 μm filtration and can be stored at room temperature.

0.85% sodium chloride solution is prepared by dissolving 0.85 g of NaCl in 100 mL of dW; 0.2 μm filtered solution can be stored at room temperature.

Citric buffer (pH 3) is prepared by mixing 46.5 mL of a 0.2 M citric acid solution (1.92 g of citric acid in 50 mL dW) with 3.5 mL of a 0.2 M sodium citrate solution (2.49 sodium citrate in 50 mL dW), adjusting the pH to 3; 0.2 μm filtered buffer can be stored at 4°C.

DAPI working solution (10 μg mL^{-1}) is prepared by dissolving 500 μL of a stock solution (1 mg DAPI in 1 mL of 0.2 μm filtered dW, stored at –20°C) in 5 mL of 0.2 μm filtered dW (the working solution is stable for 1 month if stored at –20°C).

FITC solution (100 μg mL^{-1}) is prepared by dissolving 1 mg of FITC in 0.25 mL carbonate buffer, 1.10 mL phosphate buffer, and 1.10 mL NaCl (0.85%).

Acriflavin staining solution is prepared by mixing 1 mL of a stock solution (50 mg acriflavin in 10 mL sterile dW) in 10 mL citrate buffer.

Calcofluor working solution (0.025%) is prepared by mixing 1 mL of a stock solution (1 mg Calcofluor in 1 mL dW) in 3 mL 0.2 μm filtered dW.

15.3.4 Procedures

1. In order to create the nonlinear gradient, 4.5 mL of 50% Percoll solution is centrifuged in a 15 mL falcon tube at 30,000 × g for 15 min.
2. Collect ca.1 g of sediment in a sterile tube.
3. Add 5 mL of 1.5% glutaraldehyde solution and extract cells by gently shaking the tubes 10 times. The 5 mL supernatant is then transferred to 50 mL falcon tubes (this operation is repeated 5 times, yielding a total volume of 25 mL) [NOTE 3].
4. Gently load 2 mL of the 25 mL extracted supernatant on top of a 4.5 mL silica gel gradient [NOTE 4].

5. Centrifuge samples at 5,000 × g for 15 min.
6. Discard the dense, lowermost layer holding only detritus (undesired fluorescence source).
7. Place with forceps a 2.0 µm pore size polycarbonate membrane on the filter holder (previously autoclaved), and place and fix the funnel on the filter holder with a spring clip.
8. Pour the remaining Percoll gradient inside the funnel and filter the sample at low vacuum pressure (no grater then 35 mm Hg to avoid cell damage).
9. When the sample is filtered, wash the filter with 5 mL of PFSW by vacuum (repeat washing 2–3 times).
10. Flood the filter with 4 mL of acriflavin working solution.
11. Incubate for 4 min in the dark.
12. Wash the filter with 5 mL of PFSW(repeat washing 2–3 times).
13. Pour 2 mL of the 1/10 diluted DAPI stock solution (100 µL of the stock solution in 1.8 mL dW).
14. Incubate for 15 min in the dark.
15. Wash the filter 3 times with 5 mL of PFSW.
16. Flood the filter with 2.45 mL of FITC solution.
17. Incubate for 10 min in the dark.
18. Wash the filter 3 times with 5 mL of PFSW.
19. Flood the filter with 1.5–2 mL of the optical brightener Calcofluor solution.
20. Incubate for 1 min in the dark.
21. Wash the filter 1–2 times with 5 mL of PFSW.
22. Place 1 drop of oil onto a clean microscope slide, and gently mount the filter on the top with a clean forceps (avoiding bubble formation).
23. Place another drop of immersion oil on the top of the filter and flip over a coverslip.
24. Examine the slide under an epifluorescence microscope equipped with filter for DAPI (excitation filter 360 nm, barrier filter 420 nm) and for Acriflavin/FITC (excitation filter 450–490 nm, barrier filter 520 nm) [NOTES 5 and 6].
25. Count benthic protists on at least three randomly selected areas of the polycarbonate membrane (in alternatives, replicated areas of the filter equivalent in length to the diameter of the filtered area and in height to the optical field diameter, can be analyzed) under 400× or 1000× magnification. Count at least 200 cells for each group of organisms (e.g., ciliates, nanoflagellates, thraustochytrids) [NOTE 7].
26. Sediment samples are dried in a oven at 60°C for 24–48 h until they are dried and then are weighted with an analytical balance.
27. After using, filter holders and funnels can be washed with 1N HCl.

Benthic protists abundance calculations:

$$\text{Number of protists g}^{-1} = [(\text{Protists number in counting area} \times \text{Optical field coefficient}) \times \text{dilution factor of the sediment}]/\text{sediment dry weight (g)}$$

where

Optical field coefficient = filtration area/counting area.

Dilution factor = 25/2.

15.4 NOTES

1. FITC staining resolved details of general morphology and cilia/flagella placement, whereas DAPI staining visualized well the nuclei of most of the protozoa. The combination of both provided an opportunity for some taxonomic identification of nano- and microbenthos otherwise rarely possible under epifluorescent illumination (Epstein, 1995). Acriflavine

simultaneously stains thraustochytrid sulfated polysaccharide cell walls red and their nuclei yellow-green, thus distinguishing these organisms from other similar-sized protists that are fluoresce greenish only and lack red fluorescing cell walls. Finally, to distinguish thraustochytrids from other similar-sized photosynthetic protists, natural chlorophyll fluorescence is checked under UV excitation light: photosynthetic protists are easily distinguished because of their peculiar red fluorescing cell content (Raghukumar and Shaumann 1993).

2. Fluorochrome staining solutions are prepared as described in Raghukumar and Shaumann (1993) for acriflavin, and in Sherr et al. (1993) and Epstein (1995) for DAPI and FITC. None of the fluorochromes interferes with another or has a jellying effect on Percoll.
3. Depending on the grain size of sediment (e.g., fine-silt sediment), after shaking, the suspension can be allowed to stand for a few to a maximum of 25 min before being pipetted off (Wieltschnig et al. 2003).
4. In order to avoid excess of detritus on filters, only a 2 mL subsample of a 25 mL final volume extracted is processed (ca. 1/12 of the original sample volume). According to Epstein (1995), processing of 1/15 of the original sample enables extraction of over 90% of the nano- and over 95% of the microbenthos from sand samples.
5. Heterotrophic benthic protists' fluoresce apple green; thraustochytrids' cell walls fluoresce red and their nuclei yellow-green.
6. Slides can be stored up to 3 months at −20°C. After analysis, filter holders and funnels are rinsed with dW, placed in 1 N HCl for 1 h, sonicated 3 times (3 min), and dried.
7. In order to estimate benthic protists' biomass, maximal diameter of all thraustochytrids can be measured with a micrometric eyepiece, and individual biovolume can be estimated approximating cell shape to a cylinder or a sphere. The biovolume is then converted into biomass values using specific published conversion factors for ciliates, nanoflagellates, and thraustochytrids.

Fix 1 cm^3 sediment with 5 mL of 1.5%-glutaraldehyde-PFSW solution

↓ Hand-shake the sample ca. 10 times

Move (by pipetting) the 5 mL supernatant to a Falcon tube

↓

Add to the sediments other 5 mL of 1.5%-glutaraldehyde-PFSW solution

↓ Hand-shake the sample ca. 10 times

Move (by pipetting) the 5 mL supernatant to the Falcon tube

↓ Repeat hand-shake and transfer supernatant after washing 3 times until reaching 25 mL of supernatant volume

Load by gentle pipetting, 2 mL of the supernatant to 4.5 mL Percoll gradient

↓

Centrifuge (5000 × g, 15 min)

↓

Pour supernatant into the filtration funnel by discarding the lower most dense layer

↓ Vacuum filter onto 0.2 μm-polycarbonate filter, <35 mmHg

Wash filter 3 times with 5 mL PFSW

↓ Vacuum filter, <35 mmHg

Add 4 mL Acriflavin working solution

↓ Incubate 4 min, dark, vacuum filter <35 mmHg

Wash filter 3 times with 5 mL PFSW

↓ Vacuum filter, <35 mmHg

(a)

PROTOCOL 15.1 Enumeration of benthic heterotrophic protists. *Continued*

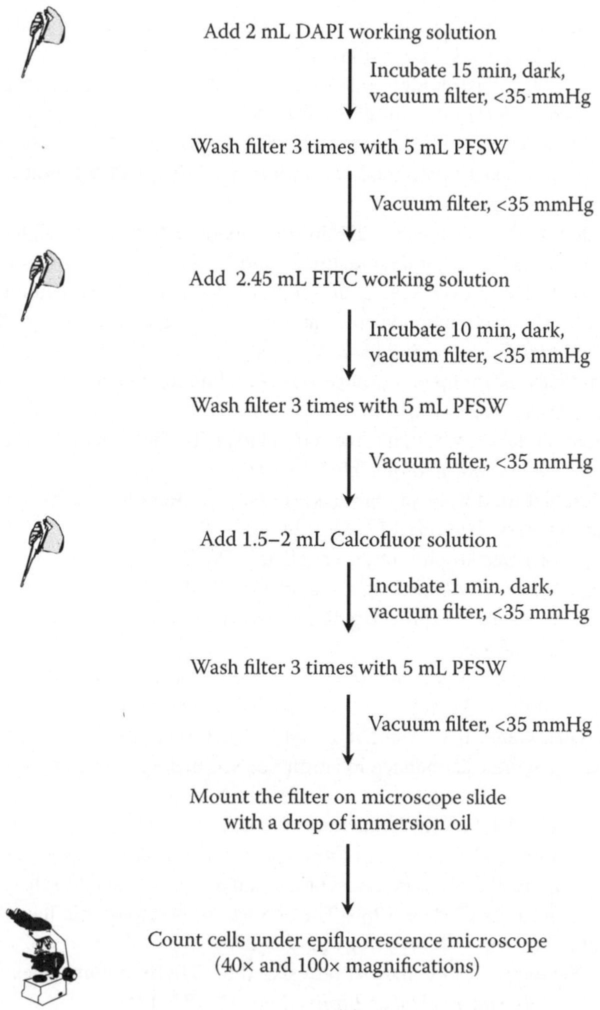

(b)

PROTOCOL 15.1 *Continued.*

REFERENCES

Alongi, D. M. 1986. Quantitative estimates of benthic protozoa in tropical marine systems using silica gel: A comparison of methods. *Estuarine, Coastal and Shelf Science* 23: 443–450.

Alongi, D. M. 1993. Extraction of protists in aquatic sediments via density gradient centrifugation. In *Aquatic microbial ecology*, ed. P. F. Kemp, B. F. Sherr, E. B. Sherr, and J. J. Cole, 109–114. Boca Raton: Lewis Publishers.

Azam, F., T. Fenchel, J. G. Field, J. S. Graf, L. A. Meyer-Reil, and F. Thingstad. 1983. The ecological role of water-column microbes in the sea. *Marine Ecology Progress Series* 10: 257–263.

Boenigk, J. and H. Arndt. 2002. Bacterivory by heterotrophic flagellate: Community structure and feeding strategies. *Antonie Van Leeuwenhoek* 81: 465–480.

Bongiorni, L. and F. Dini. 2002. Distribution and abundance of thraustochytrids in different Mediterranean coastal habitats. *Aquatic Microbial Ecology* 30: 49–56.

Bongiorni, L., S. Mirto, A. Pusceddu, R. Danovaro. 2005a. Response of benthic protozoa and thraustochytrid protists to fish-farm impact in seagrass (*Posidonia oceanica*) and soft bottom sediments. *Microbial Ecology* 50: 268–276.

Bongiorni, L., A. Pusceddu, and R. Danovaro. 2005b. Enzymatic activities of epiphytic and benthic thraustochytrids involved in organic matter degradation. *Aquatic Microbial Ecology* 41: 299–305.

Danovaro, R., D. Marrale, N. Della Croce, A. Dell'Anno, M. Fabiano. 1998. Heterotrophic nanoflagellates, bacteria and labile organic compounds in continental shelf and deep-sea sediments of the Eastern Mediterranean. *Microbial Ecology* 35: 244–255.

Epstein, S. S. 1995. Simultaneous enumeration of protozoa and micrometazoa from marine sandy sediments. *Aquatic Microbial Ecology* 9: 219–227.

Epstein, S. S. 1997a. Microbial food webs in marine sediments. I. Trophic interactions and grazing rates in two tidal flat communities. *Microbial Ecology* 34: 188–198.

Epstein, S. S. 1997b. Microbial food webs in marine sediments. II. Seasonal changes in trophic interactions in sandy tidal flat communities. *Microbial Ecology* 34: 188–198.

Fenchel, T. 1982. Ecology of heterotrophic microflagellates. IV. Quantitative occurrence and importance as bacterial consumers. *Marine Ecology Progress Series* 9: 35–42.

Fenchel, T. 1986. The ecology of heterotrophic flagellates. In *Advances in microbial ecology*, ed. K. C. Marhsall, 55–97. New York: Plenum Press.

Gaertner, A. 1968. Eine Methode des quantitativen Nachweises niederer mit Pollen koederbarer Pilze im Meerwasser und im Sediment. *Veroeff Inst. Meeresforsch Bremerhav*: 3: 75–92.

Hondeveld, B. J. M., G. Nieuwland, F. C. Van Duyl, and R. P. M. Bak. 1994. Temporal and spatial variations in heterotrophic nanoflagellate abundance in North Sea sediments. *Marine Ecology Progress Series* 109: 235–243.

Lucchesi, P. and G. Santangelo. 1995. Time behaviour of marine ciliated protozoa of sandy shores. In *Biology and ecology of shallow coastal waters. 28th European Marine Biology Symposium*, ed. A. Eleftheriou, A. D. Ansell, and C. J. Smith, 123–128. Fredensborg, Denmark: Olsen and Olsen.

Patterson, D. J., J. Larsen, and J. O. Corliss. 1989. The ecology of heterotrophic flagellates and ciliates living in marine sediments. *Progress in Protistology* 3: 185–277.

Raghukumar, S. 2002. Ecology of the marine protists, the Labyrinthulomycetes (Thraustochytrids and Labyrinthulids). *The European Journal of Protistology* 38: 127–145.

Raghukumar, S. and K. Schaumann. 1993. An epifluorescence microscope method for direct detection of and enumeration of the fungilike marine protists: The thraustochytrids. *Limnology and Oceanography* 38(1): 182–187.

Santangelo, G., L. Bongiorni, and L. Pignataro. 2000. Abundance of thraustochytrids and ciliated protozoans in a Mediterranean sandy shore determined by an improved, direct method. *Aquatic Microbial Ecology* 23: 55–61.

Santangelo, G. and P. Lucchesi. 1995. Spatial distribution pattern of ciliated protozoa in a Mediterranean interstitial environment. *Aquatic Microbial Ecology* 9: 47–54.

Sherr, B. F., E. B. Sherr, and T. Berman. 1982. Decomposition of organic detritus: A selective role for microflagellate protozoa. *Limnology and Oceanography* 27: 765–769.

Sherr, E. B., D. A. Caron, and B. F. Sherr. 1993. Staining of heterotrophicprotists for visualization via epifluorescence microscopy. In *Aquatic microbial ecology*, P. F. Kemp, B. F. Sherr, E. B. Sherr, and J. J. Cole, 213–227. Boca Raton: Lewis Publishers.

Starink, M., M. J. Bär-Gilissen, R. P. M. Bak, and T. E. Cappenberg. 1994. Quantitative centrifugation to extract benthic protozoa from freshwater sediments. *Applied and Environmental Microbiology* 60: 167–173.

Tao, S. F. and G. L. Taghon. 1997. Enumeration of protozoa and bacteria in muddy sediment. *Microbial Ecology* 33: 144–148.

Uhlig, G. 1964. Eine einfache Methode zur Extraction der vagilen, mesopsammalen Microfauna. *Helgoländer Wiss Meeresunters* 11: 178–185.

Wieltschnig, C., U. R. Fischer, A. K. T. Kirschner, and B. Velimirov. 2003. Benthic bacterial production and protozoan predation in a silty freshwater environment. *Microbial Ecology* 46: 62–72.

16 Abundance of Benthic Foraminifera

CONTENTS

16.1 INTRODUCTION

Foraminifera belong to the phylum Protozoa, unicellular organisms that are distinguishable from the other Sarcodina for the presence of granuloreticulopodia and by the possession of a test or shell that can be organic, agglutinated, or mineralized. Shelled Foraminifera create skeleton of incredible complexity that are present at all latitudes. The benthic forms exist down to deep-sea trenches, and at present occupy the marine and marginal environments with more than 3000 species (Murray 2007). Foraminifera are also used for the determination of the age and depositional environment of the strata (biostratigraphy and paleoecological investigations), in paleoclimatic studies, and in the study of the evolution of present and past oceanic environments and continental configurations (paleoceanography and paleoenvironmental reconstructions). Although, foraminifera are presently the best known and comprehensively studied calcareous microfossils, a better knowledge of the ecology and biology of modern species are essential due to the importance of these organisms in benthic communities.

The techniques used to prepare and concentrate samples for the examination of foraminiferal content vary according to the studied purpose.

16.2 THE ROSE BENGAL METHOD

Rose Bengal adheres to proteins and stains the cytoplasm of the organism (Walton 1952), exhibiting a bright pink color. This method is the most common method to study the living assemblages of foraminifera (Bernhard 2000), which is largely sufficient if the aim is to obtain a general picture of the composition of foraminiferal assemblages. However, this method is not perfectly accurate. In fact, foraminifera may be still stained even weeks or months after their death (Corliss and Emerson 1990) and, consequently, the abundance of the living assemblages may be overestimated. This is evident in hypoxic and/or anoxic sediments. Therefore, staining with Rose Bengal is suitable for broad-scale ecological and distributional surveys, but is less appropriate for accurate and short-term experimental studies (Bernhard et al. 2006).

16.2.1 Field Procedures

For quantitative studies, it is necessary to collect a sample of known area or volume of surface sediment. It is better to use the box corer or multicorer to ensure that the upper several centimeters of sediment remain undisturbed during the sampling process. When on board, the box corer or multicorer is subsampled with a small diameter core (about 3.5 cm) to assure the collection of an area of about 10 cm^2 [NOTE 1]. The core is sliced into 0.5 cm levels down to 2 cm, and in 1 cm level down to 10 cm depth; then each subsample is carefully transferred into a previously labeled container. All samples are preserved in 4% formalin buffered with sodium borate [NOTE 2], added with Rose Bengal (1 g L^{-1}) [NOTE 3], capped, and gently shaken. It is important to store the samples at 4°C until the analysis will be performed.

16.2.2 Laboratory Procedures

In the laboratory, the samples are wet sieved on 30/63 μm meshes [NOTE 4]. If the aim of the work is to study the size structure of the assemblage, a set of sieves can be used to wash the sediment to obtain several fractions (e.g., 125, 150, 250, 500 μm). It is important to collect a fixed volume of sample in order to have the number of specimens per unit volume of the sample. Subsequently, gently wash the sample with tap water until no more fine-grained sediment is left. Place the residues in labeled containers with indication of the sieved fraction and name of the sample, add buffered 4% formalin to preserve the organic material, and store at 4°C until the analysis will be performed.

The residue is placed in a gridded Petri dish and sorted in water under a binocular microscope with a magnification of 40×–100×. Only foraminifera intensively pink or red stained (except for the last few chambers) is counted, picked out, sorted, and mounted dry on micropaleontological slides to provide taxonomic reference material. It is important to analyze the residue underwater because, if dried, the protoplasm shrink, and then it is more difficult to discriminate living organisms. Agglutinated and porcellaneous taxa with not transparent test should be broken to detect the protoplasm in order to establish whether or not the protoplasm is present.

The soft-shelled foraminifera are extracted with a Pasteur pipette and stored in glycerol in a glass cavity slide, or alternatively, in Eppendorf vials with a drop of buffered 10% formalin. At the end of the analyses, all the sieve residues can be dried to analyze the dead foraminiferal fauna.

16.3 THE CELLTRACKER GREEN METHOD

One of the most practical alternatives to Rose Bengal is CellTracker Green. This method allows distinguishing accurately between dead and living foraminifera (Bernhard et al. 2006). CellTracker Green CMFDA (5-chloromethylfluorescein diacetate) is a vital fluorogenic probe that was developed to exclusively stain living cells. It is colorless and nonfluorescent until the acetate groups are cleaved by intracellular esterases, which are active in the living cell. Esterases cleave the fluorescent moiety of the probe to produce the fluorescent compound that can be microscopically viewed after excitation at the appropriate wavelengths.

16.3.1 Field Procedures

Sediment samples are collected as described previously in Section 16.2.1, "Field procedures," and mixed with an equal volume of chilled bottom seawater (*in situ* temperature). CellTracker Green (1 mg) is diluted with 1 mL DMSO (dimethyl sulfoxide) and added to the seawater-mixed sediment samples in order to obtain a final concentration of 1 μM. Samples have to be shaken and incubated at *in situ* temperature during at least 7 h in the dark for deep-sea foraminifera. Then, after incubation, the samples are fixed in 5% buffered formalin, and can be stored and analyzed later.

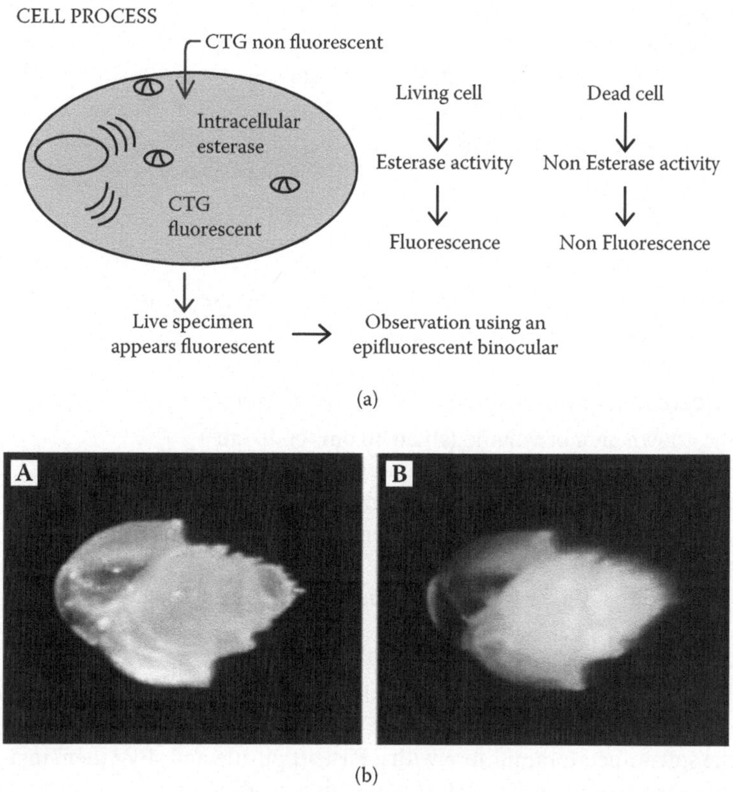

FIGURE 16.1 *(A color version of part b follows page 234.)* A live *Bulimina marginata* (Foraminifera) observed using a normal binocular (a) and an epifluorescent binocular (b). (Courtesy of E. Geslin, University of Angers.)

16.3.2 Laboratory Procedures

In the laboratory, sediment samples are washed as described for the Rose Bengal method. The residues of the sieves are wet picked using an epifluorescence microscope (Abs/Em of the CTG: 492/517 nm). Living foraminifera show green fluorescent cytoplasm, whereas dead foraminifera show no color. For porcelaneous and agglutinated foraminifera, it is often necessary to break the shell to observe the color of the cytoplasm. Green fluorescence is also frequently observed in the porcelaneous tests due to introduction of living nematodes. The fluorescent foraminifera are counted, picked out, sorted, and mounted dry on micropaleontological slides to provide taxonomic reference material. The fluorescent color is still detectable when foraminifera are dry, but it is easier to detect it when they are wet (Figure 16.1a,b).

16.3.2.1 Equipment, Instruments, and Solutions

ASTM sieves 30, 63, and 150 μm
Gridded Petri dish or picking tray
Fine sable paintbrush (#000) and/or needle
Pasteur pipette
Statistically random microsplitter
Micropaleontological slide
Glycerol
10% formalin (4% formaldehyde solution), buffered with sodium borate
Rose Bengal

Glass cavity slide
Diluted gum solution
Stereoscopic binocular microscope with incident light source

For CTG method, the same material plus

CellTracker Green CMFDA
DMSO
Micropipette (1mL and 0–100 µL)
Incubator (if *in situ* temperature is different from the room temperature)
Epifluorescent microscope with adapted filter

16.3.2.1.1 Protocol for Foram Counting: Rose Bengal Method

1. Sample the known area or volume (about 10 cm^2 or 10 cm^3).
2. Slice the core every ½ cm down to 2 cm, and every 1 cm from 2 cm down to 10 cm or more.
3. Transfer the subsamples to a labeled container.
4. Add 4% formalin buffered with sodium borate and Rose Bengal (1 g/L), and gently shake and conserve with Rose Bengal for at least 24 h.
5. Wet sieve at 30 or 63 µm, or with a set of sieves.
6. Place the residues in a gridded Petri dish.
7. Wet pick the residues under a stereomicroscope.
8. Extract the bright pink hard-shelled foraminifera and store them in a micropaleontological slide.
9. Extract the soft-walled foraminifera with a Pasteur pipette and store them in a glass cavity slide with glycerol, or in an Eppendorf vial with formalin.

16.3.2.1.2 Protocol for the Quantification of Living Foraminifera Using CellTracker Green

1. Sample the known area or volume (about 10 cm^2 or 10 cm^3).
2. Slice the core every ½ cm down to 2 cm, and every 1 cm down to 10 cm or more.
3. Store the samples at *in situ* temperature.
4. Transfer the subsamples to a labeled container.
5. Add an equal volume of chilled bottom seawater.
6. Add 1 mg CTG, diluted with 1 mL DMSO.
7. Shake the sample and incubate at *in situ* temperature for at least 7 h.
8. Fix the samples in 5% buffered formalin.
9. Wet sieved at 30 or 63 µm, or with a set of sieves.
10. Place the residues in a gridded Petri dish.
11. Wet pick the residues under an epifluorescent microscope.
12. Extract the green fluorescent hard-shelled foraminifera and store them in a micropaleontological slide.
13. Extract the green fluorescent foraminifera with a Pasteur pipette and store them in a glass cavity slide with glycerol, or in an Eppendorf vial with formalin.

Relative abundance as a percentage of each species, and absolute abundance as the number of specimens per dry gram of sediment or a known wet sediment volume are calculated. Furthermore, the sampling procedures include the size fractions analysis in order to estimate the abundance of foraminifera in different size structure and their distribution. For standing stock analyses, counts of living foraminifera is standardized to a known volume or known area (e.g., 10 cm^3 or 10 cm^2).

16.4 NOTES

1. An area of 10 cm^2 corresponds to 10 mL of 1 cm slices, and it allows the collection of a sufficient number of benthic foraminifera for statistical purposes.
2. It is possible also to use alcohol to preserve recent benthic foraminifera samples. However, soft-walled foraminifera could shrink, and they are not preserved correctly.
3. The use of Rose Bengal to distinguish living foraminifera is debated since several studies have shown that staining by Rose Bengal is still the best and cheaper method available to separate specimens that were alive or recently alive at the time of collection from dead individuals (see Bernhard 2000 and Murray and Bowser 2000 for a review).
4. The lower limit 63 μm for the examination of benthic foraminiferal assemblage has been a matter of discussion in the last few years. A few studies suggested that the inclusion of the 63 μm lower limit is a more reliable statistical base not only for paleoceanographic studies but also to investigate the biodiversity of foraminiferal association (Morigi 2009). In fact, studies performed on larger openings allow a significant loss of specimens and tend to create artificial "barren" zones in sequence dominated by small-sized species (Schröder et al. 1987). In particular, Shepherd et al. (2007) suggest that smaller sizes are important for assessments of foraminiferal assemblages in dysoxic, organic-rich environments. Moreover, a mesh net of 30 μm would allow more consistency in the comparison between foraminifera and metazoa meiofaunal variables. However, so far, only few studies have used this mesh size (Gooday et al. 1995).

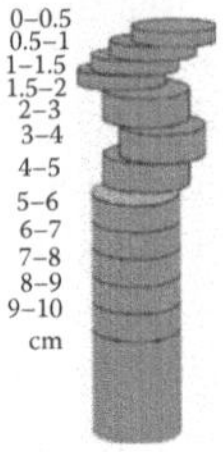

Sample known area or volume

↓

Slice the core every ½ cm down to 2 cm,
and 1 cm down to 10 cm or more

↓

Transfer sub-samples to labelled container

↓

Add 4% formalin buffered with sodium borate
and Rose Bengal (1 g/L)

↓

Shake gently and conserve with
Rose Bengal for at least 24 h

↓

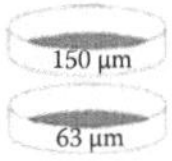

Wet-sieved at 63 μm,
or with a set of sieves

↓

Put the residues in a gridded Petri dish

↓

Wet-pick the residues under a stereomicroscope

↓

Extract the bright pink hard-shelled foraminifera
and store them in a micropaleontological slide

↓

Extract soft-wall foraminifera with a Pasteur pipette
and store them in a glass cavity slide with glycerol
or in an Eppendorf vial with formalin

PROTOCOL 16.1 Quantification of foraminifera—the Rose Bengal method.

REFERENCES

Bernhard, J. M. 2000. Distinguishing live from dead foraminifera: Methods review and proper applications. *Micropaleontology* 46: 38–46.

Bernhard, J. M., D. R. Ostermann, D. S. Williams, and J. K. Blanks. 2006. Comparison of two methods to identify live benthic foraminifera: A test between Rose Bengal and CellTracker Green with implications for stable isotope paleoreconstructions. *Paleoceanography* 21: http://hdl.handle.net/1912/1409.

Corliss, B. H. and S. Emerson. 1990. Distribution of Rose Bengal stained deep-sea benthic foraminifera from the Nova Scotian continental margin and Gulf of Maine. *Deep-Sea Research* 37: 381–400.

Gooday A. J. , M. Carstens, H. Thiel, 1995, Micro- and Nanoforaminifera from Abyssal Northeast Atlantic Sediments: A Preliminary Report. *Internationale Revue der gesamten Hydrobiologie und Hydrographie* Volume 80 Issue 2 Pages 361–383.

Morigi, C. 2009. Benthic environmental changes in the Eastern Mediterranean Sea during sapropel S5 deposition. *Palaeogeography, Palaeoclimatology, Palaeoecology* 273: 258–271.

Murray, J. W. 2007. Biodiversity of living benthic foraminifera: How many species are there? *Marine Micropaleontology* 64: 163–176.

Murray, J. W. and S. S. Bowser. 2000. Mortality, protoplasm decay, and reliability of staining techniques to recognize "living" foraminifera: A review. *Journal Foraminiferal Research* 30: 66–70.

Schröder, C. J., D. B. Scott, and F. S. Medioli. 1987. Can smaller benthic foraminifera be ignored in paleoenvironmental analysis? *Journal of Foraminiferal Research* 17(2): 101–105.

Shepherd, A. S., A. E. Rathburn, and M. E. Pérez. 2007. Living foraminiferal assemblages from the Southern California margin: A comparison of the >150, 63–150, and >63 μm fractions. *Marine Micropaleontology* 65: 54–77.

Walton, W. 1952. Techniques for recognition of living foraminifera, Contrib. Cushman Found. *Foraminiferal Research* 3: 56–60.

17 Abundance of Metazoan Meiofauna

CONTENTS

17.1 INTRODUCTION

Metazoan meiofauna represent the most abundant class of pluricellular organisms of the marine benthos, reaching abundances comprising 10^5 to 10^6 individuals per m^2 (100–1000 ind. 10 cm^{-2}) and also a biomass of 1–2 g DW m^{-2} at depth >100 m (Rex et al. 2001). Metazoan meiofauna are the dominant component of the deep-sea benthic metazoans both in terms of abundance and biomass (Vincx et al. 1994; Cook et al. 2000; Danovaro et al. 2002), and they have an important ecological role, representing the main pathway of energy transfer to higher trophic levels (as key component of the diet of juvenile demersal fish species; Leguerrier et al. 2003). Meiofauna are also characterized by a high sensitivity to environmental changes due to the short generation time, high sensitivity to any changes in environmental conditions, and lack of pelagic larval dispersion (Bongers and Ferris 1999). The total abundance and biomass of meiobenthic assemblages are known to decrease with increasing water depth, and several studies suggest that food supply is one of the major factors controlling their distribution in the deep-sea sediments (Vincx et al. 1994; Danovaro et al. 2000; Galéron et al. 2001; Danovaro et al. 2008). On a macroscale (>1000 km), deep-sea faunal distribution is assumed to depend primarily on physical parameters (temperature, hydrodynamic regimes, and near-bottom currents; Thiel 1983; Thistle and Sherman 1985; Thistle et al. 1991; Lambshead et al. 2001; Mokievsky et al. 2004) and food source availability (Schaff et al. 1992; Vanreusel et al. 1995; Relexans et al. 1996; Brown et al. 2001; Gambi and Danovaro 2006), while at smaller spatial scales (meso- and local scale), oxygen, food availability, and interactions between organisms or between organisms and the microhabitat have been suggested to be increasingly relevant (Eckman and Thistle 1988; Tyler 1995; Snelgrove et al. 1996; Levin et al. 2001). Here we report the methods for sampling and the laboratory procedures to extract and to sort and count metazoan meiofauna.

17.2 FIELD PROCEDURE

17.2.1 Sampling

In deep-sea soft sediments, multiple corer is used to collect simultaneously several meiofaunal samples (e.g., 4–12 corers; Elmgren and Radziejewska 1989). The instrument allows collection of undisturbed sediment samples of a known surface area to analyze the horizontal and vertical distribution of benthic organisms. Another common sampling instrument for soft sediments is the box corer (Fleeger et al. 1983). The weight of a box corer after sampling can range from 150 to >750 kg and collect a surface sediment of 0.02–1.0 m^2 down to a depth of 50 cm. From the box corer, meiofaunal samples could be collected using manual Plexiglas® corers (internal diameter 3.6 cm), one replicate from each different deployment. However, during sampling recovery, most box corers allow surface sediment mixing, which might alter the quality of the sample. Bett et al. (1994), Shirayama and Fukushima (1995), and Bett (2000), comparing the sampling efficiency of multiple corers and box corers in deep-sea environments, pointed out that meiofaunal abundance collected using box corers was significantly lower. However, some studies conducted in the Eastern Mediterranean demonstrated that the use of two different samplers (USNEL-type box corer and multicorers) provide identical results in terms of benthic parameters, suggesting an adequate sampling efficiency of both sampling devices (Danovaro et al. 1999; Gambi and Danovaro 2006).

17.2.2 Storage of Meiofaunal Samples at −20°C

Samples collected for quantitative analyses must be processed within few hours from collection because the short generation time and mobility of meiofauna (including the presence of some predators such as nematodes and turbellarians) might alter the state of the sample. If the vertical distribution of meiofauna is required, sediment should be immediately sliced (e.g., layers: 0–1 cm; 1–3 cm; 3–5 cm, 5–10 cm, and 10–15 cm). When sectioning is not possible, entire corers should be frozen (stored at −20°C), and can be preserved for a longer period of time. An alternative to sample freezing is fixation with buffered formalin (using $Na_2B_4O_7$, pH ca. 8.2). The quantity of formalin to add (final concentration: 4%) is calculated on the basis of the overall volume of sediment plus prefiltered seawater or distilled water containing $MgCl_2$ (80 gL^{-1}). Samples fixed with 4% formalin can be stored for a long time before being extracted and analyzed (Pfannkuche and Thiel 1988). Some soft-body meiofaunal taxa (i.e., gastrotrichs, turbellarians, and nemerteans) are subjected to major morphological alterations but are generally conserved intact. At the end, a few drops of a solution of Rose Bengal (0.5 gL^{-1}) can be added to facilitate organism sorting and counting.

17.2.2.1 Equipment and Instruments

Plexiglas corer, internal diameter 3.6 or 4.5 or 5.0 cm
Stopper for the corer
Parafilm® (American National Can Company)
Adhesive tape

17.2.2.2 Procedure

1. Keep the corer vertically [NOTE 1].
2. Eliminate the stopper from the lower opening [NOTE 2].
3. Close the lower opening with Parafilm [NOTE 3].
4. Fix Parafilm using adhesive tape [NOTE 3].
5. Eliminate the stopper from the upper opening.
6. Close the upper opening with the Parafilm.
7. Fix Parafilm using adhesive tape.
8. Write all information relative to the samples (name of the project, cruise, station, depth, replicate) using permanent ink.
9. Cover the notes with a transparent tape.

17.2.2.3 Notes

1. It is important to keep the corer vertical to avoid mixing of the sediment layers and altering the vertical distribution of meiofaunal organisms.
2. Eliminate the stopper from the lower opening and make sure that the upper opening is well closed. The upper stopper plays a key role to prevent sediments from falling down.
3. Close the corer with Parafilm and adhesive tape and keep the sample at −20°C within a few minutes of collection to avoid altering the vertical distribution of meiofaunal organisms.

17.2.3 Storage of Meiofaunal Samples Using Buffered 4% Formalin

17.2.3.1 Equipment and Instruments

Plastic jars
Pasteur pipettes
Ruler
Extruder
Spatula

17.2.3.2 Solutions

Prefiltered seawater (sieve: 0.2 μm)
Formalin
$Na_2B_4O_7$ (20 gL^{-1})
$MgCl_2$ (80 gL^{-1})
Rose Bengal (0.5 gL^{-1})

17.2.3.3 Procedure

1. Extrude the sediment from the corer, using a ruler to divide the corer in different layers [NOTE 1].
2. Transfer the sediment to a plastic jar (i.e., 150, 250, or 500 mL) [NOTES 2, 3].
3. Add a solution of prefiltered seawater with $MgCl_2$ (80 gL^{-1}) to anesthetize the organisms.
4. Add the quantity of buffered formalin ($Na_2B_4O_7$, 20 gL^{-1}) to obtain a final concentration of 4% (depending on the final volume of sediment plus the solution with $MgCl_2$ seawater) [NOTES 4, 5].
5. Add a few drops of Rose Bengal (0.5 gL^{-1}) [NOTE 6].
6. Store the samples at room temperature.

17.2.3.4 Notes

1. The number and thickness of the sediment layers should be selected according to the objectives of the study. In general, the core can be sliced in 0–1, 1–3, 3–5, 5–10, and 10–15 cm thick layers.

2. The volume of the jars is chosen on the basis of the volume (layer thickness) of the sediment.
3. Extrusion should be done carefully to move all sediment, minimizing the alteration in vertical stratification.
4. Pay attention to using formalin, follow all the safety procedures (gloves, glasses), keeping the areas aerated, that is, working under an extractor fan.
5. Shake the jar gently to mix the sediments with the 4% buffered formalin.
6. Close the stopper of the jars with Parafilm to prevent the formalin from spilling.

17.3 LABORATORY PROCEDURE

17.3.1 Extraction

Meiofaunal extraction from the sediment is carried out by means of mesh sieves:

1,000 μm as upper limit of meiofaunal size
20/30 μm as lower limit of meiofaunal size in deep-sea environments in which the size miniaturization of meiofaunal organisms is frequent and evident.

To increase extraction efficiency, before passing the sediment through the sieve, the sample can be treated with ultrasound (for 1 min 3 times, with 30 s intervals) in order to detach organisms from the grain particle surface.

17.3.1.1 Equipment and Instruments

Ultrasonic bath
Centrifuge
Sieves: 1000 and 20 μm
2 L beaker
Fresh water
50 mL tubes
Tube racks
Spoon spatulas
Pasteur
Wash bottles

17.3.1.2 Solutions

LUDOX
4% Buffered formalin
Rose Bengal (0.5 gL^{-1})

17.3.1.3 Procedure

Meiofaunal extraction from frozen samples: The sediment core is extruded and transferred into plastic jars containing prefiltered seawater. Samples can be left at room temperature and processed within a few hours. The extraction procedures are the same as reported for sediment fixed with formalin or frozen at −20°C.

Meiofauna extraction from muddy sediment fixed with formalin: The whole sample is passed through a mesh of 1000 μm, washed several times with water, and collected in a 2 L beaker. The collected material is then filtered onto a mesh of 20 μm (for deep-sea sediments) or 30 μm (for coastal and transitional environments). The residual sediment is placed in a 50 mL tube and supplemented with LUDOX (HS 30, HS 40, or TM, arranging the density to 1.15–1.18 g cm^{-3}; Heip et al. 1985; Burgess 2001; Giere 2009). Tubes are then centrifuged for 10 min at 3000 rpm (800 × g). The supernatant is filtered on a 20 μm (for deep-sea sediments) mesh and then washed again with water. The pellet sediment is again supplemented with LUDOX and centrifuged for 10 min at 3000 rpm. This procedure is repeated 3 times to ensure an extraction efficiency of >98% of meiofaunal organisms. The material collected on the top of the 20 μm mesh sieve is preserved in a 50 mL tube with 4% buffered formalin. A few drops of Bengal Rose (0.5 gL^{-1}) are added to the extracted sample to stain meiofaunal organisms. Once the extraction is completed, the pellet sediment has to be checked under binoculars to test extraction efficiency.

1. Treat the sample with ultrasound (for 1 min 3 times, with 30 s intervals).
2. Pass the whole sample through a sieve with 1000 μm mesh size and collect all the material in a beaker.
3. Pass all material through a sieve with 20 μm/30 μm mesh size [NOTES 1–3].
4. Eliminate the water as much as possible [NOTE 4].
5. Gently collect the sediment from the sieve with a spoon and place it in a 50 mL tube [NOTE 5].
6. Collect the residual sediment from the sieve with LUDOX, using a wash bottle in the 50 mL tube [NOTE 6].
7. Add LUDOX using the ratio 1:2/3 for sediment and LUDOX, respectively [NOTE 7].
8. Centrifuge the tubes for 10 min at 3000 rpm (800 × g) [NOTE 8].
9. Place the supernatant on a 20 μm/30 μm sieve and wash with water [NOTE 9].
10. Place the sieve with the sample on a Petri dish full of water.
11. Add LUDOX in the tubes and gently mix to resuspend the sediment.
12. Centrifuge the tubes for 10 min at 3000 rpm (800 × g).
13. Place the supernatant on a 20 μm/30 μm sieve and wash with water.
14. Place the sieve with the sample on a Petri dish full of water.
15. Add LUDOX in the tubes and gently mix to resuspend the sediment.
16. Centrifuge the tubes for 10 min at 3000 rpm (800 × g).
17. After centrifuging three times, collect the sample from the sieve with 4% buffered formalin solution using a wash bottle [NOTES 10–12].
18. Collect the samples in a 50 mL tube.
19. Add a few drops of Rose Bengal (0.5 gL^{-1}) [NOTE 13].
20. Check the residual sediment to verify the extraction efficiency.

17.3.1.4 Notes

1. Use the appropriate mesh sieve according to the characteristics of the ecosystems (coastal or deep-sea) in which the study on meiofauna is carried out.
2. The use of the sieve should be carried out gently and carefully to prevent damage to the organisms.
3. Collect all sediment carefully from the beaker after sieving.
4. Carefully keep out all water from the sample during the 20 μm sieving.
5. Collect all sediment carefully from the sieve and place it in a 50 mL tube.
6. Split the samples in different 50 mL tubes when the quantity of sediment is higher than 15 mL.
7. The sample should be washed very carefully because the sea salt in the sample could crystallize with LUDOX. This happens during the mixing of sediments with LUDOX or after the first centrifuge. In this case, the sediment should be immediately washed with abundant water. The sediment should be transferred to a beaker, cleaned with fresh water, and sieved again to avoid loss of the sample.
8. Centrifugation should be repeated 3 times to extract more than 98% of meiofaunal organisms.

9. During centrifugations, the sieve should be placed on a Petri dish full of water to prevent the dehydration of meiofaunal organisms collected after the first centrifugation. At the end of the last centrifugation, all the residual material has to be collected in a 50 mL tube with buffered formalin (4%).
10. Use the formalin carefully, dressing in all the safety devices (laboratory coat, gloves, glasses, respiratory protection), and work using an extractor fan.
11. The solutions containing the formalin must be collected and eliminated using standard safety procedures.
12. The effectiveness of the formalin must be accurately tested because longer treatments ensure higher preservation but can damage meiofaunal organisms.
13. Rose Bengal can stain the soft-body organisms in 15 min, whereas for hard cuticle specimens staining occur ca. 2 d.

17.3.2 Sorting

Once the extraction of organisms from the sediment is completed, it is possible to proceed with the sorting in order to identify and count meiofaunal organisms. The use of Rose Bengal is recommended to facilitate the identification of the individuals (Higgins and Thiel 1988; Giere 1993, 2009).

The examination and counting of individuals require the employment of a stereomicroscope at a minimum of 25× magnification factor. A higher magnification factor may be necessary for the smallest meiobenthic organisms and larvae (40× and 50×). Sample examination and counting are eased by using cuvettes with internal subdivisions (e.g., 200 cells in the Delfuss cuvette). The identification, to the species level, of various taxa requires the collection of individuals and preparation of permanent slides for observation under the microscope.

17.3.2.1 Equipment and Instruments

Stereomicroscope
Sieve: 20 μm/30 μm
Delfuss cuvette
Wash bottle

17.3.2.2 Solutions

Fresh water
4% buffered formalin
Rose Bengal (0.5 gL^{-1})

17.3.2.3 Procedure

A sample is transferred to the 20 μm/30 μm sieve and, after washing with fresh water to eliminate the residuals of formalin, it is transferred to the Delfuss cuvette (Figure 17.1). Before starting the sorting, wait a few minutes to allow the sedimentation of all organisms in the bottom of the cuvette. Start the analyses of the sample from the first cell on the left site of the first line of the Delfuss cuvette. At the end of the sorting, the sample is transferred to the 20 μm/30 μm sieve, and it is transferred again to the 50 mL tube with a wash bottle full of 4% buffered formalin. Finally, the sample is stained again with Rose Bengal. An example of a data sheet to annotate sorting and counting of all meiofaunal taxa, together with abiotic environmental parameters, is reported in Table 17.1.

1. Transfer the samples fixed with 4% buffered formalin to the 20 μm/30 μm sieve.
2. Wash the sample to eliminate the 4% buffered formalin.
3. Transfer the sample to a Delfuss cuvette with water using a wash bottle [NOTES 1, 2].
4. Count and identify meiofaunal organisms under a stereomicroscope (minimum magnification 25×) [NOTES 3, 4].

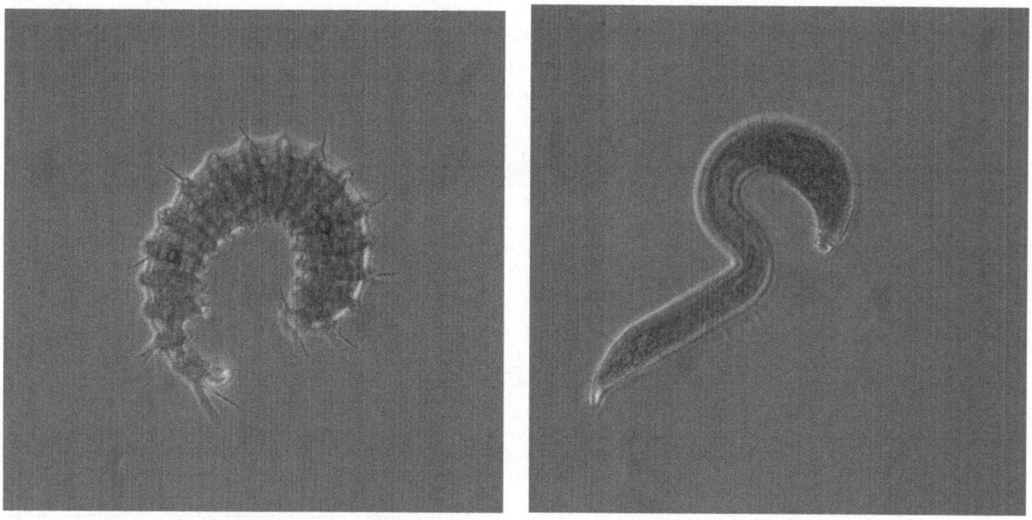

(a) (b)

FIGURE 17.1 Meiofaunal specimens extracted from the sediment.

5. Check the surface water to count and classify the specimens that can float.
6. At the end of the sorting, transfer the sample to a 20 μm/30 μm sieve [NOTE 5].
7. Transfer the samples to a 50 mL tube with 4% buffered formalin using a wash bottle [NOTE 6].
8. Add a few drops of 0.5 gL^{-1} Rose Bengal solution with a Pasteur pipette.
9. Store the samples at room temperature.

17.3.2.4 Notes

1. Wait a few minutes before starting the sorting after the transfer of the sample into the cuvette.
2. Be careful not to knock against the cuvette to prevent restarting the sorting from the beginning.
3. Check the lateral walls of the cuvette in order to prevent the loss of organisms sticking to them.
4. Use the appropriate magnification that allows observation of the entire cell.
5. Carefully collect all specimens from the tubes, the sieve, and the cuvette. Clean the sieve very well at each step during the preparation of the sample for sorting and after the end of the counting.
6. Use the formalin carefully, using all the safety devices (laboratory coat, gloves, glasses, respiratory protection), and work using an extractor fan.
7. The solutions containing the formalin must be collected and eliminated using standard safety procedures.

TABLE 17.1
An Example of the Data Sheet Utilized for the Sorting of Meiofauna

Program:

Station: Station depth:

Coordinates: Depth of sediment core (cm):

N° Replicates: Layer (cm):

TAXA	Replicate 1	Replicate 2	Replicate 3	Replicate 4	Replicate 5	Average
1. Nematoda						
2. Copepoda Nauplii						
3. Polychaeta						
4. Bivalvia						
5. Ostracoda						
6. Kinorhyncha						
7. Turbellaria						
8. Oligochaeta						
9. Tardigrada						
10. Gastrotricha						
11. Cumacea						
12. Amphipoda						
13. Isopoda						
14. Tanaidacea						
15. Acarina						
16. Nemertina						
Incertae sedis						
Others						
Number of other taxa						
Total number of taxa						

Notes

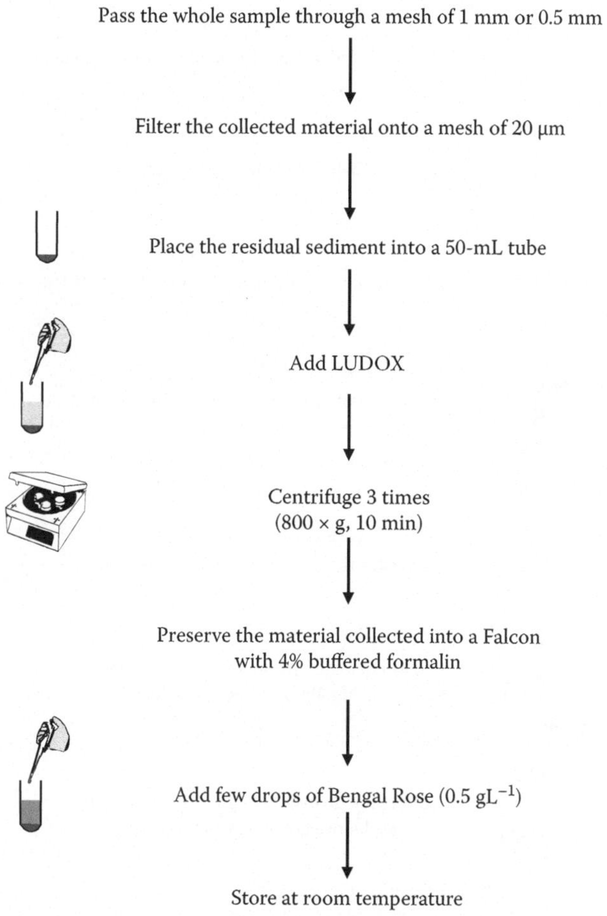

(a)

PROTOCOL 17.1 Meiofaunal extraction from sediments. *Continued*

Pass the extracted sample through a mesh of 20 μm

↓

Wash with abundant water

↓

Transfer the residual sample into a Delfuss cuvette

↓

Count and identify meiofaunal organisms under a stereomicroscope (minimum magnification × 25)

↓

At the end of the sorting, transfer the sample onto a 20 μm sieve

↓

Transfer the samples to a 50 mL tube using 4% buffered formalin

↓

Add some drops of 0.5 g L^{-1} Rose Bengal and store at room temperature

(b)

PROTOCOL 17.1 *Continued.*

REFERENCES

Bett, B. J. 2000. Comparative benthic ecology of the Rockall Trough and Faeroe–Shetland Channel. In *Atlantic margin environmental surveys of the seafloor 1996 and 1998 (Atlantic Frontier Environmental Network)* [CDROM], ed. Geotek Ltd. Section 4.3.2. London, U.K.

Bett., B., A. Vanreusel, M. Vincx, et al. 1994. Sampler bias in the quantitative study of deep-sea meiobenthos. *Marine Ecology Progress Series* 104: 197–203.

Bongers, T. and H. Ferris. 1999. Nematode community structure as a bioindicator in environmental monitoring. *Trends in Ecology and Evolution* 14: 224–228.

Brown, C. J., P. J. D. Lambshead, C. R. Smith, L. E. Hawkins, and R. Farley. 2001. Phytodetritus and the abundance and biomass of abyssal nematodes in the central, equatorial Pacific. *Deep-Sea Research Part I* 48: 555–565.

Burgess, R. 2001. An improved protocol for separating meiofauna from sediments using colloidal silica sols. *Marine Ecology Progress Series* 214: 161–165

Cook, A. A., P. J. D. Lambshead, L. E. Hawkins, N. Mitchell, and L. A. Levin. 2000. Nematode abundance at the oxygen minimum zone in the Arabian Sea. *Deep-Sea Research Part I* 47: 75–85.

Danovaro, R., D. Marrale, N. Della Croce, A. Dell'Anno, and M. Fabiano. 1999. Heterotrophic nanoflagellates, bacteria, and labile organic compounds in continental shelf and deep-sea sediments of the eastern Mediterranean. *Microbial Ecology* 35(3): 244–255.

Danovaro, R., A. Tselepides, A. Otegui, and N. Della Croce. 2000. Dynamics of meiofaunal assemblages on the continental shelf and deep-sea sediments of the Cretan Sea (NE Mediterranean): Relationships with seasonal changes in food supply. *Progress in Oceanography* 46: 367–400.

Danovaro, R., C. Gambi, and N. Della Croce. 2002. Meiofauna hotspot in the Atacama Trench (Southern Pacific Ocean). *Deep-Sea Research Part I* 49: 843–857.

Danovaro, R., C. Gambi, N. Lampadariou, and A. Tselepides. 2008. Deep-sea biodiversity in the Mediterranean Basin: Testing for longitudinal, bathymetric, and energetic gradients. *Ecography* 31: 231–244

Eckman, J. E. and D. Thistle. 1988. Small-scale spatial pattern in meiobenthos in the San Diego Trough. *Deep-Sea Research Part I* 35: 1565–1578.

Elmgren, R. and T. Radziejewska. 1989. Recommendations for quantitative benthic meiofauna studies in the Baltic. *Baltic Marine Biologists Publication* 12: 1–23.

Fleeger, J. W., W. B. Sikora, and J. P. Sikora. 1983. Spatial and long-term temporal variation of meiobenthic–hyperbenthic copepods in lake Pontchartrain, Lousiana. Estuar. *Estuarine, Coastal, and Shelf Science* 16: 441–453.

Galéron, J., M. Sibuet, A. Vanreusel, et al. 2001. Temporal patterns among meiofauna and macrofauna taxa related to changes in sediment geochemistry at an abyssal NE Atlantic site. *Progress in Oceanography* 50: 303–324.

Gambi, C. and R. Danovaro. 2006. A multiple-scale analysis of metazoan meiofaunal distribution in the deep Mediterranean Sea. *Deep-Sea Research Part I* 53: 1117–1134.

Giere, Olav. 1993. *Meiobenthology: The microscopic fauna in aquatic sediments*. Berlin: Springer-Verlag.

Giere, Olav. 2009. *Meiobenthology. The microscopic motile fauna of aquatic sediments*. Berlin: Springer-Verlag, 2nd edition.

Heip, C., M. Vincx, and G. Vranken. 1985. The ecology of marine nematodes. *Oceanography and Marine Biology Annual Review* 23: 399–489.

Higgins, R. P. and H. Thiel. 1988. *Introduction to the study of meiofauna*. Washington DC: Smithsonian Institution Press.

Lambshead, P. J. D., J. Tietjen, A. Glover, T. Ferrero, D. Thistle, and A. J. Gooday. 2001. The impact of large-scale natural physical disturbance on the diversity of deep-sea North Atlantic nematodes. *Marine Ecology Progress Series* 214: 121–126.

Leguerrier, D., N. Niquil, N. Boileau et al. 2003. Numerical analysis of the food web of an intertidal mudflat ecosystem on the Atlantic coast of France. *Marine Ecology Progress Series* 246: 17–37.

Levin, L. A., R. J. Etter, M. A. Rex et al. 2001. Environmental influences on regional deep-sea species diversity. *Annual Review of Ecology and Systematics* 32: 51–93.

Mokievsky, V. O., A. A. Udalov, and A. I. Azovsky. 2004. On the quantitative distribution of meiobenthos on the shelf of the world ocean. *Oceanology* 44: 99–109.

Pfannkuche, O. and H. Thiel. 1988. Sample processing. In *Introduction to the study of meiofauna*, ed. R. P. Higgins and H. Thiel, 134–145. London: Smithsonian Institution Press.

Relexans, J. C., J. Deming, A. Dinet, J. F. Gaillard, and M. Sibuet. 1996. Sedimentary organic matter and micro-meiobenthos with relation to trophic conditions in the tropical northeast Atlantic. *Deep-Sea Research Part I* 43: 1343–1368.

Rex, M.A., C.T. Stuart, and R.J. Etter. 2001. Do deep-sea nematodes show a positive latitudinal gradient of species diversity? The potential role of depth. *Marine Ecology Progress Series* 210: 297–298

Schaff, T. R., L. A. Levin, N. Blair, D. De Master, R. Pope, and S. Boehme. 1992. Spatial heterogeneity of benthos on the Carolina continental slope: Large (100 km)-scale variation. *Marine Ecology Progress Series* 88: 143–160.

Shirayama, Y. and T. Fukushima. 1995. Comparison of deep-sea sediments and overlying water collected using multiple corer and box corer. *Journal of Oceanography* 51: 75–82.

Snelgrove, P. V. R., J. F. Grassle, and R. F. Petrecca. 1996. Experimental evidence for aging food patches as a factor contributing to high deep-sea macrofaunal diversity. *Limnology and Oceanography* 41(4): 605–614.

Thiel, H. 1983. Meiobenthos and nanobenthos of the deep-sea. In *Deep-sea biology*, ed. G. Rowe, 167–230. New York: Wiley.

Thistle, D., S. C. Ertman, and K. Fauchald. 1991. The fauna of the Hebble site: Patterns in standing stock and sediment-dynamic effects. *Marine Geology* 99: 413–422.

Thistle, D. and K. M. Sherman. 1985. The nematode fauna of a deep-sea site exposed to strong near-bottom currents. *Deep-Sea Research Part I* 32: 1077–1088.

Tyler, P. A. 1995. Conditions for the existence of life at the deep-sea floor: An update. *Oceanography and Marine Biology. An Annual Review* 33: 221–244.

Vanreusel, A., M. Vincx, B. J. Bett, and A. L. Rice. 1995. Nematode biomass spectra at two abyssal sites in the NE Atlantic with a contrasting food supply. *Internationale Revue der gesamten Hydrobiologie* 80: 287–296.

Vincx, M., B. J. Bett, A. Dinet, et al. 1994. Meiobenthos of the deep Northeast Atlantic. *Advance in Marine Biology* 30: 2–88.

18 Macrofaunal Abundance

CONTENTS

18.1 INTRODUCTION

Macrofauna include all metazoan organisms broader than 300 μm and exclude any permanent meiofaunal taxa such as nematodes, harpacticoid copepods, and ostracods (Hessler and Jumars 1974). In the past, different mesh sizes have been used, setting the lower limit to 1.0 mm and then 0.5 mm, but the use of a smaller mesh size is recommended for studies aiming at identifying the different stages of the life cycle (Buchanan and Warwick 1974; Hartley 1982). Soft-bottom macrobenthic communities in the deep sea are mainly composed of infaunal organisms, mainly including polychaetes, amphipod and decapod crustaceans, bivalve mollusks, and echinoderms.

Macrofaunal density, biomass, and trophic structure generally decrease with increasing water depth (Thurston et al. 1994; Flach and Heip 1996; Stora et al. 1999; Cartes et al. 2002; Hughes and Gage 2004; Rex et al. 2006), so large volume of sediments are required for accurate determination of macrofaunal variables.

18.2 FIELD PROCEDURES

18.2.1 SAMPLING

Several devices have been used for soft-bottom macrobenthos sampling, including grabs, different types of box corers, and push corers handled by ROVs. Box corers have been largely utilized in quantitative studies, necessary for adequate characterizations of benthic communities since they allow the collection of a large amount sediment. In addition, they yield more precise information concerning the vertical distribution of individuals and, by collecting more integer specimens, allow more accurate biomass estimates. The most used models of box corers have been reviewed by Holme and McIntyre (1971, 1984) and by Eleftheriou and Holme (1984).

A modified USNEL box corer is generally used to collect deep-sea macrobenthos samples (Hessler and Jumars 1974; Gage and Tyler 1991). The area typically sampled by this corer is 0.25 m^2 (= 50 × 50 cm), which is a relatively good unit sampling area to provide meaningful numbers of organisms given the low densities existing in deep-sea sediments (Cosson et al. 1997). However, especially in highly oligotrophic regions, up to 3–5 independent deployments might be needed to collect a representative macrobenthos sample.

New box corers (NIOZ type) have been designed to allow the collection of samples, which close hermetically after penetration into the sediment. They thus avoid the sediment washing during retrieval, which has been hypothesized to be responsible for a significant loss of individuals (especially of meiofaunal size). Moreover, these devices maintain the substrate stratification of organisms down to the bottom profile.

An alternative sampling tool, even more efficient, is the megacorer, a hydraulically damped device that collects multiple cylindrical cores of 10–12 cm internal diameter. Megacorers could be even more effective than large oceanic box corers in the sampling of deep-sea macrofauna. Studies based on the comparison of sampling efficiencies between the box corer and megacorer suggested that the USNEL corer significantly undersampled macrobenthos (Bett 2000; Hughes and Gage 2004). The multiple corer has the advantage of causing minimal sediment disturbance (lower than that of the box corer) and allows the collection of several (pseudo) replicates (typically 10).

Whatever the type of sampler chosen, the device should ensure the collection of samples as undisturbed as possible. The sampling size is also a key element. A surface area of 0.1–1.0 m^2 that is currently utilized in shallow water sampling would not be sufficient to collect the required amount of organisms. Sediment samples should be at least 30 cm deep.

In general, the "quality" of the sediment sample is evaluated visually once on board, and only if the core sample appears to be "good" or "reasonable" that is, mostly undisturbed, is it retained for macrobenthos analysis.

After it is recovered on board, the sampler must be suitably treated in order to collect the entire sample and sieve it without losing any material. After a preliminary description of its surface, the sediment sample must be completely removed from the gear and sieved as soon as possible. If this activity is not possible, the sediment has to be preserved in 4%–10% buffered formalin.

18.2.2 Sieving

Samples must be sieved in order to eliminate water, fine sediments, and any other extraneous materials not necessary for the study. After it is recovered on board, the sampler must be suitably treated so as to collect the entire sample and sieve it without leaving out any material. After a preliminary description of its surface, the sample must be completely removed from the gear and sieved as soon as possible. It can be placed in a plastic tray and treated by having it pass through a mobile sieve according to the following steps.

First, the sample description must be recorded on an appropriate schedule, including general data concerning the sample as well as its surface features, apparent individual density, occurrence of detritus, pteropods, etc. Then, the sample must be completely removed from the gear and placed in an appropriate container. Any water occurring on the sample surface must be placed in the bowl with the sample. All sampling equipment must be carefully washed and the resulting water–sediment mixture collected after carefully sieving it with a mesh of 0.250 mm to avoid accidental introduction of other organisms.

A washing desk equipped with a wide smooth surface, on which samples can be gently washed and dissolved before being placed in the container, can be used for dissolving the sample sediment material (Figure 18.1).

A valid alternative most suitable for clay sediments is to slightly dissolve the sediment inside the bowl itself after adding 0.250 mm filtered seawater (3–4 times the volume of the sample). Water may be directly sprinkled onto the sample with a low-pressure nozzle in order to avoid any damage to animals. Small clay nodules, which can sometimes remain aggregate, will be dissolved later during sieving. Sieving consists of dissolving the sediment sample in seawater in order to separate macrofauna (Eleftheriou and Holme 1984; Kingston and Riddle 1989). A mesh of 0.5 mm is typically utilized when meiofauna also is analyzed in the same site, so that the lower size of the macrofauna is used as the upper size of the meiofauna. The sieve must have an adequate surface to avoid clogging. The choice of the appropriate surface size depends on sieve mesh, sieved volume, and type of sediment being worked on. A sieve with a diameter of 30–40 cm may be sufficient in most

FIGURE 18.1 Washing and sieving the sample on board using a mesh (size 300 μm).

cases as a good compromise between the size and robustness of the sieve. Circular sieves can be more suitable for manual sampling treatment than square ones, but square sieves are also frequently used. The depth of the sieve must enable the sample to settle and prevent any loss of material that might occur with too vigorous washing. The spray nozzle allows animals to be separated from the sediment without being damaged. Pressurized seawater is recommended, using a pump with a filter to avoid sample contamination. Care must be taken to see that the sediment does not rub against the sieve surface, in order to avoid organisms being damaged. If small-size animals pass trough the sieve mesh, it is advisable to collect the washing water in a suitable basin and sieve it again so as to recover these organisms. At the end of the sieving, the organisms can be stained with Rose Bengal (5 g L^{-1}).

After the sieving operation is completed, the remaining material (organisms, shell fragments, vegetal debris, coarse sediment, and any other matter) must be transferred to the appropriate containers. The whole sieved material must then be made to flow toward the edge (or a corner) of the sieve using a low-pressure jet; it should then be removed extremely carefully, after which the sieve must be carefully inspected to find any organisms caught in the mesh. The sieve mesh may be washed with a low-pressure jet. Once the material has been carefully and completely removed, the sieve must be thoroughly washed and cleaned with a bristle scrubbing brush to prepare it for further samples. Containers for storage must have an airtight closure and be large enough to hold the sample and treatment solutions. High density polyethylene (HDPE) containers have been found to be suitable for long-term storage. Cylindrical containers fitted with stopper and base plug have the advantage of facilitating the organisms' recovery, avoiding damage to them.

18.2.3 Fixation

Fixation of the biological sample prevents degradation of the tissues, thus preserving the biological structures for subsequent analyses. In some cases, especially for soft-body or elongated organisms, it is useful to anesthetize the animals immediately after collection and then proceed with fixation. This allows the organisms to be fixed with longer body extension, which in turn might facilitate their identification. For each taxonomic group, there are special anesthetizing/relaxation and fixation methods, depending in particular on the different chemical composition of the various organisms. A saturated solution of MgCl can be used in many cases to induce muscle relaxation.

Prior to fixing the sample, please check that the container is filled to the top so that the total volume of samples plus fixative fits into the available space. The most common preservative is formalin. A solution of 100% formalin is equivalent to 37%–40% formaldehyde (depending on purity). This has to be considered when describing the concentration of the solution utilized. In the past, 10% formalin solutions have been used, but for identification carried out within a reasonable time, a less concentrated solution is recommended (e.g., 4% formalin in prefiltered seawater). Formalin is indeed highly toxic and potentially cancerogenic, and all security measures must be applied for preventing any contamination risk in the laboratory. Since formaldehyde is an acid (pH variable from 2.5 to 5) and could cause the partial dissolution of calcified organisms, the solutions are typically buffered with sodium tetraborate (Borax, $Na_2B_4O_7$; 20 g L^{-1}). The final pH is checked using a pH meter to ensure that the value reaches a pH close to 8.0

18.2.4 Staining

Staining the organisms helps in the sorting of the samples and makes easier the identification of the smaller organisms. The most common stain is Rose Bengal, which, as in the case of formalin, must be used applying all security measures to avoid its toxicity and potential cancerogeneity. Moreover, its use may alter the color patterns of some taxa, for example, polychaetes, making more difficult their identification. Methylene blue can also be used and is far less toxic, but only at the time of identification.

18.3 LABORATORY PROCEDURES

18.3.1 Sorting

During sorting, the key steps include removal of fixatives, sample washing, subsampling (whenever necessary), preliminary sorting, fine sorting under the stereomicroscope, and separation of specimens belonging to the different high taxa.

Although sorting of fresh material is more advantageous and provides more reliable data (especially for soft-bodied specimens), it is often difficult to perform. Macrobenthos sorting is typically carried out on preserved material—after fixative solutions are removed—in a smaller-mesh sieve than the mesh size generally used for sieving on board. Sorting procedures are performed under table fume hoods to prevent the release of toxic fumes, especially during removal of fixatives such as formalin. This should be done by pouring it into a suitably sized beaker through a sieve with a smaller-sized mesh than that used for sieving the fresh sample. The sieve should be placed over a wide-necked funnel that is firmly supported by the beaker. As an alternative to the sieve–funnel combination, a funnel equipped with a mesh may be used. The container must then be rinsed well with a spray nozzle so that not even the slightest trace of the sample is left; the rising water itself must also be sieved. After inspecting and washing the sieve again, the container with the preservative solution must be emptied into the appropriate laboratory container.

The material to be sorted is placed in a small white plastic tray and analyzed under a binocular. A careful sorting can be done by moving a small quantity of the sample into an 8–10 cm Petri dish, which is then scanned under the stereomicroscope to appropriate magnification. Each organism is placed in a container of suitable size, refixed, labeled, and stored at room temperature in the dark for subsequent identification.

During the initial sorting phase, the organisms are preferably separated according to their main taxonomic groups (i.e., polychaetes, oligochaetes, nemertines, bivalves, gastropods, amphipods, echinoderms, other crustaceans, and others).

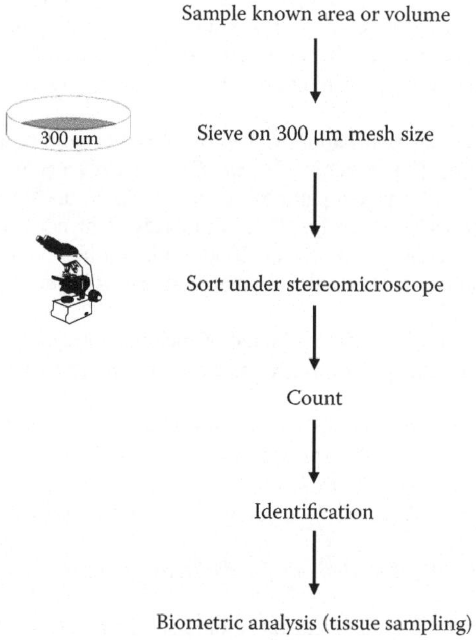

PROTOCOL 18.1 Macrofaunal counting.

REFERENCES

Bett, B. J. 2000. Comparative benthic ecology of the Rockall Trough and Faeroe-Shetland Channel. In *Atlantic margin environmental surveys of the seafloor 1996 and 1998 (AFEN)* [CDROM], ed. Geotek Ltd., Section 4.3.2. London, UKOOA.

Buchanan, J. and R. Warwick. 1974. An estimate of benthic macrofauna production in the offshore mud of the Northumberland coast. *Journal of Marine Biology Association UK* 54: 197–222.

Cartes, J. E., A. Grémare, F. Maynou, S. Villora-Moreno, and A. Dinet. 2002. Bathymetric changes in the distributions of particulate organic matter and associated fauna along a deep-sea transect down the Catalan sea slope (north-western Mediterranean). *Progress in Oceanography* 53: 29–56.

Cosson, N., M. Sibuet, and J. Gàleron. 1997. Community structure and spatial heterogeneity of the deep-sea macrofauna at three contrasting stations in the tropical northeast Atlantic. *Deep-Sea Research Part I* 44: 247–269.

Castelli, A., A. Lardicci, and D. Tagliapietra. 2004. Soft bottom macrobenthos. *Biologia Marina Mediterranea* 11:99-131

Eleftheriou, A. and N. A. Holme. 1984. Macrofauna techniques. In *Methods for the study of marine benthos*, ed. N. A. Holme and A. D. Mc Intyre, 40–121. London: Blackwell Scientific.

Ferraro, S. P., F. A. Cole, W. A. De Ben, and R. C. Swartz. 1989. Power-cost efficiency of eight macrobenthic sampling schems in Puget Sound, Washington, USA. *Canadian Journal of Fisheries Aquatic Sciences* 46: 2157–2165.

Flach, E. and C. Heip. 1996. Vertical distribution of macrozoobenthos within the sediment on the continental slope of the Goban Spur area (NE Atlantic). *Marine Ecology Progress Series* 141: 55–66.

Gage, J. D. and P. Tyler. 1991. Depth-related patterns in community composition. In *Deep-sea biology*. ed. J. D. Gage and P. Tyler, 229–247. Cambridge, U.K.: Cambridge University Press.

Galéron, J., M. Sibuet, M. L. Mahaut, and A. Dinet. 2000. Variation in structure and biomass of the benthic communities at three contrasting sites in the tropical Northeast Atlantic. *Marine Ecology Progress Series* 197: 121–137.

Gray, J. S., A. D. McIntyre, and J. Stirn. 1992. Manual of methods in aquatic environment research. Part II. Biological assessment of marine pollution with particular reference to benthos. FAO Fisheries Technical Papers 324: 1–49.

Hartley, J. P. 1982. Methods for monitoring offshore macrobenthos. *Marine Pollution Bulletin* 13: 150–154.

Hessler, R. R. and P. A. Jumars. 1974. Abyssal community analysis from replicate box cores in the central North Pacific. *Deep-Sea Research* 21: 185–209.

Holme, N. A. and A. D. McIntyre. 1971. *Methods for the study of marine benthos*. Oxford, U.K.: Blackwell Scientific..

Holme, N. A. and A. D. McIntyre. 1984. *Methods for the study of marine benthos*. Oxford, U.K.: Blackwell Scientific.

Hughes, D. J. and J. D. Gage. 2004. Benthic metazoan biomass, community structure, and bioturbation at three contrasting deep-water sites on the northwest European continental margin. *Progress in Oceanography* 63: 29–55.

Kingston, P. F. and M. J. Riddle. 1989. Cost effectiveness of benthic faunal monitoring. *Marine Pollution Bulletin* 20: 490–496.

Rex, M. A., R. J. Etter, J. S. Morris et al. 2006. Global bathymetric patterns of standing stock and body size in the deep-sea benthos. *Marine Ecology Progress Series* 317: 1–8.

Stora, G., M. Bourcier, A. Arnoux, et al. 1999. The deep-sea macrobenthos on the continental slope of the north-western Mediterranean Sea: A quantitative approach. *Deep-Sea Research Part I* 46: 1339–1368.

Thurston, M. H., B. J. Bett, A. L. Rice, and P. A. B. Jackson. 1994. Variations in the invertebrate abyssal megafauna in the North Atlantic Ocean. *Deep-Sea Research Part I* 41: 1321–1348.

19 Megafaunal Abundance

CONTENTS

19.1 INTRODUCTION

Megafauna are composed by large benthic organisms. These organisms are visible without the help of microscopes or other specific tools, and can be investigated using remote systems (video, images, etc.) or after their collection and retrieval on board. Despite their size, megafaunal organisms are still largely unexplored in deep-sea ecosystems. Megafauna are present at all depths down to the deepest trenches, but their abundance decreases significantly with depth (Figure 19.1). The slope of the regression of megafaunal abundance versus depth suggests a rate of decrease higher than that for the other faunal components inhabiting the deep-sea floor. Multiple comparisons show that slopes of two combinations—bacteria versus metazoan meiofauna, and macrofauna versus megafauna—do not differ. Within the 200–6000 m depth range reported by Rex et al. (2006), there are sequences of significant difference in the overall abundance of bacteria > meiofauna > macrofauna > megafauna.

The steeper regressions of the abundance of macrofauna and megafauna, compared to smaller-size groups, suggest that large organisms are more affected by the reduction in available energy with depth. At great depths, for certain megafaunal species, the population densities of larger organisms may become so low, due to energy restrictions, that reproduction processes could be altered (Rex 1973; Thiel 1975; Rex et al. 2005).

19.2 FIELD PROCEDURES

19.2.1 SAMPLING

Sampling deep-sea megafauna can be conducted using a variety of trawled gear and traps (Gage and Bett 2005). The type of sampling equipment used varies according to the taxa targeted and the seafloor type. The use of this trawled equipment requires a large research vessel with a winch with enough cable length to reach the seafloor. In general, the length of cable required is between 2 and 2.5 times the depth at which the sample is to be collected. The equipment is launched from the stern of the ship, lowered to the seafloor, and trawled at a speed approximately 2 knots on the seafloor for a variable length of time. As an example, it will take 12 h to sample with an OTSB14 trawl at 5000 m depth: 4 h for the trawl to reach the bottom, 4 h fishing on the bottom, and 4 h to come back to the surface.

The following are descriptions of some of the main gears used for collecting deep-sea megafauna. The two most used are the otter trawl and the Agassiz trawl. The epibenthic sled is also regularly used to collect both mega and macrofauna from the benthos and suprabenthos. Although

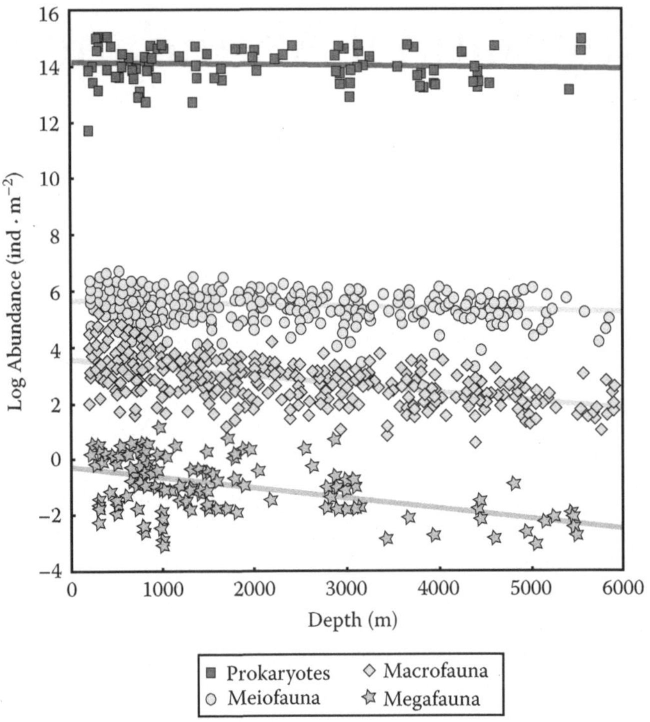

FIGURE 19.1 Bathymetric patterns of abundance of the main biological components inhabiting deep-sea sediments. Reported are the patterns of prokaryotic (squares), meiofaunal (dots), macrofaunal (diamonds), and megafaunal abundances (stars); the y-axis is Log 10 scale. (Modified from Rex M. A. et al. 2006. *Marine Ecology Progress Series* 317: 1–8.)

this equipment produces good qualitative samples, they are somehow limited in their quantitative results, even in the case of gear with odometer wheels and other bottom-tracking devices such as pingers or sensors (Bett et al. 2001). Nevertheless, an estimation of trawling time and trawl-mouth opening can be made, and benthic trawls have been used efficiently in deep-sea studies allowing for spatiotemporal comparisons in the deep-sea communities (Billett et al. 2001; Company et al. 2004; Sardà et al. 2004; Ramírez-Llodra et al. 2008). Baited traps are only used in specific cases when scavenger or predator species want to be attracted and collected. Finally, only in recent years has the use of submersibles and remotely operated vehicles (ROV) become more common, although their availability is still strongly limited by their cost.

Otter trawl: This is a fishing net with a large mouth opening (8–14 m) and two large doors (otters) that keep the mouth opened during sampling (Merrett and Marshall 1981; Sardà et al. 1998; Billett et al. 2001) (Figures 19.2 and 19.3). The head line (top rope of the mouth) is held above the seafloor by floats, while the footrope has weights. The body of the net has a decreasing mesh size from the mouth (approximately 44 mm) to the cod end (approx. 12 mm), where the sample is finally caught.

Agassiz trawl: This is a double-sided beam trawl with a main net usually of 20 mm mesh and a cod end lined with 10 mm mesh (Gage and Tyler 1991) (Figure 19.4). It is more robust than the otter trawl and is used to sample unknown grounds or when the seafloor may have rocky outcrops.

Epibenthic sled: This gear consists of a rectangular steel frame on wide runners to prevent sinking in the sediment, with one to three nets attached to it (Hessler and Sanders 1967; Rice et al. 1982; Guennegan and Martin 1985; Rice 1987). The nets can be opened and closed at the start and end of each tow, taking quantitative samples of the suprabenthonic fauna (Herring 2003). The frame slides over the sediment, and it is especially efficient in collecting epibenthic and suprabenthic fauna as

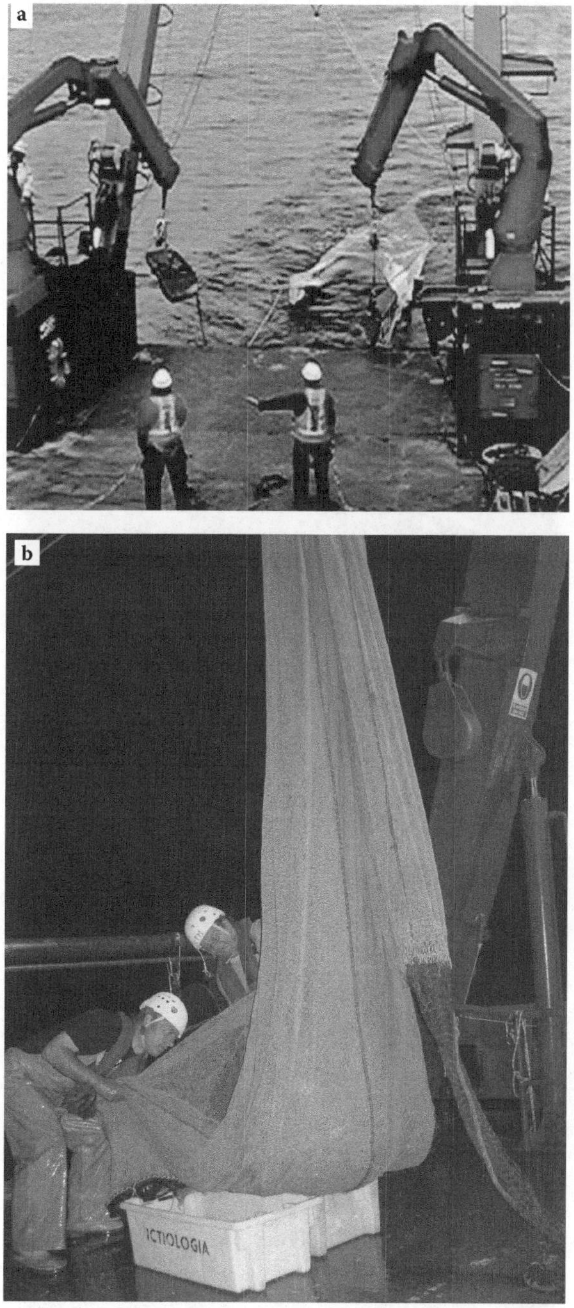

FIGURE 19.2 The otter trawl is efficient for sampling benthic megafauna on soft sediments. (a) The OTSB trawl being recovered, showing the doors and open mouth . (b) A full OTMS trawl net on deck after recovery. (Courtesy Eva Ramírez-Llodra, NOCS and ICM-CSIC. Used with permission.)

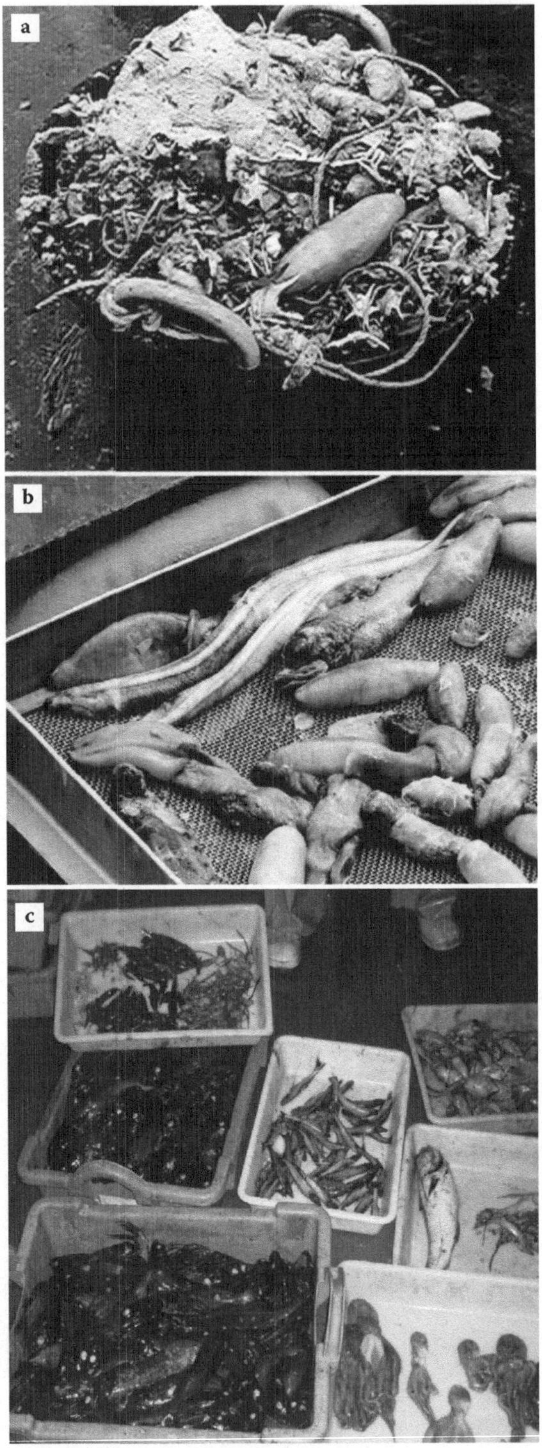

FIGURE 19.3 OTSB and OTMS samples: (a) Nonsorted sample from the NE Atlantic abyssal floor. (b) Part of the same sample on the sorting table showing deep-sea fish and holothurians. (c) Mediterranean bathyal sample showing deep-sea fish and cephalopods. (Courtesy Eva Ramírez-Llodra, NOCS and ICM-CSIC. Used with permission.)

FIGURE 19.4 Agassiz trawl. (Courtesy Eva Ramírez-Llodra, ICM-CSIC. Used with permission.)

well as organisms in the first centimeters of sediment. Some models incorporate a video camera on the frame.

Baited traps: These traps use attraction methods such as bait to collect the animals. The traps free fall to the seabed where it stays for a variable length of time (hours to days), and then it is recovered by being released on command from the surface from an acoustic release (Smith et al. 1979; Priede et al. 1994). Some models incorporate a camera. Baited traps are efficient in collecting motile animals that are dispersed on the seabed and usually able to escape trawls, such as large fish or sharks, as well as scavengers such as amphipods, crabs, or gastropods.

The deep-sea baited traps used to catch deep-sea benthic and benthopelagic megafauna can be of different shapes and sizes:

1. Cylindrical (length 150 cm, diameter 75 cm, and only one funnel opening, 23 cm in diameter)
2. Trunco-conical traps (length 150 cm, maximum diameter 75 cm, and only one funnel opening, 10 cm in diameter)
3. Cubic traps (ca. 100 cm per side and openings of all faces of size variable from 10 to 30 cm)

Smaller plastic traps (length 20 cm opening 2 cm) can be located inside the large traps to capture smaller swimming organisms (Figures 19.5–19.8).

The traps can be equipped with a bait dispenser, which is used when it is desirable to make available the bait only at a certain depth. The dispenser, a metal box fixed upside down inside the trap, was kept closed by a pin connected to a pressure device. The dispenser presents inside a coil spring and room for the bait (and eventually a chemical light). At a prefixed depth (e.g., 50 MPa), the system releases the cover, and the bait fall into the trap (Figure 19.9).

Submersibles and ROVs: The use of submersibles and ROVs has increased in recent years. These vehicles are especially important for investigation and sampling of complex topographic habitats such as, for example, hydrothermal vents, cold seeps, canyons, or seamounts, where trawling is impossible, as well as for sampling fragile ecosystems such as coldwater corals. Submersibles and ROVs have high-definition video and photographic systems used to observe and characterize the topography of the habitat as well as faunal communities. They also have articulated arms that allow

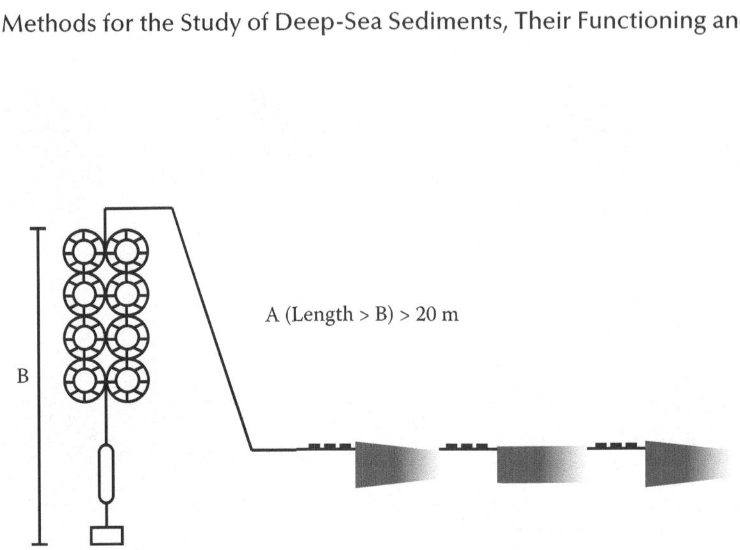

FIGURE 19.5 (A) An example of the line of baited and nonbaited traps. (B) Traps connected with a mooring consisting of a dead weight automatic releaser and buoys for the recovery of the traps after deployment. Note that the connection between traps are rigid and that three small plastic traps are attached to each connection.

FIGURE 19.6 A baited trap containing several specimens of *Centroscymnus coelolepis* (Portuguese dogfish) and the uneaten food (scomber and other flesh remains).

FIGURE 19.7 Deployment of a cubic trap. It is evident that two openings of different size are present: a CTD for recording the basic environmental variables and a camera for taking pictures of the organisms entering the cage. The bait was mainly composed of fish.

FIGURE 19.8 An example of specimens (crustaceans, *Geryon* sp. on the left side, *Aristeus antennatus* on the right side) collected by means of baited traps in the Eastern Mediterranean Sea at >2000 m depth.

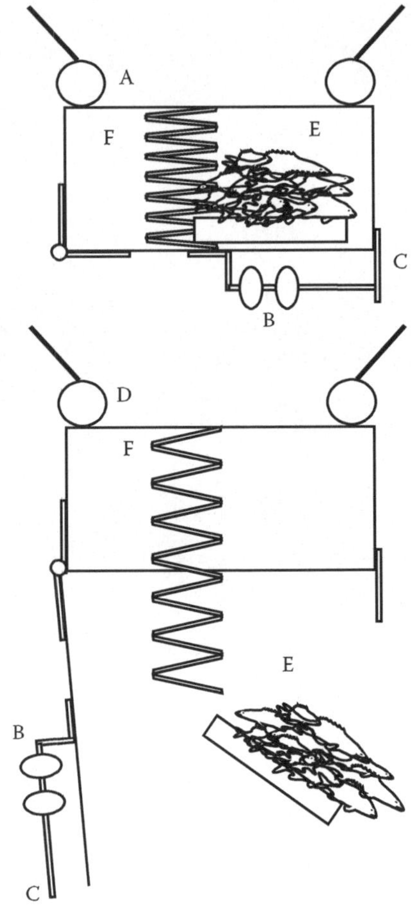

FIGURE 19.9 Schematic view of the bait dispenser: (A) dispenser closed; (B) pressure device; (C) pin; (D) dispenser open; (E) bait and chemical light; (F) spring.

sampling the fauna, rocks, and sediment, deploying equipment in specific locations and conducting detailed experiments on the seafloor. Submersibles (Figure 19.10A) are independent of the mother ship once launched. The inhabited space of deep-rates submersibles is a titanium sphere of about 2 meters in diameter with three small portholes, with space for three people (pilots and scientists). Submersibles that dive to a maximum depth of 1000 meters can have an acrylic sphere that provides a panoramic view to the pilot and scientist. Because submersibles obtain their energy supply from batteries, the autonomy of a submersible is around 8 h. The advantage of using ROVs is that they are connected to the mother ship by an optical cable that is used to send power from the ship to the ROV, and all the data gathered by the ROV to the ship, theoretically allowing the ROV to work in the water nonstop. The ROV is piloted from a control van on the ship, with TV screens showing the images captured by the ROV's cameras (Figure 19.10B).

19.2.2 Sorting and Preservation

The sorting of the otter trawl and Agassiz trawl is done on deck. The sample is placed in buckets and trays, and sorted by species or group. When the sample is clean (i.e., no sediment), the animals are separated by species on a sorting table. Large animals can be separated directly, while the remaining sample of small and broken animals has to be sieved through a series of meshes to avoid losing the small specimens (including macrofauna). When the trawl collects sediment (i.e.,

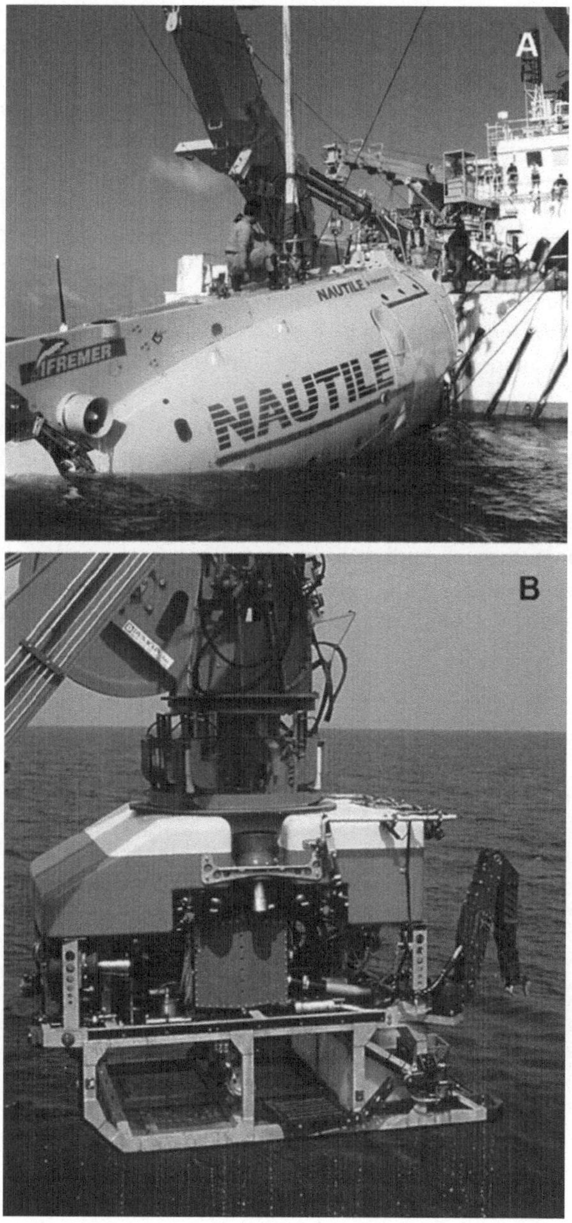

FIGURE 19.10 Deep-water vehicles: (A) The French manned submersible Nautile (courtesy of Daniel Desbruyères, Ifremer); (B) the British ROV Isis. (Courtesy of Hermes Luppi, NOCS. Used with permission.)

the net gets slightly caught on the bottom during sampling), the sample has to be sieved using decreasing mesh sizes to extract the fauna from the fine deep-sea sediment. The Epibenthic sled samples the suprabenthos on three containers (upper, medium, and lower) at the end of each net.

Before detaching each container, the net has to be softly washed down with a hose to ensure no specimens remain in the net and that all specimens are in the container. The containers, marked as upper, medium, and lower, are opened over a tray and taken to the boat's laboratory. The sample is washed down in a 300 μm sieve, and it can be either preserved all together or the main species/groups can be sorted out and preserved separately.

Once the animals are sorted, the samples are preserved. The preservation method depends on the analyses to be conducted. If the sample is to be used for morphological taxonomy, biodiversity, biogeography, life-cycle studies, or histology, the animals can be fixed in 5%–10% borax-buffered formaldehyde in seawater and, after 3–5 days, transferred into 80% alcohol for preservation (Gage and Tyler 1991; Billett et al. 2001). When the samples are to be used for biochemical or molecular analyses, the animals will be frozen at −80°C or −20°C. These samples can also be preserved with 90% ethanol. Subsamples of several organs can also be taken for a variety of analyses. For example, subsamples of gonad, gut, and muscle can be preserved frozen for biochemical studies such as lipids, pigments, or stable isotopes, and gonads can be dissected out for reproductive studies. Muscle tissue is usually used for molecular analyses.

Abundance and biomass are standardized by area sampled in order to allow for quantitative comparisons between samples, areas, and studies. The distance fished on the seafloor is calculated from the navigation data on board (latitude and longitude) and the touchdown and lift-up times of the net. This distance is transformed into the fished area by taking into account the trawl mouth opening. The abundance and biomass data are divided by area fished (square kilometer, km^2) providing the area corrected data in individuals per square kilometer ($N \cdot Km^{-2}$) and kilograms (or grams) per square kilometer ($Kg \cdot Km^{-2}$) (Company et al. 2004; Ramírez-Llodra et al. 2008).

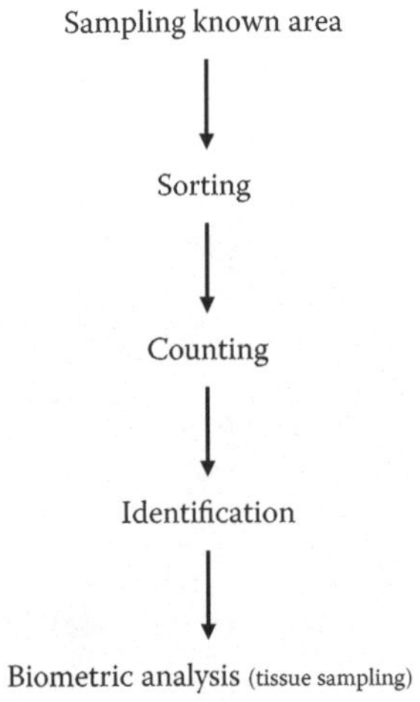

PROTOCOL 19.1 Quantification of megafaunal abundance.

REFERENCES

Bett, B. J., M. G. Malzone, B. E. Narayanaswamy, and B. D. Wigham. 2001. Temporal variability in phytodetritus and megabenthic activity at the seabed in the deep Northeast Atlantic. *Progress in Oceanography* 50: 349–368.

Billett, D. S. M., B. J. Bett, A. L. Rice et al. 2001. Long-terms change in the megabenthos of the Porcupine Abyssal Plain. *Progress in Oceanography* 50: 325–348.

Company, J. B., P. Maiorano, A. Tselepides et al. 2004. Deep-sea decapod crustaceans in the western and central Mediterranean Sea: Preliminary aspects of species distribution, biomass and population structure. *Scientia Marina* 68 (Suppl. 3): 73–86.

Gage, J. D. and P. A. Tyler. 1991. *Deep-sea biology: A natural history of organisms at the deep-sea floor.* Cambridge, U.K.: Cambridge University Press.

Gage J. and B. Bett. 2005. Deep-sea benthic sampling. In *Ecology of Marine Benthos*, ed. A. Eleftheriou and A. McIntyre, 273–325. Oxford, U.K: Blackwell.

Guennegan, Y. and V. Martin. 1985. Techniques de prélèvement. In *Peuplements Profonds du Golfe de Gascogne: Campagne BIOGAS*, ed. L. Laubier and C. Monniot, 571–602. Brest, France: IFREMER.

Herring, P. 2003. *The biology of the deep ocean.* Oxford, U.K: Oxford University Press.

Hessler, R. R. and H.L. Sanders. 1967. Faunal diversity in the deep-sea. *Deep-Sea Research* 14: 65–78.

Merret, N. R. and N. B. Marshall. 1981. Observations on the ecology of deep-sea bottom-living fishes collected off northwest Africa (08°–27°N). *Progress in Oceanography* 9: 185–244.

Priede, I. G., P. M. Bagley, A. Smith, S. Creasey, and N. R. Merrett. 1994. Scavenging deep demersal fishes of the Porcupine Seabight, North-East Atlantic: Observations by baited camera, trap and trawl. *Journal of the Marine Biological Association of the United Kingdom* 74: 481–498.

Ramírez-Llodra, E., M. Ballesteros, J. B. Company, L. Dantart, and F. Sardà. 2008. Spatio-temporal variations of biomass and abundance in bathyal non-crustacean megafauna in the Catalan Sea (North-western Mediterranean). *Marine Biology* 153: 297–309.

Rex, M. A. 1973. Deep-sea species diversity: Decreased gastropod diversity at abyssal depths. *Science* 181: 1051–1053.

Rex, M. A. 1981. Community structure in the deep-sea benthos. *Annual Review of Ecological Systems* 12: 331–353.

Rex, M. A., C. R. McClain, N. A. Johnson et al. 2005. A source-sink hypothesis for abyssal biodiversity. *The American Naturalist* 165: 163–178.

Rex, M. A., R. J. Etter, J. S. Morris et al. 2006. Global bathymetric patterns of standing stock and body size in the deep-sea benthos. *Marine Ecology Progress Series* 317: 1–8.

Rice, A., R. G. Aldred, E. Darlington, and R. A. Wild. 1982. The quantitative estimation of the deep-sea megabenthos: A new approach to an old problem. *Oceanologica Acta* 5: 63–72.

Rice, A. 1987. Benthic transect photography. Report 248. Institute of Oceanographic Sciences, Deacon Laboratory.

Sardà, F., J. E. Cartes, J. B. Company, and A. Albiol. 1998. A modified trawl used to sample deep-sea megabenthos. *Fisheries Science* 64: 492–493.

Sardà, F., G. D'Onghia, C.-Y. Politou, and A. Tselepides. 2004. Mediterranean deep-sea biology. *Scientia Marina* 68 (Suppl. 3): 204.

Smith, K. L., G. A. White, M. B. Laver, R. R. McConnaughey, and J. P. Meador. 1979. Free vehicle capture of abyssopelagic animals. *Deep-Sea Research* 26: 57–64.

Thiel, H. 1975. The size structure of the deep-sea benthos. *Internationale Revue der Gesamten Hydrobiologie* 60: 575–606.

Section IV

Deep-Sea Benthic Diversity

20 Extraction and Purification of DNA from Marine Sediments Suitable for the Analysis of Prokaryotic Diversity

CONTENTS

20.1 INTRODUCTION

In the last decades, molecular methods have been increasingly used for investigating the biodiversity and richness of benthic prokaryotic assemblages in a variety of aquatic environments. Most of these methods, with the exception of those based on staining and subsequent microscopic visualization of intact cells (i.e., FISH and CARD-FISH) require, as the initial step, the isolation and purification of genomic DNA from prokaryotes in order to proceed with subsequent molecular analyses (e.g., PCR targeting 16S rRNA genes). As such, DNA extraction from environmental matrices is the key step in any molecular-based diversity analysis.

There are two different approaches for recovering the prokaryotic DNA from sediment samples. One approach is based on the direct *in situ* lysis of cells within the sediments, whereas the other is based on physical/chemical separation of cells from the sediment (i.e., cell extraction) and their subsequent lysis (Steffan et al. 1988; Gabor et al. 2003). The *in situ* lysis can be performed by means of different physical (e.g., bead mill homogenization, sonication, and/or freeze-thawing) or chemical procedures (e.g., utilizing detergents such as SDS or Sarkosyl), or a combination of both (Hurt et al. 2001; Luna et al. 2006). These treatments can be carried out by using time-consuming laboratory protocols (such as the Zhou protocol; Zhou et al. 1996) or by means of commercial kits (such as the UltraClean Soil DNA Isolation kit [MoBio] or the Bio 101 DNA extraction kit [QBIOgene]). These kits enable the fast recovery of PCR-ready nucleic acids from a large number of samples that are generally collected for ecological investigations. In addition, recent studies highlighted that estimates of bacterial diversity in marine sediments obtained using the commercial kit MoBio and those obtained using the Zhou protocol are not significantly different (Luna et al. 2006). Conversely, the procedure based on cell extraction, when applied to marine sediments, provides lower DNA yields than *in situ* lysis. This is also reflected in a lower bacterial diversity, mainly because procedure generally allows a low recovery of cells from the sediment (20%–30% of the total cells; Lindahl 1996).

In this chapter, we present three different procedures that can be used for DNA extraction and purification from marine sediments: (1) *in situ* lysis using the UltraClean Soil DNA Isolation kit [MoBio], (2) *in situ* lysis using the Zhou protocol, (3) cell extraction and subsequent cellular lysis according to the protocol by Steffan et al. (1988).

20.2 FIELD PROCEDURES

Undisturbed sediment samples can be collected manually by scuba divers or using specific devices (i.e., multicorer or, if available, by ROV). A sediment subsample (ca. 5–10 cm^3) is then transferred into sterile Falcon tubes (50 mL) by using a sterile spatula or sterile syringe. Samples are then stored at −80°C or alternatively at −20°C until DNA extraction if conducted within a few weeks from sampling.

20.3 LABORATORY PROCEDURES—PROTOCOL 20.1. DNA EXTRACTION USING THE ULTRACLEAN SOIL DNA ISOLATION KIT [MOBIO]

20.3.1 Equipment and Instruments

Water bath or thermal block
Centrifuges (must be suitable for 2-mL test tubes and for 50 mL test tubes)
Vortex
Laminar air flow (LAF)

20.3.2 Solutions and Reagents

Wash Solution 1: 50 mM Tris-HCl, pH 8.3; 200 mM NaCl; 5 mM Na_2EDTA; 0.05% Triton X-100.
Wash Solution 2: 50 mM Tris-HCl, pH 8.3; 200 mM NaCl; 5 mM Na_2EDTA.
Wash Solution 3: 10 mM Tris-HCl, pH 8.3; 0.1 mM Na_2EDTA.
Ultrapure water: Reagent-grade water, previously autoclaved, filtered (through 0.2 μm) and stored in sterile 2-mL Eppendorf tubes at 4°C.
UltraClean Soil DNA Isolation Kit (MoBio, catalog #12800-100). The kit contains: Bead Solution tubes, solution S1, solution IRS, solution S2, solution S3, solution S4, and solution S5. Moreover, spin columns are also included for DNA purification.

20.3.3 Procedure

1. Using a sterile spatula, insert 2.5 g of wet sediment into a 50 mL test tube and add 5 mL of Wash Solution 1 (see Notes).
2. Vortex for 2 min.

3. Centrifuge for 2 min at 6000 × g.
4. Discard the supernatant.
5. Add 5 mL of Wash Solution 2.
6. Mix to resuspend the sediment pellet.
7. Centrifuge for 2 min at 6000 × g.
8. Discard the supernatant.
9. Add 5 mL of Wash Solution 3 [NOTE 1].
10. Mix to resuspend the sediment pellet.
11. Centrifuge for 2 min at 6000 × g.
12. Discard the supernatant.
13. Transfer 1 g of the sediment sample into the 2 mL Bead Solution tube of the UltraClean™ Soil DNA Isolation Kit (MoBio, catalog #12800-100).
14. Add 60 μL of solution S1.
15. Mix for a few seconds.
16. Add 200 μL of solution IRS.
17. Vortex the samples for 10 min.
18. Incubate for 5 min at 70°C [NOTE 2].
19. Vortex for 1 min.
20. Incubate again for 5 min at 70°C.
21. Vortex for 1 min.
22. Centrifuge for 1 min at 10,000 × g.
23. Transfer 400–500 μL of the supernatant into a clean Eppendorf tube (provided by the manufacturer).
24. Add 250 μL of solution S2.
25. Vortex for a few seconds.
26. Incubate for 5 min at 4°C.
27. Centrifuge for 1 min at 10,000 × g.
28. Transfer the entire supernatant into a clean Eppendorf tube (provided by the manufacturer).
29. Add 1.3 mL of solution S3.
30. Vortex for a few seconds.
31. Load approximately 700 μL onto a spin column.
32. Centrifuge for 1 min at 10,000 × g.
33. Discard the flow through (i.e., liquid passing the filter of the column).
34. Load approximately 700 μL onto a spin column.
35. Centrifuge for 1 min at 10,000 × g.
36. Discard the flow through (i.e., liquid passing the filter of the column).
37. Add the remaining supernatant to the spin filter.
38. Centrifuge for 1 min at 10,000 × g.
39. Discard the flow through (i.e., liquid passing the filter of the column).
40. Add 300 μL of solution S4 to the spin column.
41. Centrifuge for 1 minute at 10,000 × g.
42. Place spin filters in a new Eppendorf tube.
43. Add 50 μL of solution S5 to the center of the spin filter.
44. Centrifuge for 30 s at 10,000 × g.
45. Discard the spin filter.
46. DNA in the tube can be stored at −80°C or alternatively at −20°C [NOTE 3].

20.3.4 Notes

1. The initial washing steps presented here (i.e., washing with Wash Solution 1, 2, and 3) are not included in the manufacturers instructions. Based on our experience of large sets of

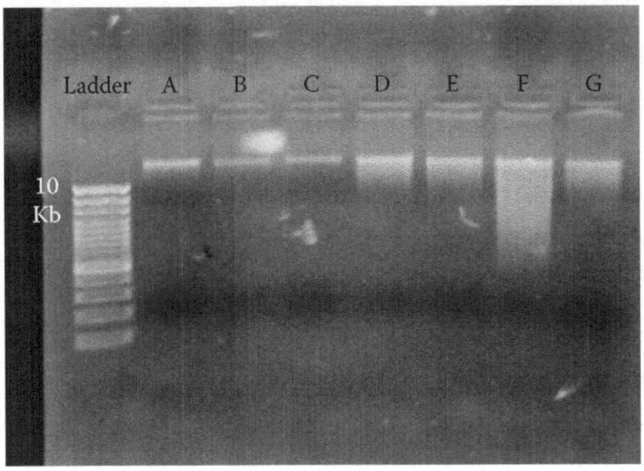

FIGURE 20.1 Agarose gel electrophoresis of genomic DNA extracted from marine sediments.

deep-sea samples, these steps, applied for the first time by Fortin et al. (2004), improve the success of DNA amplification by PCR in marine sediments.

2. Sample incubation for 5 min at 70°C is not included in the instructions provided by the manufacturer. In our experience, this step increases the DNA extraction yield from marine sediments.
3. We generally check the quality and molecular weight of extracted DNA on 1X TBE-agarose gel prior to its quantification (Figure 20.1). The DNA extraction procedure described here will generally yield a single band of high-molecular-weight DNA (Figure 20.1, lanes A, B, and C). In some samples, extracted DNA with the MoBio protocol can present a variable degree of shearing (as evident from DNA smearing on the gel; Figure 20.1, lanes D, E, G and, particularly, F). As shearing can negatively affect subsequent molecular analyses, we suggest in these cases that DNA extraction be repeated, reducing vortexing time in the MoBio protocol to 2 min. In our experience, this reduction in vortexing time can significantly reduce the mechanical shearing of DNA.

20.4 LABORATORY PROCEDURES—PROTOCOL 20.2. DNA EXTRACTION USING THE ZHOU ET AL. (1996) PROTOCOL

20.4.1 Equipment and Instruments

Water bath
Horizontal shaker
Centrifuges (must be suitable for 50 mL test tubes)
Vortex
Laminar air flow (LAF)

20.4.2 Solutions and Reagents

DNA extraction buffer (100 mM Tris-HCl pH 8.0; 100 mM sodium EDTA, pH 8.0; 100 mM sodium phosphate, pH 8.0; 1.5 M NaCl; 1% CTAB)
Proteinase K solution (10 mg/mL)
SDS (20%)
Phenol/chloroform/isoamyl alcohol (25:24:1) (v/v)

Chloroform/isoamyl alcohol (24:1) (v/v)
Isopropanol
70% Ethanol (v/v)
Ultrapure water: Reagent-grade water, previously autoclaved, filtered (through 0.2 μm) and stored in sterile 2-mL Eppendorf tubes at 4°C

20.4.3 Procedure

1. In a Falcon test tube (50 mL), mix ca. 2.5 g of wet sediment with 6.75 mL of DNA extraction buffer and 50 μL of proteinase K (10 mg/mL) [NOTE 1].
2. Shake horizontally at 225 rpm (30 min, 37°C).
3. After shaking, add 0.75 mL of 20% SDS.
4. Incubate in a water bath at 65°C for 2 h, with gentle "end over end" inversions every 20 min.
5. Centrifuge at 6000 × g (10 min, room temperature).
6. Transfer the supernatant into sterile 50-mL Falcon tubes.
7. Extract the sediment pellet two more times by adding 2.25 mL of DNA extraction buffer and 0.25 mL of 20% SDS, vortexing for 10 s, incubating at 65°C for 10 min and centrifuging as described earlier.
8. Combine the supernatants from all these three cycles (depending on water content, it will be ca. 12–14 mL).
9. Add an equal volume of phenol/chloroform/isoamyl alcohol (25:24:1).
10. Mix thoroughly and centrifuge at 6000 × g for 10 min.
11. Collect the aqueous phase and transfer it to a new tube.
12. Add one volume of chloroform/Isoamyl alcohol (24:1).
13. Mix thoroughly and centrifuge at 6000 × g for 10 min.
14. Collect the aqueous phase and transfer it to a new tube.
15. Add 0.6 volumes of room-temperature isopropanol.
16. Incubate for 1 h at room temperature.
17. Centrifuge at 16,000 × g for 20 min.
18. Discard the supernatant and wash the pellet by adding 5 mL of cold 70% ethanol.
19. Centrifuge at 16,000 × g for 10 min at 4°C.
20. Discard the supernatant, then dry the pellet using a speed-vac (if available) or by placing it under laminar air flow (LAF).
21. Resuspend in 50 μL ultrapure water.
22. Store at −80°C [NOTE 2].

20.4.4 Notes

1. The initial washing steps described (i.e., before using MoBio kit) can also be applied to sediment samples processed using the Zhou protocol (1996).
2. The Zhou protocol is not really effective in removing coextracted substances (such humic acids), which are known inhibitors of the PCR reaction. The presence of these substances is generally evident from the yellow-brownish color of extracted DNA. Therefore, crude DNA extracts should be purified before any other molecular analyses. This can be achieved using commercially available DNA purification systems (e.g., Wizard Clean-up System, Promega, Madison, WI) or other spin column-based technologies. Other purification methods include agarose gel electrophoresis (Moré et al., 1994) or Sephadex G-200 column chromatography (Rochelle et al. 1992; Kuske et al. 1998), used individually or in combination.

20.5 LABORATORY PROCEDURES—PROTOCOL 20.3. DNA EXTRACTION USING THE STEFFAN ET AL. (1988) PROTOCOL

20.5.1 Equipment and Instruments

UltraTurrax mixer
Centrifuges (must be suitable for 50 mL test tubes and for 2 mL Eppendorf test tubes)
Vortex
Incubator
LAF

20.5.2 Solutions and Reagents

Na-phosphate buffer (0.1M, pH 8.0)
Acid-washed polyvinylpolypirrolydone (PVPP), prepared as described by Holben et al. (1988)
SDS 10%
0.1% sodium hexametaphosphate–0.1% sodium pyrophosphate
Chrombach buffer (0.33 M Tris-hydrochloride, 1 mM EDTA, pH 8.0)
DNase I (Sigma or similar)
Lysozime (50 mg/mL)
Proteinase K solution (10 mg/mL)
NaCl 5 M
CTAB 10%
Phenol/chloroform/isoamyl alcohol (25:24:1) (v/v)
Chloroform/isoamyl alcohol (24:1) (v/v)
Na-Acetate (10 M)
Isopropanol
70% ethanol
Ultrapure water: Reagent-grade water, previously autoclaved, filtered (through 0.2 μm) and stored in sterile 2 mL Eppendorf at 4°C

20.5.3 Procedure

20.5.3.1 Phase 1. Separation of Cells from the Sediment Matrix

1. Add to 2.5 g of wet sediment 7.5 mL of Na-phosphate buffer and 0.5 g acid-washed PVPP.
2. Homogenize samples using an UltraTurrax mixer for three 1-min cycles, with 1 min of ice-cooling between each cycle.
3. Add 100 μL of SDS (0.1% final concentration).
4. Homogenize the samples again for 10 s.
5. Chill the samples on ice and allow the foaming to settle for 5 min.
6. Centrifuge the samples at 500 x g for 5 min at 4°C to allow only sediment separation.
7. Transfer the supernatant into a sterile Falcon tube (50 mL).
8. Wash the sediment pellets two times more by adding each time 5 mL of Na-phosphate buffer and homogenizing as described earlier (but now without further additions of SDS).
9. Combine all resulting supernatants and centrifuge for 30 min at 10,000 x g (at 4°C) to collect cells.
10. After centrifugation, wash cellular pellets two times with 5 mL of 0.1% sodium hexametaphosphate–0.1% sodium pyrophosphate at 4°C by shaking manually for 1 min and then centrifuging for 30 min at 10,000 x g at 4°C.
11. Suspend the cellular pellet in 5 mL of Chrombach buffer and centrifuge again as described for previous washings.

12. Transfer the resulting pellet into a 2 mL Eppendorf tube.
13. Suspend in 500 μL of ultrapure water.

20.5.3.2 Phase 2. DNA Extraction from the Cell Suspension

1. Add to the pellet 10 U of DNase I (Sigma) and 20 mM $MgCl_2$ (provided by the manufacturer) and incubate at 37°C for 2 h (this procedure is carried out to avoid any contamination by extracellular DNA).
2. After incubation, wash the pellets two times with 2 mL of Na-phosphate buffer (in order to eliminate DNase) and resuspend in 500 of μL of Na-phosphate buffer.
3. Add 70 μL of lysozime and incubate at 37°C for 30 min.
4. Add 30 μL of SDS and incubate at 50°C for 1 h.
5. Add 8 μL of Proteinase K and incubate at 50°C for 30 min.
6. Add 100 μL of NaCl (5 M) and 80 μL of CTAB 10% and incubate at 65°C for 10 min.
7. Add 700 μL of phenol/chloroform/isoamyl alcohol (25:24:1) and mix for 15 s.
8. Centrifuge at 14,000 × g for 2 min.
9. Transfer the supernatant into a new Eppendorf tube and add 700 μL of chloroform/isoamyl alcohol (24:1).
10. Mix for 15 s.
11. Centrifuge at 14,000 × g for 2 min.
12. Transfer the supernatant into a new Eppendorf tube and add 140 μL of Na-acetate (10 M) and 600 μL of isopropanol.
13. Centrifuge at 13,000 × g for 20 min.
14. Discard the supernatant and add 500 μL of 70% ethanol.
15. Centrifuge at 13,000 × g for 5 min.
16. Discard the supernatant and dry the pellet under LAF to eliminate all traces of ethanol.
17. Add 50 μL of ultrapure water.
18. Store at −80°C.

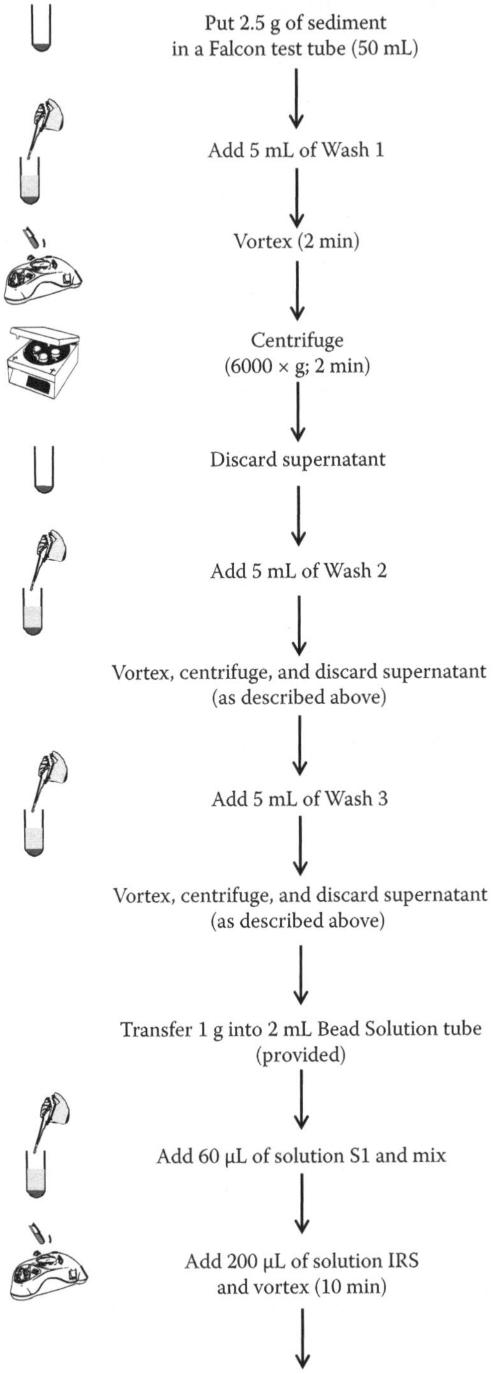

(a)

PROTOCOL 20.1 Protocol for DNA extraction using the UltraClean Soil DNA Isolation kit [MoBio].

Continued

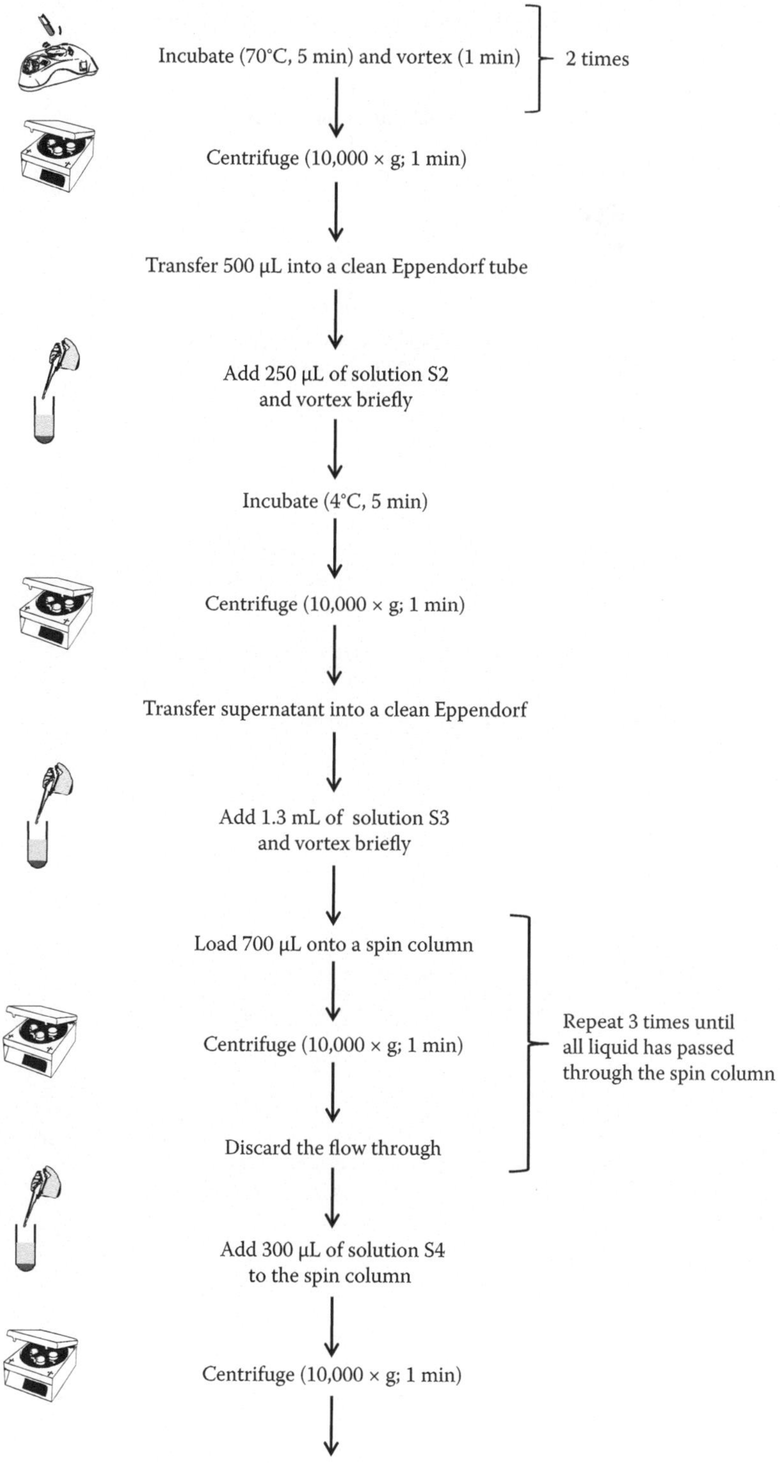

(b)

PROTOCOL 20.1 *Continued.*

Place the spin filter into a clean Eppendorf tube

↓

Add 50 μL of solution S5

↓

Centrifuge (10,000 × g; 30 sec)

↓

Discard the spin and store extracted DNA
(–20°C or –80°C)

(c)

PROTOCOL 20.1 *Continued.*

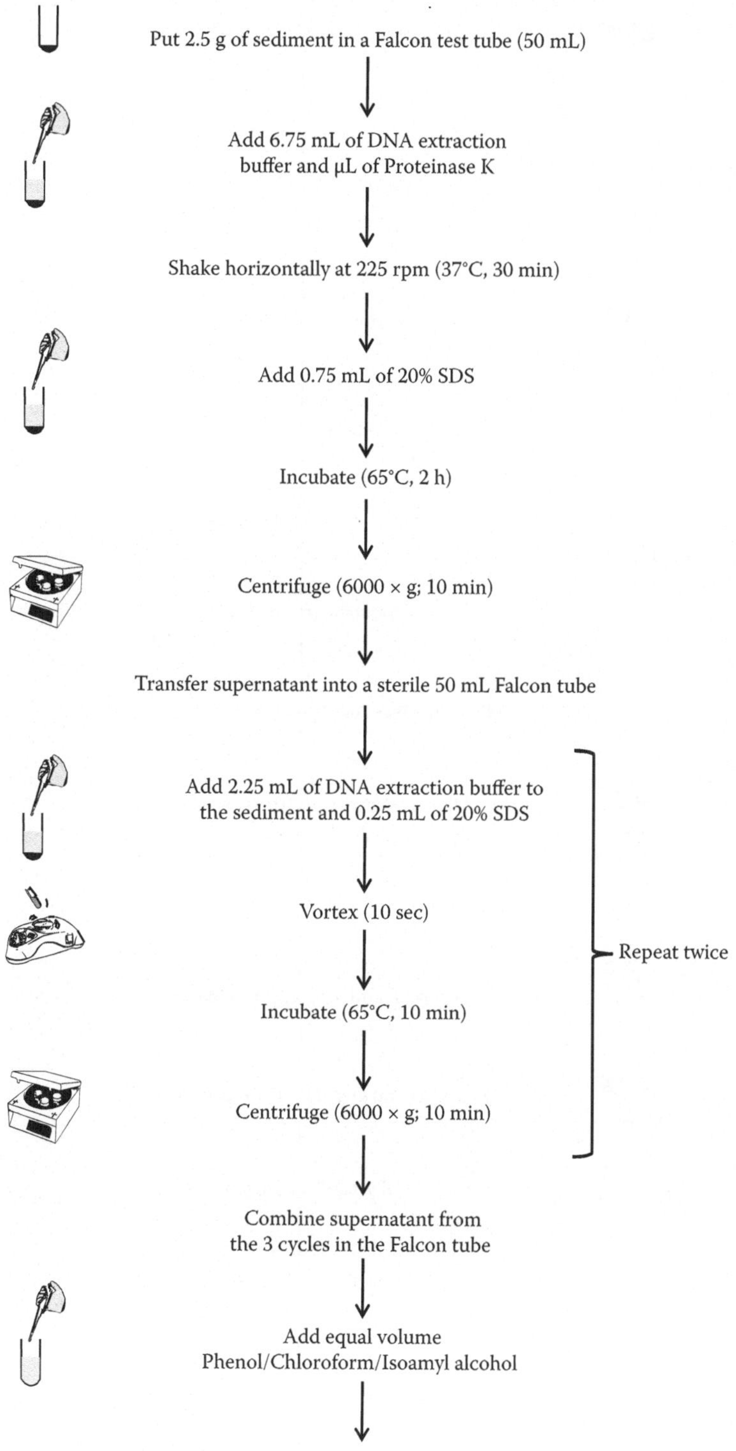

(a)

PROTOCOL 20.2 Protocol for DNA extraction (Zhou et al. 1996). *Continued*

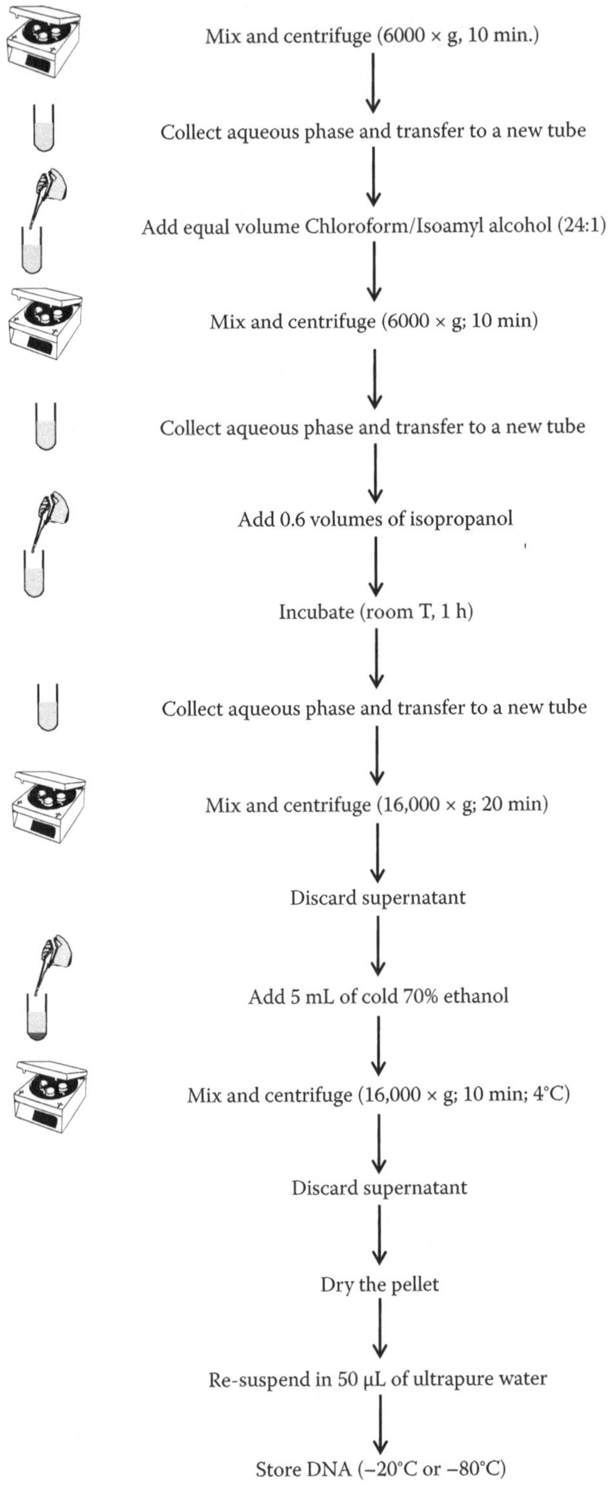

(b)

PROTOCOL 20.2 *Continued.*

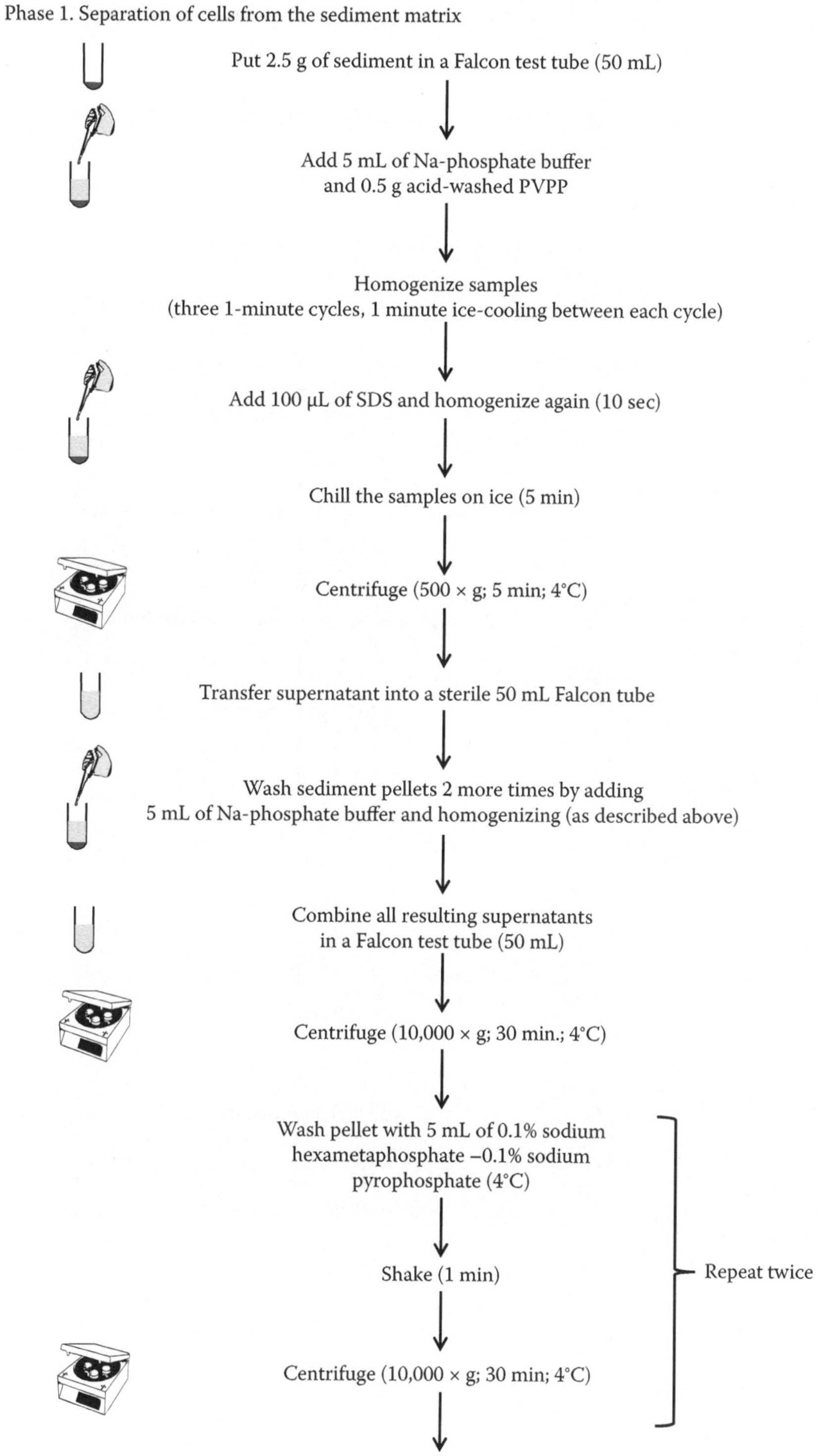

(a)

PROTOCOL 20.3 Protocol for DNA extraction (Steffan et al. 1988).

Continued

Suspend the pellet in 5 mL of Chrombach buffer

↓

Centrifuge (10,000 × g; 30 min, 4°C)

↓

Phase 2. DNA extraction from the cell suspension

Add the pellet with 10 U of Dnase I and 20 mM $MgCl_2$

↓

Incubate (37°C, 2 h)

↓

Wash pellet 2 times with 2 mL Na-phosphate buffer

↓

Re-suspend in 500 μL of Na-phosphate buffer

↓

Add 70 μL of lysozime

↓

Incubate (37°C, 30 min.)

↓

Add 30 μL of SDS

↓

Incubate (50°C, 1 h)

↓

Add 8 μL of Proteinase K

↓

Incubate (50°C, 30 min)

↓

Add 100 μL of NaCl and 80 μL of CTAB 10%

↓

(b)

PROTOCOL 20.3 *Continued.*

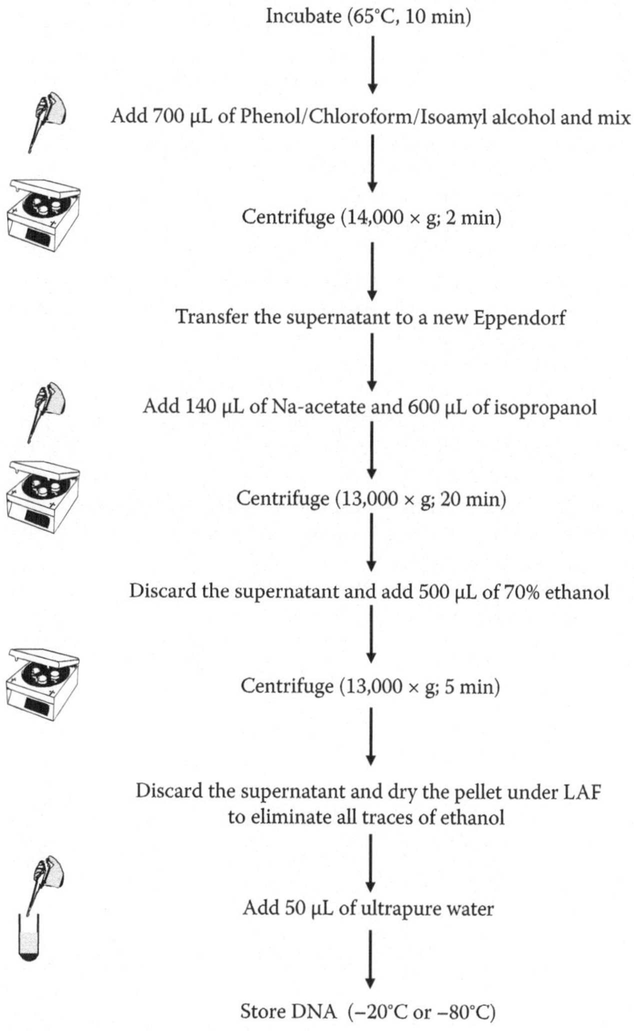

(c)

PROTOCOL 20.3 *Continued.*

REFERENCES

Fortin, N., D. Beaumier, K. Lee, and C. W. Greer. 2004. Soil washing improves the recovery of total community DNA from polluted and high organic content sediments. *Journal of Microbiological Methods* 56: 181–191.

Gabor, E. M., E. J. de Vries, and D. B. Janssen. 2003. Efficient recovery of environmental DNA for expression cloning by indirect extraction methods. *FEMS Microbiology Ecology* 44: 153–163.

Holben, W. E., J. K. Jansson, B. K. Chelm, and J. M. Tiedje. 1988. DNA probe method for the detection of specific microorganisms in the soil bacterial community. *Applied and Environmental Microbiology* 54: 703–711.

Hurt, R. A., X. Qiu, L. Wu et al. 2001. Simultaneous recovery of RNA and DNA from soils and sediments. *Applied and Environmental Microbiology* 67: 4495–4503.

Lindahl, V. 1996. Improved soil dispersion procedures for total bacteria counts, extraction of indigenous bacteria and cell survival. *Journal of Microbiological Methods* 25: 279–286.

Luna, G. M., A. Dell'Anno, and R. Danovaro. 2006. DNA extraction procedure: A critical issue for bacterial diversity assessment in marine sediments. *Environmental Microbiology* 8: 308–320.

Moré, M. I., J. B. Herrick, M. C. Silva, W. C. Ghiorse, and E. L. Madsen. 1994. Quantitative cell lysis of indigenous microorganisms and rapid extraction of microbial DNA from sediment. *Applied and Environmental Microbiology* 60: 1572–1580.

Kuske, C. R., K. L. Banton., D. L. Adorada, P. C. Stark, K. K. Hill, and P. J. Jackson. 1998. Small-scale DNA sample preparation method for field PCR detection of microbial cells and spores in soil. *Applied and Environmental Microbiology* 64: 2463–2472.

Rochelle, P. A., J. C. Fry, R. J. Parkes, and A. J. Weightman. 1992. DNA extraction for 16S rRNA gene analysis to determine genetic diversity in deep sediment communities. *FEMS Microbiology Letters* 100: 59–66.

Steffan, R. J., J. Goksoyr, A. K. Bej, and R. M. Atlas. 1988. Recovery of DNA from soils and sediments. *Applied and Environmental Microbiology* 54: 2908–2915.

Zhou, J., M. A. Bruns, and J. M. Tiedje. 1996. DNA recovery from soils of diverse composition. *Applied and Environmental Microbiology* 62: 316–322.

21 Archaeal Diversity Analysis Using 16S rDNA T-RFLP (Terminal-Restriction Fragment Length Polymorphisms)

CONTENTS

21.1 INTRODUCTION

The analysis of microbial diversity using molecular methods has increased significantly the ability to study the structure and dynamics of marine microbial assemblages (Rappè and Giovannoni 2003). The importance of Archaea in marine ecosystems (Fuhrman and Ouverney 1998; Church et al. 2003) is enhancing the scientific interest on the investigation of the patterns of Archaeal diversity, which in the deep-sea sediments is almost completely unexplored. While 16S rDNA-based surveys have initially provided evidence for the widespread occurrence of Archaea in oxygenated coastal superficial waters (DeLong 1992), more recent studies have indicated that pelagic Archaea are abundant members of meso- and bathypelagic picoplankton (Karner et al. 2001; Béjà et al. 2002). Other studies have revealed that Archaea also account for a significant fraction of prokaryotic assemblages in coastal and deep-sea sediments (Vetriani et al. 1999; Braker et al. 2001). Molecular studies have consistently revealed that marine archaeal populations are diverse, complex, and widespread (Kim et al. 2005; Chaban et al. 2006).

Among the available methods to study the diversity of Archaeal assemblages in marine ecosystems, PCR-based fingerprinting techniques represent probably one of the most convenient approaches in spite of some well-known limitations (e.g., the inability to produce a complete inventory of all taxa present, and the exclusion of rare ones; Danovaro et al. 2006; Danovaro et al. 2007; Fuhrman and Hagstrom 2008). Techniques known as Denaturing Gradient Gel Electrophoresis (DGGE), T-RFLP, and Automated Ribosomal Intergenic Spacer Analysis (ARISA) are now routinely used in microbial ecology to estimate the number of dominant prokaryotic taxa (richness) and, most importantly, to investigate the spatial and temporal dynamics of prokaryotic assemblages (Vetriani et al. 1999; Luna et al. 2004; Luna et al. 2006; Danovaro et al. 2006; Schwarz et al. 2006). While ARISA is used for the study of bacterial assemblages (Danovaro et al. 2006), T-RFLP is preferred for the study of Archaeal diversity (Braker et al. 2001; Vetriani et al. 2003) when compared to other fingerprinting methods (such as DGGE) because it has the advantages of (1) providing a more accurate discrimination among the different taxa of a given assemblage, and (2) allowing semiquantitative estimations of the relative importance of different ribotypes (Liu et al. 1997; Marsh 1999; Blackwood et al. 2003). When using universal primers for Archaea, each peak within the electropherogram is assumed to correspond to an Archaeal taxon (genus or species).

In this chapter we present the detailed procedure for applying the 16S rDNA T-RFLP fingerprinting technique in marine sediments in order to estimate benthic Archaeal ribotype richness and community composition.

21.2 FIELD PROCEDURES

Undisturbed sediment samples are typically collected using a mini- or a multicorer or, in case available, an ROV (remotely operated vehicle). A subsample (ca. 5–10 cm^3) from the superficial layer (i.e., 0–1 cm) is then aseptically transferred from the core into sterile Falcon test tubes (50 mL) using a sterile metal or plastic spatula or either a 10-mL sterile syringe from which the tip has been removed. The tube is then frozen at –20°C or (preferably) at –80°C and kept stored until DNA extraction analyses.

21.3 LABORATORY PROCEDURES

21.3.1 Equipment and Instruments

Water bath or thermal block
Centrifuges (must be suitable for 2 mL test tubes and for 50 mL test tubes)
Vortex
Thermalcycler
Transilluminator
UV-VIS spectrophotometer
Complete apparatus for horizontal electrophoresis (including casting trays, electrophoretic chamber, and power supply)
ABI 310 or 3100 sequencer (if not available, the capillary electrophoresis of digests can be carried out by an external subcontractor)
Microwave

21.3.2 Solutions and Reagents

Wash Solution 1: 50 mM Tris-HCl, pH 8.3; 200 mM NaCl; 5 mM Na_2EDTA; 0.05% Triton X-100
Wash Solution 2: 50 mM Tris-HCl, pH 8.3; 200 mM NaCl; 5 mM Na_2EDTA
Wash Solution 3: 10 mM Tris-HCl, pH 8.3; 0.1 mM Na_2EDTA

Ultrapure water: Reagent-grade water, previously autoclaved, filtered (through 0.2 μm) and stored into sterile 2 mL Eppendorf tubes at 4°C or −20°C
UltraClean™ Soil DNA Isolation Kit (MoBio, catalog #12800-100)
MasterTaq® kit (Eppendorf AG, Germany)
Agarose
TBE 10× (can be purchased from BIORAD or similar, or, alternatively, prepared as follows (for 1 L): 108 g TRIS Base, 55 g Boric Acid, 40 mL EDTA 0.5M pH 8, reagent-grade water to volume)
TBE 1×: dilute TBE 10× (100 mL TBE 10× + 900 mL of reagent-grade water)
Ethidium Bromide (10 mg mL^{-1} in ultrapure water)
Loading dye 6X (Promega, Fermentas, or similar)
Molecular weight 100 bp (Fermentas or similar)
Wizard® PCR cleanup system (Promega)
Alu I restriction enzyme (Promega, Fermentas, or other)
Formamide
Internal size standard (GS1000ROX; Applied Biosystems)

21.3.3 Procedure

The application of the 16S rDNA gene T-RFLP method to sediment samples involves the following steps:

1. DNA extraction and purification (with the aim of extracting genomic DNA from benthic microrganisms, which must be of sufficient purity to carry out subsequent PCR-based analyses).
2. DNA quantification (to accurately quantify the concentration of DNA extracts and to utilize known amounts of DNA in the PCR reaction).
3. PCR amplification of Archaeal 16S rRNA genes (to amplify the 16S rDNA using universal Archaeal primers, thus providing sufficient amount of this gene sequences for being analyzed. As the PCR is carried out using a fluorescently labeled primer, it will yield a mixture of 16S rDNA amplicons of the same size, having a fluorescent label on one end).
4. Agarose gel electrophoresis of PCR products (to check the success of the amplification reaction).
5. Cleanup of PCR products (to purify the amplicons from salts, Taq, primers, excess of dNTPs, and all other reagents utilized during the PCR reaction before proceeding with enzymatic digestion).
6. Enzymatic digestion of amplicons (to digest the purified 16S rDNA amplicons with tetrameric restriction enzymes in order to provide different terminal restriction fragments for each Archaeal ribotype).
7. Capillary electrophoresis of digests (to resolve and accurately size the generated terminal restriction fragments).
8. Analysis and interpretation of results (to transform the observed results into estimates of Archaeal richness and community structure).

21.3.3.1 DNA Extraction and Purification

Extract the DNA following the described procedure (NOTE 1).

21.3.3.2 DNA Quantification (Using a Spectrophotometer)

Use only quartz cuvettes (path 1 cm, internal volume 1 mL).
Set up the spectrophotometer at 260 nm.

If working with a dual-channel spectrophotometer, place in both the "reference" and "sample" cuvette 1 mL of ultrapure water and then operate the procedure to reset the instrument.
Discard the water from the "sample" cuvette and then place in this cuvette 990 μL of reagent-grade water and 10 μL of each DNA sample (i.e., dilute the sample to a factor of 100).
Measure the absorbance at 260 nm.
Discard the liquid from the "sample" cuvette and repeat the procedure for all extracted DNA samples.
At the end of the procedure, DNA concentration in each DNA extract is calculated as follows:

$$DNA\ (ng\ \mu L^{-1}) = ABS_{260} \cdot 50 \cdot 100 \qquad (21.1)$$

where 100 is the dilution factor of the DNA within the cuvette (i.e., 10 μL of DNA into 990 μL of reagent-grade water), and 50 is the conversion factor from absorbance to DNA concentration (i.e., an absorbance of 1 at 260 nm contains approximately 50 μg/mL (or 50 ng/μL) of double-stranded DNA). The purity of the DNA can be checked by measuring the ratio of ABS at 260 to ABS at 280 nm: a ratio greater than 1.8–2 is considered as an indication of DNA purity from proteins [NOTE 2].

21.3.3.3 PCR Amplification of the Archaeal 16S rRNA Genes

Dilute the DNA extracts to ca. 5 ng μL^{-1}.
Prepare four 200 μL sterile PCR test tubes for each DNA sample (i.e., each sediment sample must be amplified in four replicates), plus one each for the positive and negative PCR controls.
For PCR amplification, it is recommended that the MasterTaq® kit (Eppendorf AG, Germany) be used, which increases the success of PCR amplification on DNA extracted from marine sediments by reducing the effects of residual, potentially PCR-inhibiting contaminants.
Each PCR tube will contain the following ingredients:

- 1 μL of Taq Polymerase (the stock solution must be 1 Unit μL^{-1})
- 5 μL of 10× Taq Buffer with Mg^{2+}
- 10 μL 5× of TaqMaster PCR Enhancer
- 1 μL of 10 mM dNTPs
- 0.5 μL of the 100 μM forward primer "ARCH109f" (5′-ACK GCT CAG TAA CAC GT-3′) [NOTE 3]
- 0.5 μL of the 100 μM HEX-labeled reverse primer "ARCH915r" (5′-GTG CTC CCC CGC CAA TTC CT-3′)
- 31 μL of ultrapure water
- 1 μL of sample DNA

It is recommended that a single master mix be prepared containing all the ingredients (with the exception of DNA) for all the test tubes to be analyzed. The master mix should be prepared for a larger number of samples in order to compensate for errors in pipetting (i.e., if the number of sediment samples is 10, consider = 40 [each sample in four replicates] + 2 [the two PCR controls] + 2 = 44).

Dispense 49 μL of master mix in each tube and then add template DNA (1 μL, containing ca. 5 ng of DNA).
In the negative control, add 1 μL of ultrapure water.
In the positive control, add 1 μL containing ca. 1–5 ng of *Methanococcus jannaschii* (ATCC 43067D).
Cycling conditions in the thermalcycler: 30 PCR-cycles, consisting of 94°C for 1 min, 52°C for 1 min, and 72°C for 2 min, preceded by 3 min of denaturation at 94°C, followed by a final extension of 10 min at 72°C.

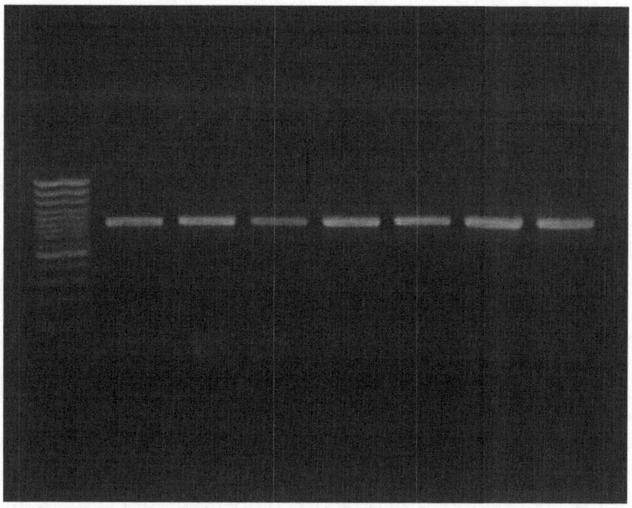

FIGURE 21.1 Typical bands obtained from the PCR amplification of the Archaeal 16S rRNA genes using the primer set "ARH109-915" on DNA extracted from sediment samples. The size of the expected amplicon is ca. 820 bp.

21.3.3.4 Agarose Gel Electrophoresis of PCR Products

To prepare the agarose gel,

Weigh 0.35 g of agarose.

Add 35 mL of TBE 1×.

Dissolve the agarose in the microwave.

Let the liquid cool (down to ca. 60°C–70°C) and add 1 μL of ethidium bromide.

Place the liquid in the casting tray containing a sample comb and let it solidify at room temperature.

Fill the electrophoretic chamber with TBE 1×.

Insert the gel (still in its plastic tray) horizontally into the electrophoresis chamber and cover with TBE 1× buffer.

Once the gel is ready, mix, on a piece of parafilm, 5 μL aliquots of each PCR product with 1 μL of loading dye.

Load the mixture in each well.

In the first well of each gel, place 4 μL of molecular weight marker.

Place the lid and power leads on the apparatus and apply current (100 V, 25 min; note that these settings are only indicative and should be adjusted for each instrument). Check that the current is really flowing by observing bubbles coming off the electrodes. DNA will migrate toward the positive electrode, which is usually red colored.

At the end of the electrophoretic run, transfer the gel on the transilluminator and check the presence of the expected amplicon. One discrete band should be visible on the gel for each PCR product (Figure 21.1) [NOTE 4].

21.3.3.5 Cleanup of PCR Products

Before purification, pool together two independent PCR replicates for each sediment sample. This process is carried out in order to minimize stochastic PCR biases (Polz and Cavanaugh 1998). According to this step, two T-RFLP replicates will then be digested and subsequently analyzed for each sediment sample [NOTE 5].

Purify pooled PCR products using the Wizard PCR cleanup system by carefully following the procedure predescribed by the manufacturer.

Quantify the eluted DNA spectrophotometrically as described earlier.

21.3.3.6 Enzymatic Digestion of Amplicons

For each duplicate purified PCR, digest between 200 and 300 ng of purified 16S rDNA amplicons in 20 μL reactions, according to the following recipe:

- 10 μL of DNA
- 2 μL of 10× buffer (supplied by the manufacturer)
- 1 μL of Alu I enzyme (containing 10 units, i.e., the stock solution must be 10 unit μL^{-1}) [NOTE 6]
- 0.2 μL of BSA (bovine serum albumin, supplied by the manufacturer)
- 6.8 μL reagent-grade water

It is recommended that a single master mix be prepared, containing all the ingredients (with the exception of the amplicons), for all the test tubes to be digested. The master mix should be prepared for two more samples in order to compensate for errors in pipetting.

Incubate in the dark at 37°C for 3 h.

Inactivate the restriction enzymes by heating to 70°C for 20 min (this step can be carried out in the thermalcycler).

Keep each tube frozen at −20°C until analysis.

21.3.3.7 Capillary Electrophoresis

Capillary analysis must be performed in an ABI Prism 310 or 3100 Genetic Analyzer (Applied Biosystems).

Mix 1–2 μL of digest (corresponding to ca. 15–30 ng of DNA) with 14 μL of internal size standard (GS1000ROX) in deionized formamide [NOTE 7].

Denature at 94°C for 2 min and immediately chill into ice.

Analyze fragments in GeneScan mode using 47 cm × 50 μm capillaries, POP-4 polymer, 40 s injection time, 15 kV injection voltage, 15 kV run voltage, and 60°C capillary temperature (these parameters can however be varied according to fluorescence signal intensity).

Determine T-RFLP fragments in the range 35–800 (ca.) bp using Genescan analytical software 2.02 (ABI; Figure 21.2) or similar (Peak Scanner Software, ABI).

21.3.3.8 Analysis and Interpretation of Results

Extract Genescan numerical outputs from each electropherogram and transfer to Microsoft Excel.

Remove, from each profile, “nonreproducible” and “shoulder” peaks, as described for the ARISA technique.

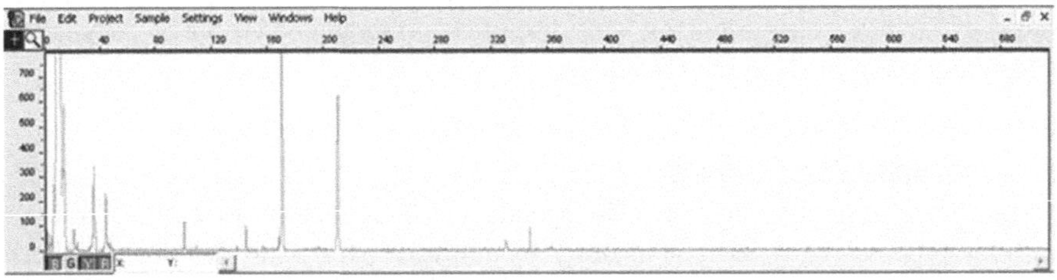

FIGURE 21.2 A typical electropherogram obtained from T-RFLP analysis of Archaeal benthic assemblages. Depending on the sample, from a few to more than 50 peaks, each corresponding to a different Archaeal ribotype, can be detected.

To discriminate Archaeal ribotypes or OTUs (operational taxonomic units) from baseline fluorescence, the total integrated height of each profile must be calculated (by summing the height of each peak in the T-RFLP electropherogram). Then calculate the relative percentage of each peak (as a percentage over the total integrated height), considering only the peaks accounting for more than 0.16% of the total integrated height (Luna et al. 2006) [NOTE 8].

After this procedure, Archaeal ribotype richness is defined as the number of remaining OTUs.

Archaeal diversity indices (such as the Shannon–Wiener index and others) or the Pielou index (evenness) can be calculated assuming that the number of OTU represents the species number, and that the height associated with each peak represents the relative abundance of each Archaeal OTU.

However, caution must be used, as the calculation of diversity indices using the outputs of fingerprinting techniques can be biased by several factors (Bent and Forney 2008).

Data from T-RFLP profiles can be utilized for analyses of ß-diversity and comparisons among samples in terms of species composition and community structure by using specific software (such as Primer; Clarke 1993).

21.3.4 Notes and Problems

1. DNA extraction and purification: the same as reported for Chapter 20, "Extraction and Purification of DNA from Marine Sediments Suitable for the Analysis of Prokaryotic Diversity."
2. DNA quantification (using a spectrophotometer): the same as reported in Chapter 23, "Benthic Bacterial Diversity Analysis Based on ARISA (Automated Ribosomal Intergenic Spacer Analysis)."
3. PCR amplification of the Archaeal 16S rRNA genes. Other authors (DeLong 1992; Moeseneder et al. 2001; Braker et al. 2001) have used, for studies carried out in both marine and terrestrial environments, another widely utilized universal primer set (ARCH21f–ARCH958r). Recently, we have found that, when applied to marine sediments, this primer set produces estimates of richness and diversity significantly lower than those obtained using the set ARCH109f–ARCH915r (Luna et al. 2009). As such, we recommend that the ARCH109f–915r primer set be used for ecological investigations in marine sediments.
4. Agarose gel electrophoresis of PCR products: (1) Ethidium bromide is potentially carcinogenic; follow all rules of personal safety to prevent risks; (2) one single band having the size of ca. 820 bp must be visible on the gel. If more than one band will appear (Figure 21.3), optimize PCR reactions (by changing the annealing temperature) or gel-purify the band having proper size (using a commercial kit) before proceeding with the digestion.
5. Cleanup of PCR products. Egert and Friedrich (2003) have recently reported the occurrence of "pseudo-terminal restriction fragments" ("pseudo-TRFs") that form during PCR amplification of mixed templates. If not identified, these additional pseudo-TRFs can lead to an overestimation of the diversity of the sample. In this case it is suggested that PCR amplicons be treated with mung bean nuclease prior to restriction digest, and that T-RFLP analysis minimizes the occurrence of these pseudo-T-RFs. This process can be carried out before proceeding with the cleanup.
6. Enzymatic digestion of amplicons. Other tetrameric enzymes (such as Hha I, Hae III, Msp I, Taq I, and Rsa I) have been utilized in previous T-RFLP-based studies of Archaea. In a recent study, we have found that Alu I is the enzyme providing the highest richness estimates when applied to marine sediments (Luna et al. 2009); as such, the use of this enzyme is recommended when screening benthic Archaeal assemblages.

   Other authors (Winter et al. 2004) have reported higher incubation times during restriction digestions (i.e., from 4 to overnight) to assure a complete digestion of the amplicons.

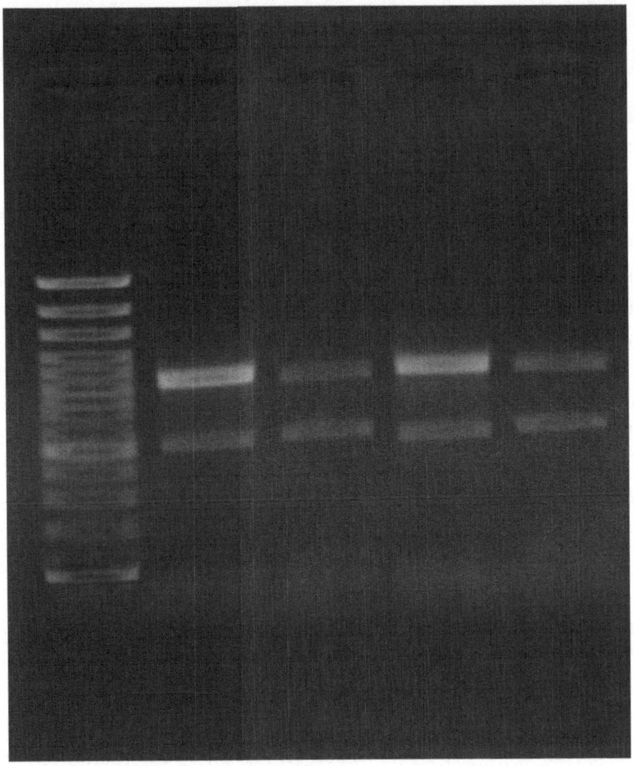

FIGURE 21.3 Example of unspecific PCR products. The more evident band corresponds to the size of the amplified sequence, whereas the other is an unspecific product. In this case, optimization of PCR conditions, or, alternatively, gel purification (by excising the band of proper size and recovering the PCR product with a gel purification kit, such as Qiagen, MoBio, or Promega), is required before proceeding with digestion.

However, in on our experience, we have found that 3 h are sufficient to degrade the amounts of DNA typically used in each reaction.

7. Capillary electrophoresis. The GS1000-ROX internal size standard can sometimes display unoptimal performances; as alternative, it is possible to utilize other standards, such as MapMarker® 1000 X-Rhodamine (Bioventures) or GS500-ROX. GS500-ROX is commonly considered the more precise marker for sizing terminal restriction fragments; however, the use of this marker may lead to missing fragments longer than 500 bp, thus providing underestimations of the true diversity.
8. Analysis and interpretation of results. The same as reported in the Chapter 23, "Benthic Bacterial Diversity Analysis based on ARISA (Automated Ribosomal Intergenic Spacer Analysis)," using 0.16% as the threshold limit for Archaea.

Extract DNA from the sediment sample
(as described above in Chapter 20)

↓

Put 990 μL of reagent grade water and
10 μL of DNA in the spectrophotometer cuvette

↓

Measure the absorbance at 260 nm
to quantify DNA concentration

↓

Dilute the DNA extracts to ca. 5 ng μL^{-1}

↓

Prepare four 200-μL sterile PCR tubes (for each DNA sample)
plus one for the positive and one for the negative control

↓

Prepare the appropriate PCR master mix

↓

Dispense 49 μL of master mix in each tube and then add 1 μL of DNA

↓

Add 1 μL of ultrapure water in the negative control and 1 μL of
Methanococcus jannaschii DNA in the positive control

↓

Put the tubes in the thermal cycle
and run the appropriate program

↓

Prepare the agarose gel, as follows:

↓

(a)

PROTOCOL 21.1 Determination of benthic Archaeal diversity by T-RFLP. *Continued*

Transfer the gel on the trans-illuminator and check the presence of the expected amplicon

↓

Pool together 2 PCR test tubes for each sample

↓

Purify pooled PCR products using the Wizard PCR clean-up system

↓

Quantify the DNA spectrophotometrically (as described above)

↓

Digest 200–300 ng of purified DNA in 20-μL reactions, each containing 10 μL of DNA, 2 μL of buffer, 1 μL of Alu I, 0.2 μL of BSA and 6.8 μL reagent grade water

↓

Incubate in the dark at 37°C for 3 h

↓

Heat to 70°C for 20 min

↓

Keep each tube frozen at –20°C until capillary analysis

↓

To perform capillary analysis, mix 1–2 μL of digest with 14 μL of appropriate Internal Size Standard in deionized formamide

↓

Denature at 94°C for 2 min and immediately chill into ice

↓

Analyze fragments in GeneScan mode

↓

Extract GeneScan file into Excel and analyze data

(b)

PROTOCOL 21.1 *Continued.*

REFERENCES

Béjà, O., E. V. Koonin, L. Aravind et al. 2002. Comparative genomic analysis of Archaeal genotypic variants in a single population and in two different oceanic provinces. *Applied and Environmental Microbiology* 68: 335–345.

Bent, S. J. and L. J. Forney. 2008. The tragedy of the uncommon: Understanding limitations in the analysis of microbial diversity. *The ISME Journal* 2: 689–695.

Blackwood, C. B., T. Marsh, S. H. Kim, and E. Y. Paul. 2003. Terminal restriction fragment length polymorphism data analysis for quantitative comparison of microbial communities. *Applied and Environmental Microbiology* 69: 926–932.

Braker, G., H. L. Ayala-del-Rio, A. H. Devol, A. Fesefeldt, and J. M. Tiedje. 2001. Community structure of denitrifiers, bacteria, and archaea along redox gradients in Pacific Northwest marine sediments by terminal restriction fragment analysis of amplified nitrite reductase (nirS) and 16S rRNA genes. *Applied and Environmental Microbiology* 67: 1893–1901.

Chaban, B., S. Y. Ng, and K. F. Jarrell. 2006. Archaeal habitats—from the extreme to the ordinary. *Canadian Journal of Microbiology* 52: 73–116.

Church, M. J., E. F. DeLong, H. W. Ducklow, M. B. Karner, C. M. Preston, and D. M. Karl. 2003. Abundance and distribution of planktonic Archaea and Bacteria in the Western Antarctic Peninsula. *Limnology and Oceanography* 48: 1893–1902.

Clarke, K. R. 1993. Non parametric multivariate analyses of changes in community structure. *Australian Journal of Ecology* 18: 117–143.

Danovaro, R., G. M. Luna, A. Dell'Anno, and B. Pietrangeli. 2006. Comparison of two fingerprinting techniques, terminal restriction fragment length polymorphism and automated ribosomal intergenic spacer analysis, for determination of bacterial diversity in aquatic environments. *Applied and Environmental Microbiology* 72: 5982–5989.

Danovaro, R., G. M. Luna, A. Dell'Anno, and B. Pietrangeli. 2007. Measuring species richness based on microbial community fingerprints: The emperor has no clothes. *Applied and Environmental Microbiology* 73: 2399–2401.

DeLong, E. F. 1992. Archaea in coastal marine environments. *The Proceedings of the National Academy of Sciences* 89: 5685–5689.

DeLong, E. F. 1998. Everything in moderation: Archaea as "non-extremophiles." *Current Opinion in Genetics and Development* 8: 649–654.

Egert, M. and M. W. Friedrich. 2003. Formation of pseudo-terminal restriction fragments, a PCR-related bias affecting terminal restriction fragment length polymorphism analysis of microbial community structure. *Applied and Environmental Microbiology* 69: 2555–2562.

Fuhrman, J. A. and A. Hagstrom. 2008. Bacterial and archaeal community structure and its patterns. In *Microbial ecology of the oceans*, ed. D. L. Kirchman, 45–90. Hoboken, New Jersey: John Wiley & Sons.

Fuhrman, J. A. and C. C. Ouverney. 1998. Marine microbial diversity studied via 16S rRNA sequences: Cloning results from coastal waters and counting of native archaea with fluorescent single cell probes. *Aquatic Ecology* 32: 3–15.

Karner, M. B., E. F. DeLong, and D. M. Karl. 2001. Archaeal dominance in the mesopelagic zone of the Pacific Ocean. *Nature* 409: 507–510.

Kim, B. S., H. M. Oh, H. Kang, and J. Chun. 2005. Archaeal diversity in tidal flat sediment as revealed by 16S rDNA analysis. *The Journal of Microbiology* 43: 144–151.

Liu, W. T., T. L. Marsh, H. Cheng, and L. J. Forney. 1997. Characterization of microbial diversity by determining terminal restriction fragment length polymorphisms of genes encoding 16S rRNA. *Applied and Environmental Microbiology* 63: 4516–4522.

Luna, G. M., A. Dell'Anno, and R. Danovaro. 2006. DNA extraction procedure: A critical issue for bacterial diversity assessment in marine sediments. *Environmental Microbiology* 8: 308–320.

Luna, G. M., A. Dell'Anno, L. Giuliano, and R. Danovaro. 2004. Bacterial diversity in deep Mediterranean sediments: Relationship with the active bacterial fraction and substrate availability. *Environmental Microbiology* 6: 745–753.

Luna, G. M., K. Stumm, A. Pusceddu, and R. Danovaro. 2009. Archael diversity in deep-sea sediments estimated by means of different terminal-restriction fragment length polymorphisms (T-RFLP) protocols. *Current Microbiology* 59: 356–361.

Marsh, T. L.1999. Terminal-Restriction Fragment Length Polymorphism (T-RFLP): An emerging method for characterizing diversity among homologous populations of amplicons. *Current Opinion Microbiology* 2: 323–327.

Moeseneder, M. M., C. Winter, J. M. Arrieta, and G. J. Herndl. 2001. Terminal-restriction length polymorphism (T-RFLP) screening of a marine archaeal clone library to determine the different phylotypes. *Journal of Microbiological Methods* 44: 159–172.

Polz, M. F. and C. M. Cavanaugh. 1998. Bias in template-to-product ratios in multitemplate PCR. *Applied and Environmental Microbiology* 64: 3724–3730.

Rappè, M. S. and S. J. Giovannoni. 2003. The uncultured microbial majority. *Annual Review of Microbiology* 57: 369–374.

Schwarz, J. I., W. Eckert, and R. Conrad. 2006. Community structure of Archaea and Bacteria in a profundal lake sediment Lake Kinnert (Israel). *Systematic and Applied Microbiology* 30: 239–254.

Vetriani, C., H. W. Jannasch, B. J. MacGregor, D. A. Stahl, and A. L. Reysenbach. 1999. Population structure and phylogenetic characterization of marine benthic Archaea in deep-sea sediments. *Applied and Environmental Microbiology* 65: 4375–4384.

Vetriani, C., H. V. Tran, and L. J. Kerkhof. 2003. Fingerprinting microbial assemblages from the oxic/anoxic chemocline of the Black Sea. *Applied and Environmental Microbiology* 69: 6481–6488.

Winter, C., A. Smit, G. J. Herndl, and M. G. Weinbauer. 2004. Impact of virioplankton on archaeal and bacterial community richness as assessed in seawater batch cultures. *Applied and Environmental Microbiology* 70: 804–813.

22 Benthic Bacterial Diversity Based on Cloning and Sequencing of 16S rRNA Genes

CONTENTS

22.1 INTRODUCTION

Sequence analysis of PCR-amplified and subsequently cloned 16S ribosomal RNA genes (also defined as 16S rDNA) is a widely utilized approach to assess microbial diversity and community composition in marine sediments (Bowman and McCuaig 2003; Polymenakou et al. 2005; Hunter et al. 2006; Stevens et al. 2007). Despite not being immune, analogously to all other PCR-based methods, of known potentially occurring biases (von Wintzingerode et al. 1997), this method allows an accurate description of prokaryotic biodiversity in marine sediments, and identification of the taxonomic identity of Bacteria (or Archaea) present in the investigated sample. Compared to fingerprinting techniques described here (such as automated ribosomal intergenic spacer analysis [ARISA] or terminal-restriction fragment length polymorphism [T-RFLP]), this method has the advantage of providing more detailed information in terms of species composition and allowing a phylogenetic-based identification of bacteria in the sediment sample.

However, despite having higher resolution compared to fingerprinting methods, cloning and sequencing also suffers from the inability of detecting the rare components of the assemblage, which has so far been investigated only using 454 sequencing (Sogin et al. 2006). While the use of 454 technology is today too expensive to be applied to a large set of samples, approaches based on the use of fingerprinting methods of a large set of samples, followed by clone library analysis of a subset, are currently reputed as an optimal balance between the resources required and the information gained (Bent and Forney 2008; Fuhrman and Hagstrom 2008).

The cloning technique can be described as follows: a DNA fragment is inserted into a circular DNA molecule (cloning vector), generally by means of a ligase enzyme (this process is named "ligation"). Then, ligation products are brought into competent *Escherichia coli* bacterial strains (a process named "transformation"). The competent cells are plated onto an agar medium and, once incubated, start duplicating the recombined cloning vectors (e.g., the plasmids) during growth, thus passing copies to the succeeding generations. As a consequence, the colonies grown will be genetically identical and will contain the recombined DNA fragments. When applied to a heterogeneous mixture of PCR products (i.e., amplicons possessing the same length but different nucleotide sequence), this method will allow the separation of the different products constituting the mixture by introducing each one into a single line of *E. coli*. Separated products can then be analyzed, for instance, by sequencing or other molecular methods.

The method can also be applied, if used in combination with Archaeal universal primers, to study the biodiversity of Archaea. However, this approach is time consuming and more expensive than genetic fingerprinting methods (Denovaro et al. 2007). As such, the choice of one of the two approaches (or their combination) will depend upon the specific objectives of each investigation.

22.2 FIELD PROCEDURES

Undisturbed sediment samples are typically collected using a mini- or a multicorer or, if available, a remotely operated vehicle (ROV). A subsample (ca. 5–10 cm^3) from the superficial layer (i.e., 0–1 cm) is then aseptically transferred from the core into sterile Falcon test tubes (50 mL) using a sterile metal or plastic spatula or a 10 mL sterile syringe from which the tip has been removed. The tube is then frozen at –20°C for a limited time, generally a few weeks, or at −80°C until DNA extraction.

22.3 LABORATORY PROCEDURES

The application of the 16S rDNA cloning method to sediment samples involves the following steps in succession:

1. DNA extraction and purification (with the aim of extracting genomic DNA from benthic microorganisms, which must be of sufficient purity for subsequent PCR-based analyses).
2. DNA quantification (needed to quantify the concentration of DNA extracts and to utilize known amounts of DNA in the PCR reaction).
3. PCR amplification of Bacterial 16S rRNA genes (to amplify the 16S rDNA present in the sample by using universal bacterial primers, thus providing a sufficient amount of this gene for cloning).
4. Agarose gel electrophoresis of PCR products (after PCR amplification, to check the success of the amplification reaction and the quality of the obtained amplicon).
5. Cloning (to ligate PCR products into a vector and transform them into chemically competent cells of *E. coli* using the TOPO-TA cloning kit).
6. Screening of clones by PCR and RFLP (to identify unique clones by performing the analysis called RFLP).
7. Sequencing of clones (which allows partial sequencing of the cloned 16S rRNA genes).
8. Analysis and interpretation of results to (1) perform phylogenetic analyses with the aim of finding the identity of the clones, and (2) obtain estimates of bacterial richness and community structure in the sample.

22.3.1 Equipment and Instruments

Water bath or thermal block
Centrifuges (must be suitable for 2 mL test tubes and 50 mL test tubes)
Vortex
Thermalcycler
Transilluminator
UV-VIS spectrophotometer
Complete apparatus for horizontal electrophoresis (including casting trays, electrophoretic chamber, and power supply)
ABI 310 or 3100 sequencer
Horizontal shaking apparatus
Incubator (37°C)

22.3.2 Solutions and Reagents

Wash Solution 1: 50 mM Tris-HCl, pH 8.3; 200 mM NaCl; 5 mM Na_2EDTA; 0.05% Triton X-100
Wash Solution 2: 50 mM Tris-HCl, pH 8.3; 200 mM NaCl; 5 mM Na_2EDTA
Wash Solution 3: 10 mM Tris-HCl, pH 8.3; 0.1 mM Na_2EDTA
Ultrapure water: Reagent-grade water, previously autoclaved, filtered (through 0.2 μm) and stored in sterile 2 mL Eppendorf tubes at 4°C or −20°C
UltraClean™ Soil DNA Isolation Kit (MoBio, catalog #12800-100)
MasterTaq® kit (Eppendorf AG, Germany)
Primers (such as "27f," "1492r," "M13forward," and "M13reverse"; can be purchased by SIGMA, MWG, or others after providing the nucleotide sequence)
Agarose
TBE 10× (can be purchased from BIORAD or similar, or, alternatively, prepared as follows [for 1 L]: 108 g TRIS Base; 55 g Boric Acid; 40 mL EDTA 0.5 M, pH 8; reagent-grade water to volume)
TBE 1×: dilute TBE 10× (100 mL TBE 10× + 900 mL of reagent-grade water)
Ethidium bromide (10 mg mL^{-1} in ultrapure water)
Loading dye 6× (Promega, Fermentas, or similar)
Molecular weight 100 bp (Fermentas or similar)
Wizard® PCR cleanup system (Promega)
TOPO-TA cloning kit for sequencing (Invitrogen)
Ampicillin (or alternatively, Kanamycin)
Tryptone
Yeast extract
Sodium chloride
Bacteriological agar
Restriction enzyme Hae III (Promega or similar)
BigDye Terminator v3.1 cycle-sequencing kit (optional)

22.3.3 Procedure

22.3.3.1 DNA Extraction and Purification

Extract the DNA following the procedure reported in Chapter 20 of this book. [NOTE 1]

22.3.3.2 DNA Quantification (Using a Spectrophotometer)

Use only quartz cuvettes (path 1 cm, internal volume 1 mL).

Set up the spectrophotometer at 260 nm.
If working with a dual-channel spectrophotometer, place in both the "reference" and "sample" cuvettes, 1 mL of ultrapure water and then operate the procedure to reset the instrument.
Discard the water from the "sample" cuvette and place in it 990 μL of reagent-grade water and 10 μL of each DNA sample (i.e., dilute the sample to a factor of 100).
Measure the absorbance at 260 nm.
Discard the liquid from the "sample" cuvette and repeat the procedure for all extracted DNA samples.
At the end of the procedure, DNA concentration in each DNA extract is calculated as follows:

$$DNA\ (ng\ \mu L^{-1}) = ABS_{260} \cdot 50 \cdot 100 \quad (22.1)$$

where 100 is the dilution factor of the DNA within the cuvette (i.e., 10 μL of DNA in 990 μL of reagent-grade water) and 50 is the conversion factor from absorbance to DNA concentration (i.e., an absorbance of 1 at 260 nm contains approximately 50 μg/mL [or 50 ng/μL] of double-stranded DNA). The purity of the DNA can be checked by measuring the ratio of 260: 280 nm; a ratio greater than 1.8–2 is considered an indication of DNA purity from proteins [NOTE 2].

22.3.3.3 PCR Amplification of the Bacterial 16S rRNA Genes

Dilute the DNA extracts to ca. 5 ng μL^{-1}.
Prepare four 200 μL sterile PCR test tubes for each DNA sample (i.e., each sediment sample must be amplified in four replicates), plus one for the positive and one for the negative PCR controls.
For PCR amplification, it is recommended that the MasterTaq® kit (Eppendorf AG, Germany) be used, which increases the success of PCR amplification on DNA extracted from marine sediments by reducing the effects of residual, potentially PCR-inhibiting contaminants.
Each PCR tube will contain the following ingredients:

- 1 μL of Taq Polymerase (the stock solution must be 1 Unit μL^{-1})
- 5 μL of 10× Taq Buffer with Mg^{2+}
- 10 μL of 5× TaqMaster PCR Enhancer
- 1 μL of 10 mM dNTPs
- 0.5 μL of the 100 μM forward primer "27f" (5′-AGA GTT TGA TCM TGG CTC AG-3′)
- 0.5 μL of the 100 μM reverse primer "1492r" (5′-GGT TAC CTT GTT ACG ACT T-3′) [NOTE 3]
- 31 μL of ultrapure water
- 1 μL of sample DNA

Prepare a single master mix containing all the ingredients (with the exception of DNA) for all the test tubes to be analyzed. The master mix should be prepared for a larger number of samples in order to compensate for errors in pipetting (i.e., if the number of sediment samples is 10, consider 40 [each sample in four replicates] + 2 [the two PCR controls] + 2 = 44). However, compared to fingerprinting techniques (where more than one sample can be processed each time), the cloning and sequencing procedure, given its complexity, will be carried out by running only 1–2 sediment samples per time.

Dispense 49 μL of master mix in each tube and then add template DNA (1 μL, containing ca. 5 ng of DNA).
In the negative control, add 1 μL of ultrapure water.

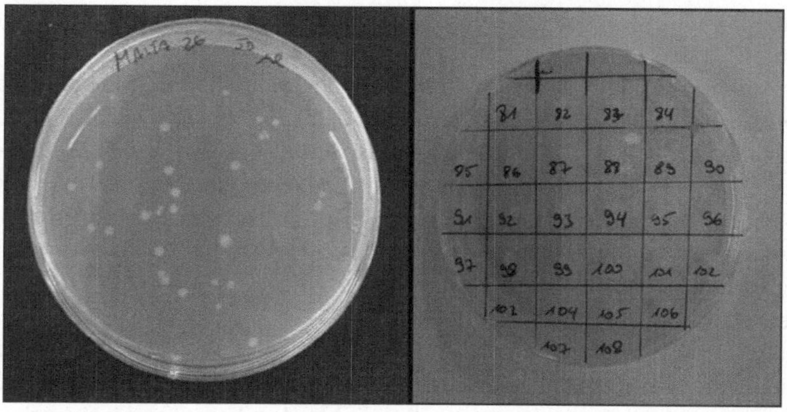

FIGURE 22.1 Left: example of *E. coli* colonies grown on LB agar medium, containing 50 μg/mL ampicillin, following a TOPO-TA reaction. Right: positive transformants can subsequently be numbered and organized into new LB agar plates. (The back side of a plate is shown.)

In the positive control, add 1 μL containing ca. 1–5 ng of genomic DNA of *E. coli* (SIGMA or other).

Cycling conditions in the thermalcycler: 30 PCR-cycles, consisting of 94°C for 1 min, 55°C for 1 min, and 72°C for 2 min, preceded by 3 min of denaturation at 94°C and followed by a final extension of 10 min at 72°C.

22.3.3.4 Agarose Gel Electrophoresis of PCR Products

To prepare the agarose gel:

Weigh 0.35 g of agarose.

Add 35 mL of TBE 1×.

Boil to dissolve.

Let the liquid cool (down to ca. 60°–70°C) and add 1 μL of ethidium bromide.

Place the liquid in the casting tray containing a sample comb and let it solidify at room temperature.

Fill the electrophoretic chamber with TBE 1×.

Insert the gel (still in its plastic tray) horizontally into the electrophoresis chamber and cover with TBE 1× buffer.

Once the gel is ready, mix, on a piece of parafilm, 5 μL aliquots of each PCR-product with 1 μL of loading dye.

Load the mixture in each well.

In the first well of each gel, place 4 μL of molecular weight marker.

Place the lid and power leads on the apparatus and apply current (100 V, 25 min; note that these settings are only indicative, and should be adjusted for each instrument). Ensure that the current is really flowing by observing bubbles coming off the electrodes. DNA will migrate toward the positive electrode, which is usually colored red.

At the end of the electrophoretic run, transfer the gel to the transilluminator and check the presence of the desired amplicon. One discrete band, having the size of ca. 1500 bp, should be visible on the gel for each PCR product [NOTE 4].

22.3.3.5 Cloning

Pool together the four PCR replicates. This will allow the minimization of biases caused by PCR-generated artifacts (Polz and Cavanaugh 1998).

Perform the TOPO-TA reaction by carefully following the instructions provided by the manufacturer [NOTE 5].

At the end of the procedure, plate 10 to 50 μL aliquots from the transformation vial on LB agar plates containing 50 μg/mL of kanamycin or 50 μg/mL ampicillin. LB plates will be prepared in advance (e.g., the day before) using tryptone, yeast extract, sodium chloride, and bacteriological agar, according to the recipe and procedure described in the TOPO-TA kit instructions. Be sure to plate different volumes to ensure that at least some plates have well-spaced colonies. For plating small volumes, add 20 μL of SOC medium to allow an even spreading of the cell suspension on the plates.

Invert the plate and incubate them overnight at 37°C.

The day after, check the plates for growth. An efficient cloning reaction should produce several hundred of colonies. Colonies of interest (e.g., positive transformants containing the 16S rDNA insert) can then subsequently be numbered and organized into new agar plates (Figure 22.1).

22.3.3.6 Screening of Clones by PCR and RFLP

To confirm the presence of the 16S rDNA insert on the clones, select well-spaced clones on the plates and analyze them by PCR using external (vector) primers (e.g., "M13forward" [5′-GTA AAA CGA CGG CCA G-3′] and "M13reverse" [5′-CAG GAA ACA GCT ATG AC-3′]. At this stage, it is important not to use the 27f–1492r bacterial primer set described earlier, to avoid the coamplification of *E. coli* 16S rDNA.

Perform the PCR as described earlier (Section 22.3.3.3, "PCR Amplification of the Bacterial 16S rRNA Genes"), but carrying out only 20 amplification cycles (to minimize the introduction of artifacts).

As DNA template, touch each grown colony with a sterile tip (or an autoclaved toothpick) and immerse it in the test tube containing the amplification mix. This will yield sufficient amount of material for amplification (given the sensitiveness of PCR, the colony material should not be visible on the tip).

At the end of the PCR, analyze positive transformants (i.e., clones carrying an insert of the size) by agarose gel electrophoresis (as described in Section 22.3.3.4 Agarose Gel Electrophoresis of PCR products).

To identify unique clones, perform RFLP analysis as follows:

- Digest aliquots (5 μL) of individual PCR products with a four-cutting restriction enzyme (we generally use Hae III) for 3 h according to the instructions supplied by the manufacturer, in a final volume of 20 μL.
- At the end of the 3 h, inactivate the enzyme (20 min at 70°C) and perform an electrophoresis (as described in Section 22.3.4, but using a 2% agarose gel [2.5 h, 80 V]).
- The resulting RFLP patterns are then observed (Figure 22.2) and used to classify clones into operational taxonomic units (OTU). Clones having different restriction patterns (such as, for instance, those indicated by the white arrows in Figure 22.2) are considered different ribotypes of OTU.

It is often necessary to analyze hundreds of clones to screen adequately the diversity of a library (see NOTE 6).

22.3.3.7 Sequencing of Clones [NOTE 7]

Purify and sequence all unique clones on a capillary sequencer using the primer 27f (5′-AGA GTT TGA TCM TGG CTC AG-3′). If the sequencer is available in the lab, it is necessary to purchase the appropriate kit for sequencing (e.g., the BigDye Terminator v3.1 Cycle Sequencing Kit [Applied Biosystems]). Otherwise, it is possible to send

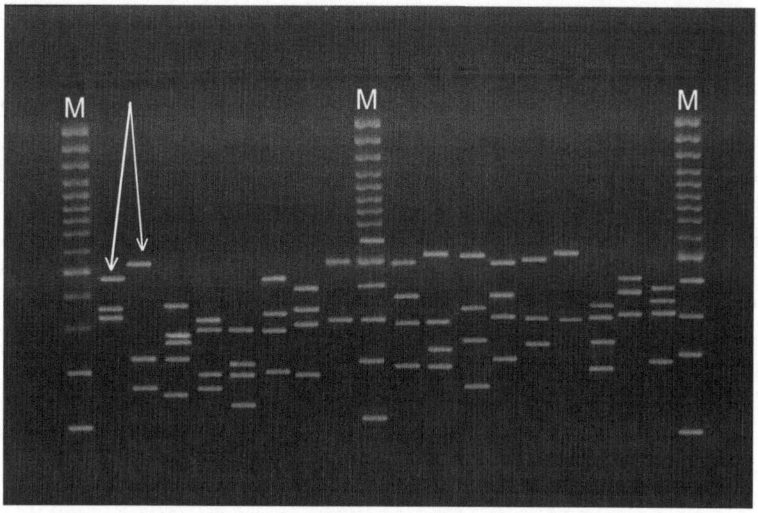

FIGURE 22.2 Examples of RFLP patterns obtained on 16S rDNA clones from marine sediments. All the analyzed clones display different restriction patterns. M: molecular weight ladder (Fermentas, 100 bp).

aliquots of each PCR-amplified insert to an external biotech company performing the sequencing [NOTE 7].

22.3.3.8 Analysis and Interpretation of the Results

With the results from the RFLP patterns, the RFLP-based distribution of clones in different OTUs can be used to estimate species richness with the help of a rarefaction calculator software. Most researchers utilize a program available from the Web: http://www2.biology.ualberta.ca/jbrzusto/rarefact.php. Species richness is generally estimated using the nonparametric estimator Chao (Chao 1984). Once the sequences are available, the software DOTUR (Distance-based OTU and richness Determination; Schloss and Handelsman 2005) can be used to obtain estimates of richness and diversity indices. To use DOTUR, it will first be necessary to align the obtained sequences. A simple and widely used alignment program is CLUSTALW, available at the Web site http://align.genome.jp/.

Aligned sequences then need to be processed with the program PHYLIP: http://evolution.genetics.washington.edu/phylip.html, which generates a "DNA-distance" matrix used as an input file to DOTUR.

Phylogenetic analyses can be performed using the software ARB (http://www.arb-home.de/) to assign 16S rRNA sequences to a taxonomical hierarchy in order to identify the bacteria present in the sample at the species, genus, or lower taxonomic resolution. Using this software, sequences are aligned using the integrated aligner tool and the fast aligner option followed by manual alignment of the sequences to closely related sequences in the ARB database. For comparisons of libraries, the statistical significance of the differences between two clone libraries can be tested using the software LIBSHUFF (http://whitman.myweb.uga.edu/libshuff.html). For phylogenetic analyses, it is also possible to utilize the programs provided by the Ribosomal Database Project (http://rdp.cme.msu.edu/).

22.3.4 Notes and Problems

1. DNA extraction and purification: The same as reported for Chapter 20, "Extraction and Purification of DNA from Marine Sediments Suitable for the Analysis of Prokaryotic Diversity."

2. DNA quantification (using a spectrophotometer): The same as reported in Chapter 23, "Benthic Bacterial Diversity Analysis Based on ARISA (Automated Ribosomal Intergenic Spacer Analysis)."
3. PCR amplification of the Bacterial 16S rRNA genes. Other sets of primers than those described in this chapter can be used as universal primers for 16S rRNA bacterial gene amplification. For a complete list of available primer sets, see the papers by Marchesi et al. (1998) and Baker et al. (2003). For the analysis of Archaeal diversity, the same cloning procedure described here can be adopted by using Archaeal universal primers such as the sets "A21f–958r" or "A109f–915r" described in Chapter 21, "Archaeal Diversity Analysis using 16S rDNA T-RFLP."
4. Agarose gel electrophoresis of PCR products: (1) Ethidium bromide is possibly carcinogenic; follow all rules of personal safety to avoid any risks; (2) one single band (in this case, having a size of ca. 1500 bp) must be visible on the gel. For cloning, it is very important that unspecific bands, if present, are eliminated by optimizing PCR reactions or by gel-purification (using a commercial kit) before proceeding with the cloning protocol.
5. Cloning: It is important to use fresh (i.e., maximum 1–2 d dated) PCR products. The ligation protocol relies upon the fact that Taq DNA polymerase has a nontemplate-dependent terminal transferase activity, which adds a single 3′-deoxyadenosine to the ends of PCR products. The linearized vector supplied in this kit has single, overhanging 3′-deoxythymidine (T) residues, which allows PCR inserts to ligate efficiently with the vector. As the terminal 3′-deoxyadenosine residue has a tendency to fall off over time, it is extremely important that fresh PCR products be used for the cloning reaction.
6. Screening of clones by PCR and RFLP. In many cases, it may be necessary to screen more than several hundreds clones to describe the whole diversity of the library and thus obtain realistic estimates of the true benthic bacterial diversity; only rarely are 100 clones or less sufficient to reach saturation (Figure 22.3).
7. Sequencing of clones: In some cases, "bad" sequences can be obtained (Figure 22.4).

When sequencing clones, the most common causes for "bad," multiple-peaks sequences include:

1. "Double pick" of two different colonies (the *E. coli* colonies on the library plates were too close, and the PCR-generated fragment is thus a mix of two different 16S rDNA).
2. Poor quality of the PCR template (the PCR fragment has not been purified before sequencing).
3. A wrong primer-annealing temperature has been used in the sequencing reaction.

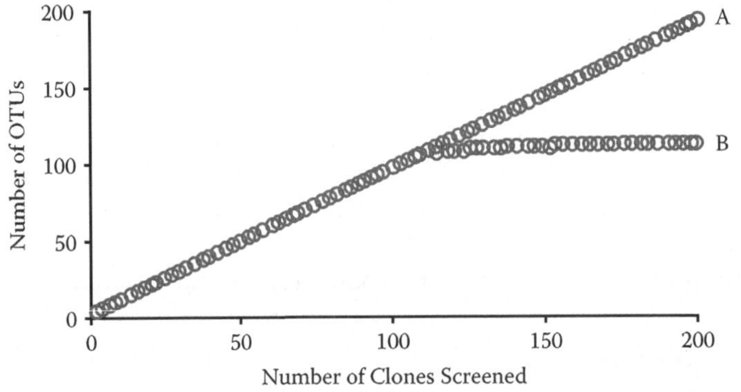

FIGURE 22.3 Examples of rarefaction analysis of 16S rDNA RFLP patterns in two sediment clone libraries. The total number of analyzed clones is plotted against the number of unique OTUs identified by RFLP.

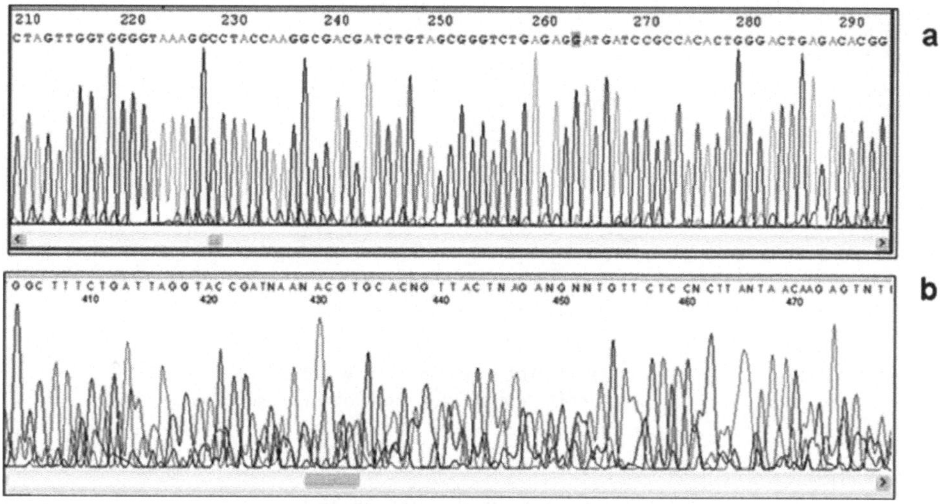

FIGURE 22.4 ***(A color version of this figure follows page 234.)*** Examples of a "good" (a) and a "bad" (b) sequence of a 16S rDNA clones obtained from marine sediments. The "bad" sequence presents multiple peaks.

Extract DNA from the sediment sample
(as described above)

↓

Put 990 μL of reagent grade water and
10 μL of DNA in the spectrophotometer cuvette

↓

Measure the absorbance at 260 nm
to quantify DNA concentration

↓

Dilute the DNA extracts to ca. 5 ng μL

↓

Prepare four 200-μL sterile PCR tubes (for each DNA sample),
plus one for the positive and one for the negative control

↓

Prepare the appropriate PCR master mix

↓

Dispense 49 μL of master mix in each tube
and then add 1 μL of DNA

↓

Add 1 μL of ultrapure water in the negative control and
1 μL of *Escherichia coli* DNA in the positive control

↓

Put the tubes in the thermalcycler and
run the appropriate program

↓

(a)

PROTOCOL 22.1 Determination of benthic bacterial diversity by clone libraries. *Continued*

Prepare the agarose gel, as follows:

↓

Weigh 0.35 g of agarose

↓

Add 35 mL of TBE 1X

↓

Dissolve the agarose in the microwave

↓

Let the liquid cool and
add 1 μL of Ethidium Bromide

↓

Put the liquid into the casting tray
containing a sample comb and let it solidify

↓

Fill the electrophoretic chamber with TBE 1X

↓

Insert the gel horizontally into the electrophoresis chamber
and cover with TBE 1X buffer

↓

Once the gel is ready, mix 5-μL aliquots
of each PCR product with 1 μL of loading dye

↓

Load the mixture in each well
(leave one well empty for the molecular weight marker)

↓

(b)

PROTOCOL 22.1 *Continued.*

Put 4 μL of molecular weight marker in the empty well

↓

Place the lid and power leads on the apparatus and apply current (100 V, 25 min)

↓

Transfer the gel on the trans-illuminator and check the presence of the expected amplicon

↓

Pool together the four PCR test tubes for each sample

↓

Purify pooled PCR products using the Wizard PCR clean-up system

↓

Quantify the DNA spectrophotometrically as described above

↓

Perform the TOPO-TA reaction as indicated by the manufacturer

↓

Plate 10 to 50 μL aliquots from the transformation vial on LB agar plates containing 50 μg/mL of kanamycin or ampicillin

↓

Invert the plate and incubate them overnight at 37°C

↓

(The day after) Check the plates for growth

↓

(c)

PROTOCOL 22.1 *Continued.*

Select well-spaced clones on the plates and analyze them by PCR using "M13forward" and "M13reverse" primers

↓

Perform the PCR (as described above), with 20 amplification cycles

↓

Analyze amplicons by agarose gel electrophoresis (as described above)

↓

To identify unique clones, perform RFLP analysis as follows:

↓

Digest 5 μL of PCR products with *Hae III* for 3 h according to the instructions supplied by the manufacturer in a final volume of 20 μL

↓

Inactivate the enzyme (20 min at 70°C)

↓

Perform agarose gel electrophoresis (as described above), but using a 2% agarose gel (2.5 h, 80 V)

↓

Observe RFLP patterns and classify clones into operational taxonomic units (OTU)

↓

Purify and sequence unique clones on a capillary sequencer using the primer 27f. If the sequencer is available in the lab, use the kit for sequencing (e.g., the BigDye Terminator v3.1 Cycle Sequencing Kit [Applied Biosystems]). Otherwise, send aliquots of PCR-amplified inserts to an external biotech company

↓

Once the sequences are available, perform data analyses using appropriate software

(d)

PROTOCOL 22.1 *Continued.*

REFERENCES

Baker, G. C., J. J. Smith, and D. A. Cowan. 2003. Review and re-analysis of domain-specific 16S primers. *Journal of Microbiological Methods* 55: 541–555.

Bent, S. J. and L. J. Forney. 2008. The tragedy of the uncommon: Understanding limitations in the analysis of microbial diversity, *The ISME Journal* 2: 689–695.

Bowman, J. P. and R. D. McCuaig. 2003. Biodiversity, community structural shifts, and biogeography of prokaryotes within Antarctic continental shelf sediment. *Applied and Environmental Microbiology* 69: 2463–2483.

Chao, A. 1984. Nonparametric estimation of the number of classes in a population. *Scandinavian Journal of Statistics* 11: 265–270.

Danovaro, R., G. M. Luna, A. Dell'Anno, and B. Pietrangeli. 2007. Measuring species richness based on microbial community fingerprints: The emperor has no clothes. *Applied and Environmental Microbiology* 73: 2399–2401.

Fuhrman, J. A. and A. Hagstrom. 2008. Bacterial and Archaeal community structure and its patterns. In *Microbial ecology of the oceans*, ed. D. L. Kirchman, 45–90. Hoboken, New Jersey: John Wiley & Sons.

Hunter E. M., H. J. Mills, and J. E. Kostka. 2006. Microbial community diversity associated with carbon and nitrogen cycling in permeable shelf sediments. *Applied and Environmental Microbiology* 72: 5689–5701.

Marchesi J. R., T. Sato, A. J. Weightman et al. 1998. Design and evaluation of useful bacterium-specific PCR primers that amplify genes coding for bacterial 16S rRNA *Applied and Environmental Microbiology* 64: 795–799.

Polymenakou, P. N., S. Bertilsson, A. Tselepides, and E. G. Stephanou. 2005. Bacterial community composition in sediments from the Eastern Mediterranean Sea: A comparison of four 16S rDNA clone libraries. *Microbial Ecology* 50: 447–462.

Polz, M. F. and C. M. Cavanaugh. 1998. Bias in template-to-product ratios in multitemplate PCR. *Applied and Environmental Microbiology* 64: 3724–3730.

Schloss, P. D. and J. Handelsman. 2005. Introducing DOTUR, a computer program for defining operational taxonomic units and estimating species richness. *Applied and Environmental Microbiology* 71: 1501–1506.

Sogin, M. L., H. G. Morrison, J. A. Huber et al. 2006. Microbial diversity in the deep sea and the underexplored "rare biosphere." *The Proceedings of the National Academy of Sciences (USA)* 103: 12115–12120.

Stevens, H., T. Brinkhoff, B. Rink, J. Vollmers, and M. Simon. 2007. Diversity and abundance of Gram positive bacteria in a tidal flat ecosystem. *Environmental Microbiology* 9: 1810–1822.

von Wintzingerode, F., U. B. Göbel, and E. Stackebrandt. 1997. Determination of microbial diversity in environmental samples: Pitfalls of PCR-based rRNA analysis. *FEMS Microbiology Ecology* 21: 213–229.

23 Benthic Bacterial Diversity Analysis Based on ARISA (Automated Ribosomal Intergenic Spacer Analysis)

CONTENTS

23.1 INTRODUCTION

In the last few years, the analysis of bacterial ribotype richness has become increasingly important in microbial ecology as it allows identification of hot spots of microbial biodiversity, the study of the spatio-temporal dynamics of prokaryotic assemblages, the identification of factors influencing their distribution, and the investigation of bacterial biogeography. Among the wide array of molecular techniques available for investigating prokaryotic diversity (Amann et al. 2001), genetic fingerprinting techniques allow the estimation of prokaryotic richness and community composition of the dominant components of the assemblage in a rapid and rather accurate way. PCR-based fingerprinting techniques, despite limitations related to their weakness in addressing species identity and their inability to detect rare taxa within the assemblage (Bent and Forney 2008; Fuhrman and Hagstrom 2008), represent a low-cost, rapid tool for comparative studies of the biodiversity of prokaryotes in different aquatic ecosystems. This holds true particularly in the framework of ecological contexts in which it is useful for a large-scale comparison of putatively different habitats and/or ecosystems (Figure 23.1).

Techniques such as T-RFLP (Terminal-Restriction Fragment Length Polymorphism; Marsh 1999) applied to the 16S rRNA gene and ARISA (Automated Ribosomal Intergenic Spacer Analysis; Fisher and Triplett 1999) are now increasingly used in the framework of marine ecological studies because they allow, compared to other time-consuming methods (such as cloning and sequencing; Hewson and Fuhrman 2004), rapid processing of large numbers of aquatic samples (Hewson and Fuhrman

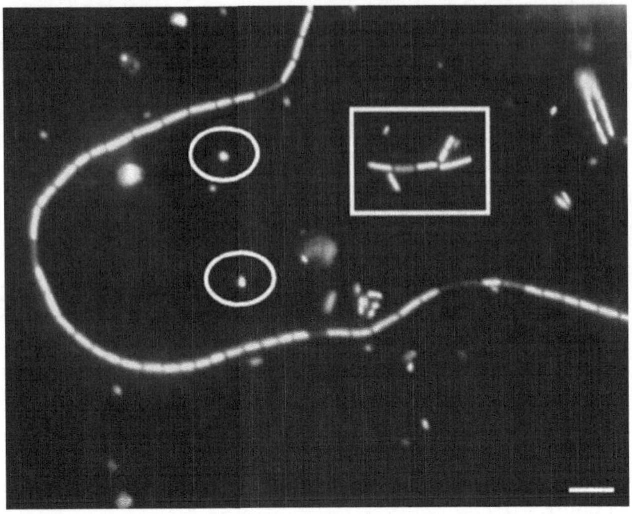

FIGURE 23.1 Epifluorescence microscopy image of a deep-sea sediment sample filtered onto a Whatman 0.2 mm anodisc filter after staining with fluorochrome SYBR Green I. The picture shows several prokaryotic cells displaying different morphology. Circles highlight cells having a spherical, cocci-like morphology; the rectangle indicates bacilli-like cells. The scale bar is 5 μm.

2004; Luna et al. 2004). ARISA and T-RFLP have been suggested to be more sensitive than other fingerprinting methods such as DGGE or SSCP (Marsh 1999; Fisher and Triplett 1999), especially for their ability to detect also less abundant taxa within assemblages. In addition, in a recent study, it has been demonstrated that ARISA, compared to T-RFLP (which is based on the analysis of the 16S rRNA gene), provides a more accurate estimation of richness and community composition due to the higher resolution power of the 16S-23S ITS region (Danovaro et al. 2006). ARISA is based on the extraction of DNA from benthic microbes and the subsequent amplification of the intergenic region between the 16S and 23S rRNA genes in the rRNA operon (known as the ITS1 region), which is characterized by a significant variability in length and nucleotide sequence among different bacterial genotypes (Daffonchio et al. 2003). As for T-RFLP, the output of each ARISA analysis is an electropherogram characterized by the presence of a series of peaks that allow quantification of the number and relative abundance of the different taxa of the bacterial assemblage. This cultivation-independent method is considered at present as the best fingerprinting method available for obtaining quick evaluations of bacterial richness and community structure when processing large numbers of sediment samples (Hewson and Fuhrman 2004; Luna et al. 2004). Advantages and disadvantages of the ARISA approach are discussed by Fuhrman and Hagstrom (2008).

23.2 FIELD PROCEDURE

Undisturbed sediment samples are typically collected using a mini- or a multicorer. A subsample (ca. 5–10 cm^3) from the sediment cores (i.e., superficial layer 0–1 cm) is then aseptically transferred into sterile Falcon test tubes (50 mL) using a sterile metal or plastic spatula, or using a 10 mL sterile syringe from which the tip has been removed. The tube is then frozen at −20°C or (preferably) at −80°C and kept stored until DNA extraction.

23.3 LABORATORY PROCEDURE

23.3.1 Equipment and Instruments

Water bath or thermal block

Centrifuges (must be suitable for 2 mL and 50 mL test tubes)
Vortex
Thermalcycler
Transilluminator
UV-VIS spectrophotometer
Complete apparatus for horizontal electrophoresis (including casting trays, electrophoretic chamber, and power supply)
ABI (Applied Biosystems) 310 or 3100 sequencer (if not available, the capillary electrophoresis of purified PCR products can be carried out by an external subcontractor)

23.3.2 Solutions and Reagents

Wash Solution 1: 50 mM Tris-HCl, pH 8.3; 200 mM NaCl; 5 mM Na_2EDTA; 0.05% Triton X-100
Wash Solution 2: 50 mM Tris-HCl, pH 8.3; 200 mM NaCl; 5 mM Na_2EDTA
Wash Solution 3: 10 mM Tris-HCl, pH 8.3; 0.1 mM Na_2EDTA
Ultrapure water: Reagent-grade water, previously autoclaved, filtered (through 0.2 μm) and stored into sterile 2 mL Eppendorf tubes at 4°C or –20°C
UltraClean™ Soil DNA isolation kit (MoBio, catalog #12800-100)
MasterTaq® kit (Eppendorf AG, Germany)
Agarose
TBE (Tris-Borate-EDTA)10× (can be purchased from BIORAD or similar or, alternatively, prepared as follows [for 1 L]: 108 g TRIS Base; 55 g Boric Acid; 40 mL EDTA 0.5 M, pH 8; reagent-grade water to volume)
TBE 1×: Dilute TBE 10× (100 mL TBE 10× + 900 mL of reagent-grade water)
Ethidium bromide (10 mg mL^{-1} in ultrapure water)
Loading dye 6× (Promega, Fermentas, or similar)
Molecular weight 100 bp (Fermentas or similar)
Wizard® PCR cleanup system (Promega)
Formamide
Internal size standard (GS2500-ROX; Applied Biosystems)

23.3.3 Procedure

The application of the ARISA method to sediment samples involve the following steps:

1. DNA extraction and purification (with the aim of extracting genomic DNA from benthic microrganisms, which must be of sufficient purity for subsequent PCR-based analyses).
2. DNA quantification (needed to quantify the concentration of DNA extracts and use known amounts DNA in the PCR reaction).
3. PCR amplification of the bacterial ITS1 region (to amplify the ITS1 bacterial region, thus providing sufficient amount of DNA for analysis).
4. Agarose gel electrophoresis of PCR products (after PCR amplification, to check the success of the amplification reaction).
5. Cleanup of PCR products (to purify the ITS1 amplicons from salts, Taq, primers, excess of dNTPs, and all other reagents used during the PCR reaction).
6. Capillary electrophoresis of purified PCR products (allows resolution and sizing of all the different ITS1 regions amplified in the sample).
7. Analysis and interpretation of results (to transform the observed results into estimates of bacterial richness and community structure).

23.3.3.1 DNA Extraction and Purification

Extract the DNA following the procedure reported Chapter 20 of this book [NOTE 1].

23.3.3.2 DNA Quantification (Using a Spectrophotometer)

Use only quartz cuvettes (path 1 cm, internal volume 1 mL).

Set up the spectrophotometer at 260 nm.

If working with a dual-channel spectrophotometer, place in both the "reference" and "sample" cuvettes 1 mL of ultrapure water and then operate the procedure to reset the instrument.

Discard the water from the "sample" cuvette and place in this cuvette 990 μL of reagent-grade water and 10 μL of each DNA sample (i.e., dilute the sample to a factor of 100).

Measure the absorbance at 260 nm.

Discard the liquid from the "sample" cuvette and repeat the procedure for all extracted DNA samples.

At the end of the procedure, DNA concentration in each DNA extract is calculated as follows:

$$DNA\ (ng\ \mu L^{-1}) = ABS_{260} \cdot 50 \cdot 100 \tag{23.1}$$

where

100 is the dilution factor of the DNA within the cuvette (i.e., 10 μL of DNA into 990 μL of reagent-grade water).

50 is the conversion factor from absorbance to DNA concentration (i.e., an absorbance of 1 at 260 nm contains approximately 50 μg/mL [or 50 ng/μL] of double-stranded DNA).

The purity of the DNA can be checked by measuring the ratio 260:280 nm; a ratio greater than 1.8–2 is considered an indication of DNA purity from proteins [NOTE 2].

23.3.3.3 PCR Amplification of the Bacterial ITS1 Region

Dilute the DNA extracts to ca. 5 ng μL^{-1}.

Prepare 200 μL sterile PCR test tubes for each DNA sample (i.e., each sediment sample must be amplified in four replicates), plus one for the positive and one for the negative PCR controls.

For PCR amplification, it is recommended that the MasterTaq® kit (Eppendorf AG, Germany) be used, which increases the success of PCR amplification on DNA extracted from marine sediments by reducing the effects of residual, potentially PCR-inhibiting contaminants.

Each PCR tube will contain the following ingredients:

- 1 μL of Taq polymerase (the stock solution must be 1 Unit μL^{-1})
- 5 μL of 10× Taq Buffer with Mg^{2+}
- 10 μL 5× of TaqMaster PCR enhancer
- 1 μL of 10 mM dNTPs
- 0.5 μL of the 100 μM forward primer "16S-1392F" (5′-GYACACACCGCCCGT-3′)
- 0.5 μL of the 100 μM HEX-labeled reverse primer "23S-125R" (5′-GGGTTBCCCCATTCRG-3′) [NOTE 3].
- 31 μL of ultrapure water
- 1 μL of sample DNA

Preparation of a single master mix is recommended, which contains all the ingredients (with the exception of DNA) for all the test tubes to be analyzed. The master mix should be prepared for a larger number of samples in order to compensate for errors in pipetting (i.e., if the number of sediment samples is 10, consider 40 [each sample in four replicates] + 2 [the two PCR controls] + 2 = 44).

Dispense 49 μL of master mix in each tube and then add template DNA (1μL, containing ca. 5 ng of DNA).

In the negative control, add 1 μL of ultrapure water.

In the positive control, add 1 μL containing ca. 1–5 ng of *Escherichia coli* genomic DNA (SIGMA or similar).

Cycling conditions in the thermalcycler: 30 PCR-cycles, consisting of 94°C for 1 min, 55°C for 1 min, and 72°C for 2 min, preceded by 3 min of denaturation at 94°C, followed by a final extension of 10 min at 72°C.

23.3.3.4 Agarose Gel Electrophoresis of PCR Products

To prepare the agarose gel,

Weigh out 0.35 g of agarose.

Add 35 mL of TBE 1×.

Boil to dissolve.

Let the liquid cool (down to ca. 60–70°C), and add 1 μL of ethidium bromide [NOTE 4].

Place the liquid in the casting tray containing a sample comb and let it solidify at room temperature.

Fill the electrophoretic chamber with TBE 1×.

Insert the gel (still in its plastic tray) horizontally into the electrophoretic chamber and cover with TBE 1× buffer.

Once the gel is ready, mix, on a piece of parafilm, 5 μL aliquots of each PCR product with 1 μL of loading dye.

Load the mixture in each well.

In the first well of each gel, place 4 μL of molecular weight marker.

Place the lid and power leads on the apparatus and apply current (100 V, 25 min). Check that the current is really flowing by observing bubbles coming off the electrodes. The DNA will migrate toward the positive electrode, which is usually colored red.

At the end of the electrophoretic run, transfer the gel on the transilluminator and check the presence of the desired amplicon. As ITS1 is variable in length among different bacterial species, several bands should be visible on the gel.

23.3.3.5 Cleanup of PCR Products

Before purification, pool together two independent PCR replicates for each sediment sample. This process is carried out in order to minimize stochastic PCR biases (Polz and Cavanaugh 1998). According to this step, two ARISA replicates will be analyzed for each sediment sample.

Purify pooled PCR products using the Wizard PCR cleanup system by carefully following the procedure described by the manufacturer [NOTE 5].

Quantify the eluted DNA spectrophotometrically as described earlier.

23.3.3.6 Capillary Electrophoresis of Purified PCR Products

Capillary analysis must be performed in an ABI Prism 310 or 3100 Genetic Analyzer (Applied Biosystems).

Dilute purified amplicons to ca. 5 ng μL^{-1} and mix 1 μL with 14 μL of internal size standard (GS2500-ROX) in deionized formamide [NOTE 6].

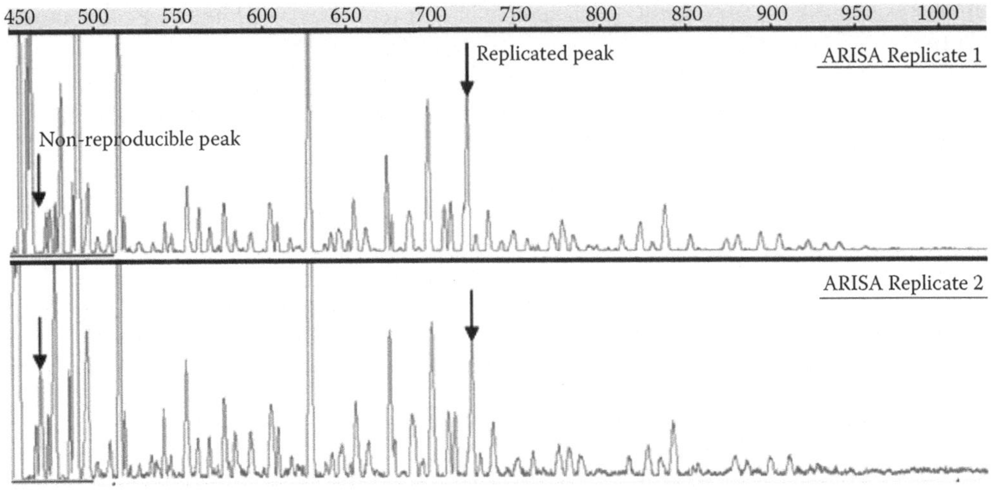

FIGURE 23.2 A typical replicated electropherogram obtained from the ARISA analysis of a sediment sample. The arrows on the right side indicate an example of replicated peaks (which are considered true bacterial operational taxonomic units [OTUs]), while arrows on the left side indicate an example of a nonreproducible peak (i.e., a peak that is present only in one replicate), which is excluded from diversity estimates.

Denature at 94°C for 2 min and immediately chill into ice.

Analyze fragments in GeneScan mode using 47 cm × 50 μm capillaries, POP-4 polymer, 40 s injection time, 15 kV injection voltage, 15 kV run voltage, and 60°C capillary temperature (these parameters can however be varied according to fluorescence signal intensity).

Determine ARISA fragments in the range 390–1400 bp using Genescan analytical software 2.02 (ABI) or similar (Peak Scanner Software, ABI).

23.3.3.7 Analysis and Interpretation of Results

Extract Genescan numerical outputs from each electropherogram and transfer to Microsoft Excel.

Remove from each profile those peaks that are not present in both replicates (i.e., "nonreproducible" peaks; Figure 23.2), as they are assumed to be possible PCR artifacts.

Remove peaks that are less than 1.5 bp apart from a larger peak, which are considered "shoulders" peaks (Figure 23.3).

For discriminating bacterial OTUs (operational taxonomic units) from baseline fluorescence, the total integrated height of each profile (by summing the height of each peak in the ARISA electropherogram) must be calculated. Then calculate the relative

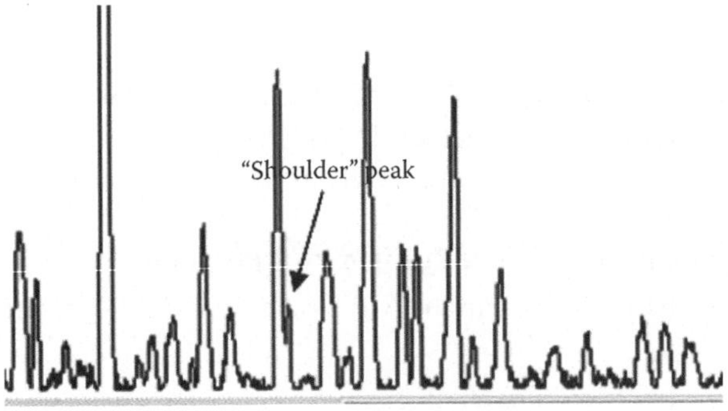

FIGURE 23.3 The arrows indicate a typical "shoulder" peak, which is eliminated from the analysis.

percentage of each peak (as a percentage over the total integrated height); consider only those peaks accounting for more than 0.11% of the total integrated height (Danovaro et al. 2006) [NOTE 7].

After this procedure, bacterial richness is defined as the number of remaining OTUs.

Bacterial diversity indices (such as the Shannon–Wiener index and others) or the Pielou index (evenness) can be calculated assuming that the number of the OTU represents the species number, and the peak height associated with each peak represents the relative abundance of each bacterial OTU. However, caution must be used, as the calculation of diversity indices using the outputs of fingerprinting techniques can be biased by several factors (Bent and Forney 2008).

In addition, data from ARISA profiles can be used for analyses of ß-diversity and comparisons of samples in terms of species composition and community structure by using specific software (such as Primer; Clarke 1993).

23.3.4 Notes

1. DNA extraction and purification: The same as reported for Chapter 20, "Extraction and Purification of DNA from Marine Sediments Suitable for the Analysis of Prokaryotic Diversity."
2. DNA quantification using a spectrophotometer: Some sediments may be characterized by low bacterial abundances and, consequently, low DNA content (e.g., deep-sea sediments). This can result in DNA concentrations below the detection limit of the spectrophotometer. In this case, more sensitive DNA quantification procedures, such as the fluorometric method described by Corinaldesi et al. (2005), must be used.
3. PCR amplification of the bacterial ITS1 region: None.
4. Agarose gel electrophoresis of PCR products: Ethidium bromide has suspected carcinogenic effects; follow all rules of personal safety for protection from risks.
5. Cleanup of PCR products: None.
6. Capillary electrophoresis of purified PCR products: Some authors reported unoptimal performances of the GS2500-ROX internal standard for sizing larger fragments and have utilized other standards, such as Map Marker 1000 X-Rhodamine (Bioventures; Hartmann et al. 2005), or custom-made ROX-labeled standard from Bioventures (i.e., a 1500-bp standard [Fuhrman et al. 2006] or a 200–2000 bp standard [Fisher and Triplett 1999]).
7. Analysis and interpretation of results: Different methods are reported in the literature to define the fluorescence threshold within each electropherogram, that is, to define the amount of fluorescence that allows distinguishing true bacterial peaks from instrumental fluorescent noise. They include 50 fluorescence units (Hartmann et al. 2005), 100 fluorescence units (Kent et al. 2006), or five times the fluorescence baseline (Fuhrman et al. 2006). We suggest using the one reported in this chapter (i.e., 0.11% of the total integrated height), whose rationale is described in Luna et al. (2006) and Danovaro et al. (2006).

Extract DNA from the sediment sample
(as described above in Chapter 20)

↓

Put 990 μL of grade reagent water and
10 μL of DNA in the spectrophotometer cuvette

↓

Measure the absorbance at 260 nm to quantify DNA concentration

↓

Dilute the DNA extracts to ca. 5 ng μL^{-1}

↓

Prepare four 200-μL sterile PCR tubes (for each DNA sample),
plus one for the positive and one for the negative control

↓

Prepare the appropriate PCR master mix

↓

Dispense 49 μL of master mix in each tube
and then add μL of DNA

↓

Add 1 μL of ultrapure water in the negative control and 1 μL of
Escherichia coli DNA in the positive control

↓

Put the tubes in the thermalcycler and run the appropriate program

↓

(a)

PROTOCOL 23.1 Determination of bacterial diversity by ARISA. *Continued*

Prepare the agarose gel, as follows:

↓

Weigh 0.35 g of agarose

↓

Add 35 mL of TBE 1X

↓

Dissolve the agarose in the microwave

↓

Let the liquid cool and add 1 μL of Ethidium Bromide

↓

Put the liquid into the casting tray
containing a sample comb and let it solidify

↓

Fill the electrophoretic chamber with TBE 1X

↓

Insert the gel horizontally into the electrophoresis
chamber and cover with TBE 1X buffer

↓

Once the gel is ready, mix –5 μL aliquots
of each PCR product with 1 μL of loading dye

↓

Load the mixture in each well (leave one
well empty for the molecular weight marker)

↓

Put 4 μL of molecular weight marker in the empty well

↓

(b)

PROTOCOL 23.1 *Continued.*

Place the lid and power leads on the apparatus
and apply current (100 V, 25 min)

↓

Transfer the gel on the trans-illuminator
and check the presence of the expected amplicon

↓

Pool together 2 PCR test tubes for each sample

↓

Purify pooled PCR products using the
Wizard PCR clean-up system

↓

Quantify the DNA spectrophotometrically
(as described above)

↓

Keep each tube frozen at –20°C until capillary analysis

↓

To perform capillary analysis, mix 1–2 μL of digest with
14 μL of appropriate Internal Size Standard in deionized formamide

↓

Denature at 94°C for 2 min
and immediately chill into ice

↓

Analyze fragments in GeneScan mode

↓

Extract GeneScan file into Excel and analyze data

(c)

PROTOCOL 23.1 *Continued.*

REFERENCES

Amann, R., B. M. Fuchs, and S. Behrens. 2001. The identification of microorganisms by fluorescence *in situ* hybridisation. *Current Opinion in Biotechnology* 12: 231–236.

Bent, S. J. and L. J. Forney. 2008. The tragedy of the uncommon: Understanding limitations in the analysis of microbial diversity. *The ISME Journal* 2: 689–695.

Clarke, K. R. 1993. Non parametric multivariate analyses of changes in community structure. *Australian Journal of Ecology* 18: 117–143.

Corinaldesi, C., R. Danovaro, and A. Dell'Anno. 2005. Simultaneous recovery of extracellular and intracellular DNA suitable for molecular studies from marine sediments. *Applied and Environmental Microbiology* 71: 46–50.

Daffonchio, D., A. Cherif, L. Brusetti et al. 2003. Nature of polymorphisms in 16S-23S rRNA gene intergenic transcribed spacer fingerprinting of Bacillus and related genera. *Applied and Environmental Microbiology* 69: 5128–5137.

Danovaro, R., G. M. Luna, A. Dell'Anno, and B. Pietrangeli. 2006. Bacterial diversity in aquatic environments: Comparison of two fingerprinting techniques (T-RFLP and ARISA). *Applied and Environmental Microbiology* 72: 5982–5989.

Fisher, M. M. and E. W. Triplett. 1999. Automated approach for ribosomal intergenic spacer analysis of microbial diversity and its application to freshwater bacterial communities. *Applied and Environmental Microbiology* 65: 4630–4636.

Fuhrman, J. A. and A. Hagstrom. 2008. Bacterial and Archaeal community structure and its patterns. In *Microbial ecology of the oceans*, ed. D. L. Kirchman, 45–90. Hoboken, New Jersey: John Wiley & Sons.

Fuhrman, J. A., I. Hewson, M. S. Schwalbach, J. A. Steele, M. V. Brown, and S. Naeem. 2006. Annually reoccurring bacterial communities are predictable from ocean conditions. *Proceedings of the National Academy of Sciences* 103: 13104–13109.

Hartmann, M., B. Frey, R. Kolliker, and F. Widmer. 2005. Semi-automated genetic analyses of soil microbial communities: Comparison of T-RFLP and RISA based on descriptive and discriminative statistical approaches. *Journal of Microbiological Methods* 61: 349–360.

Hewson, I. and J. A. Fuhrman. 2004. Richness and diversity of bacterioplankton species along an estuarine gradient in Moreton Bay, Australia. *Applied and Environmental Microbiology* 70: 3425–3433.

Kent, A. D., S. E. Jones, G. H. Lauster, J. M. Graham, R. J. Newton, and K. D. McMahon. 2006. Experimental manipulations of microbial food web interactions in a humic lake: Shifting biological drivers of bacterial community structure. *Environmental Microbiology* 8: 1448–1459.

Luna, G. M., A. Dell'Anno, and R. Danovaro. 2006. DNA extraction procedure: A critical issue for bacterial diversity assessment in marine sediments. *Environmental Microbiology* 8: 308–320.

Luna, G. M., A. Dell'Anno, L. Giuliano, and R. Danovaro. 2004. Bacterial diversity in deep Mediterranean sediments: Relationship with the active bacterial fraction and substrate availability. *Environmental Microbiology* 6: 745–753.

Marsh, T. L. 1999. Terminal-Restriction Fragment Length Polymorphism (T-RFLP): An emerging method for characterizing diversity among homologous populations of amplicons. *Current Opinion in Microbiology* 2: 323–327.

Polz, M. F. and C. M. Cavanaugh. 1998 Bias in template-to-product ratios in multitemplate PCR. *Applied and Environmental Microbiology* 64: 3724–3730.

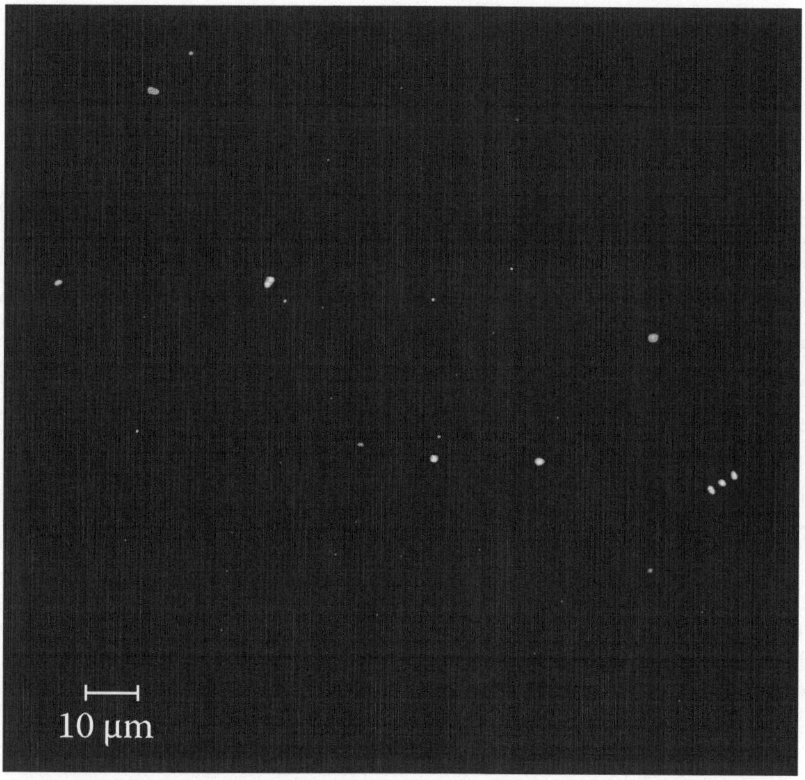

FIGURE 11.1 A typical view of marine microbial community stained with SYBR.

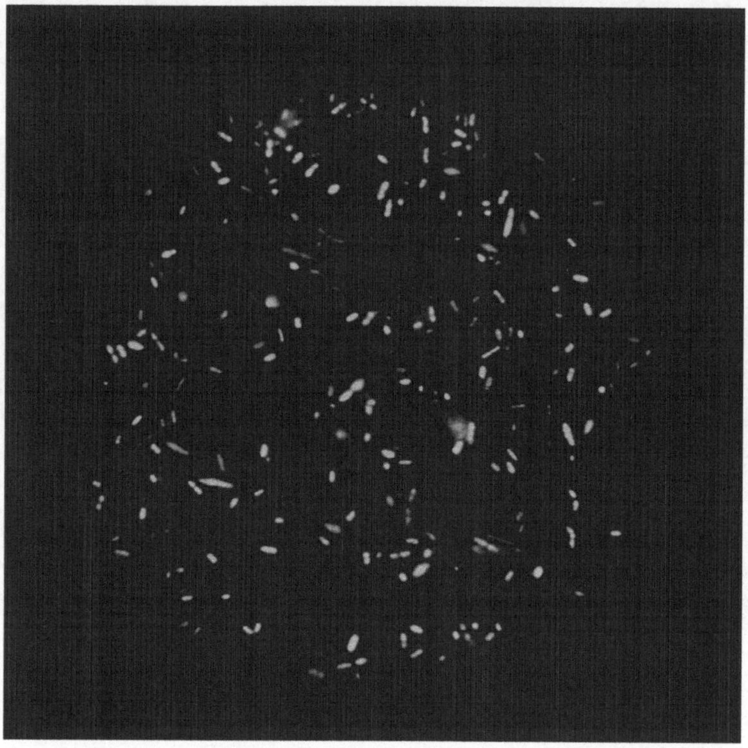

FIGURE 12.1 Visualization of prokaryote cells with SYBR Green I staining. Prokaryotic cells appear green/yellow.

FIGURE 12.2 Visualization of the prokaryote cells with AO staining. (Left) Sediment particles appear orange/red and prokaryotic cells appear green/yellow. Sometimes cells could appear orange/red, these should be counted anyway identified on the base of their prokaryotic shape and dimension. (Right) Visualization of the prokaryotic cells with a strong background. The background is red, and the cells show a strong color.

FIGURE 13.1 Microscopic evaluation of the bacteria (probe: EUB338).

FIGURE 13.2 Example of a high background sample.

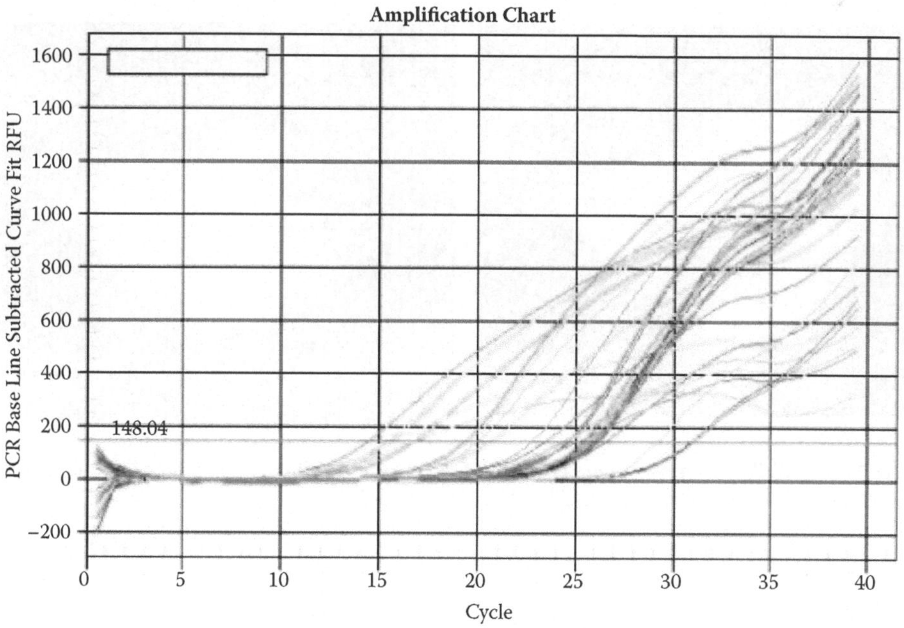

FIGURE 14.3 Real-time amplification chart: Pattern of amplification with increasing number of cycles.

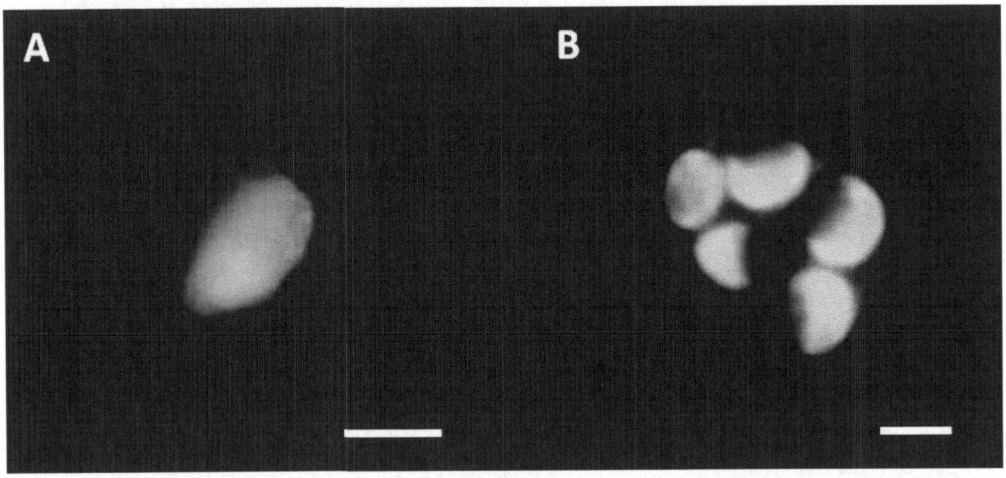

FIGURE 15.1 Ciliate (B) stained with DAPI and FITC (A) and a cluster of thraustochytrid cells stained with Acriflavin; scale bars are 10 and 5 µm in A and B, respectively.

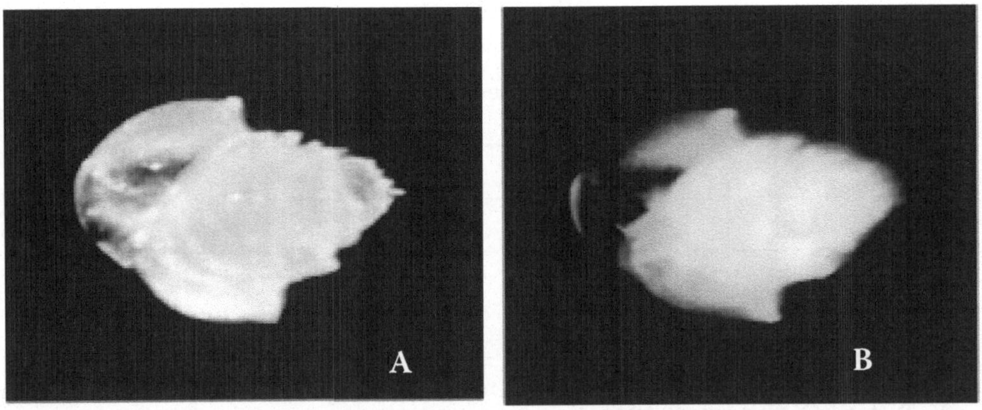

FIGURE 16.1 A live *Bulimina marginata* (Foraminifera) observed using a normal binocular (a) and an epifluorescent binocular (b). (Courtesy of E. Geslin, University of Angers.)

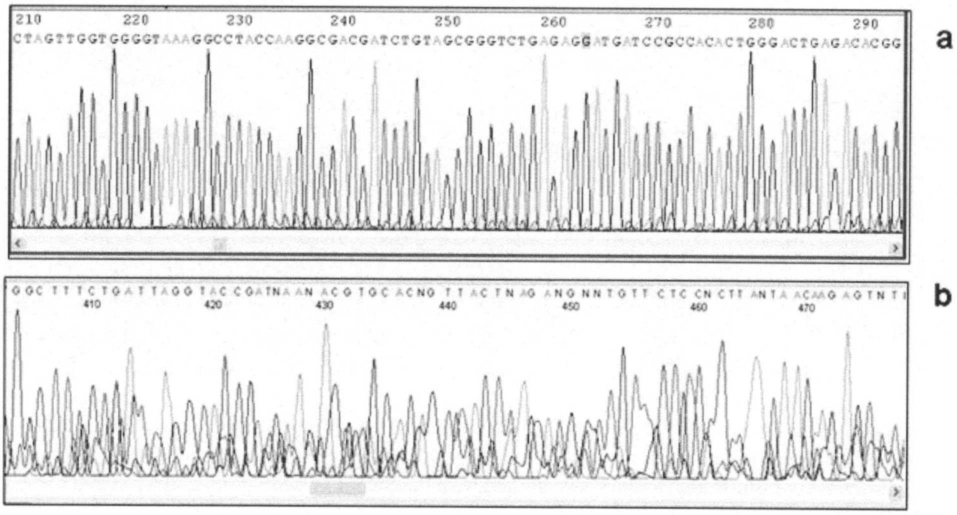

FIGURE 22.4 Examples of a "good" (a) and a "bad" (b) sequence of a 16S rDNA clones obtained from marine sediments. The "bad" sequence presents multiple peaks.

FIGURE 31.1 An image showing the release of DNA from prokaryotic cells. Extracellular DNA is a source of bioavailable N and P, which sustains deep-sea heterotrophic metabolism. (Courtesy of M. Tangherlini. Used with permission.)

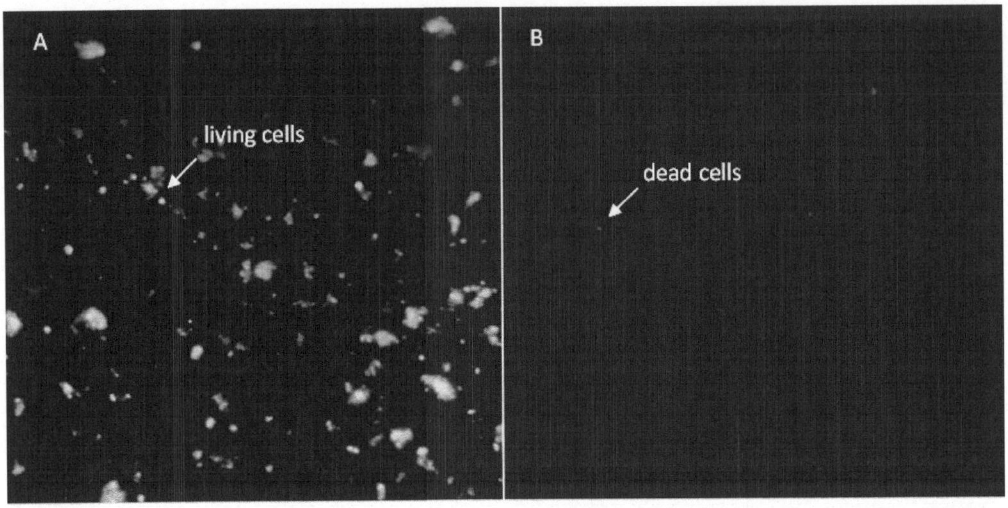

FIGURE 34.1 Marine prokaryotes examined using epifluorescence microscopy (magnification 1000×; Zeiss Axioskop 2, BP 490 filter set for SYBR Green I for the yellow-green florescence and the BP 545 filter for propidium iodide). Figure illustrates prokaryotes stained with the SYBR Green I (total cells) (A) and prokaryotes stained with propidium iodide (dead cells) (B).

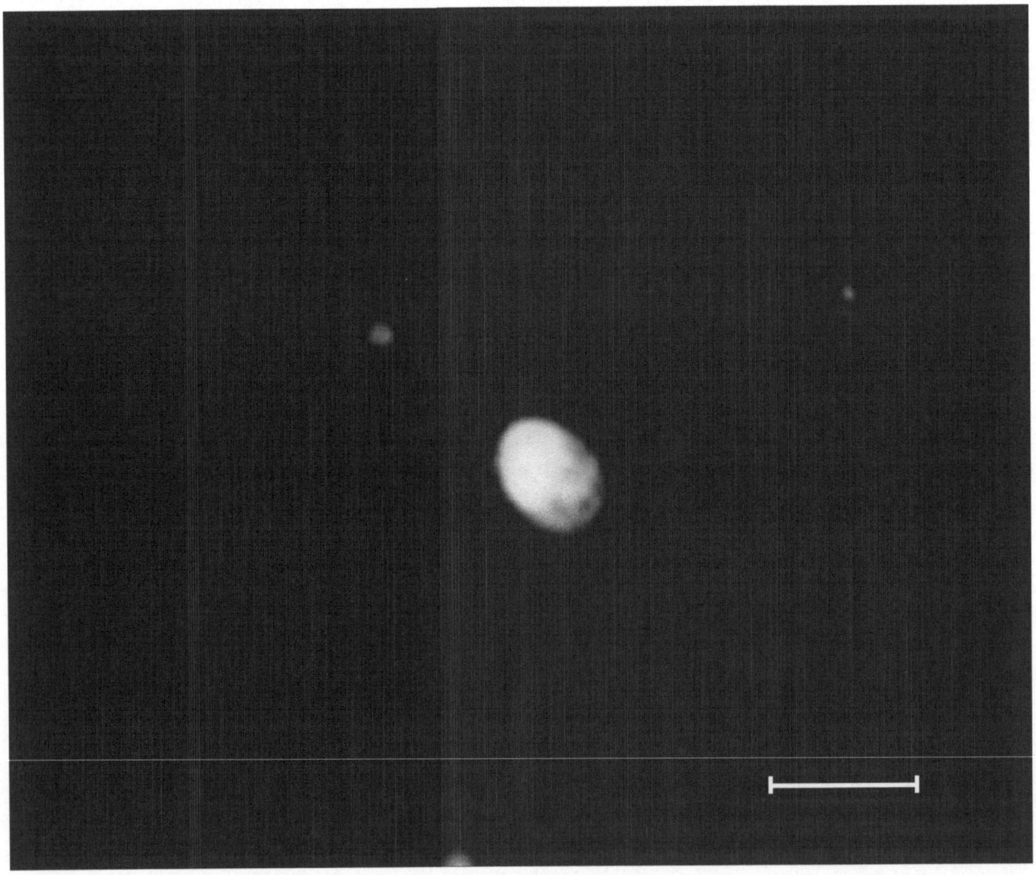

FIGURE 40.1 Nanoflagellate and prokaryote cells stained with DAPI. Scale bar is 5 μm.

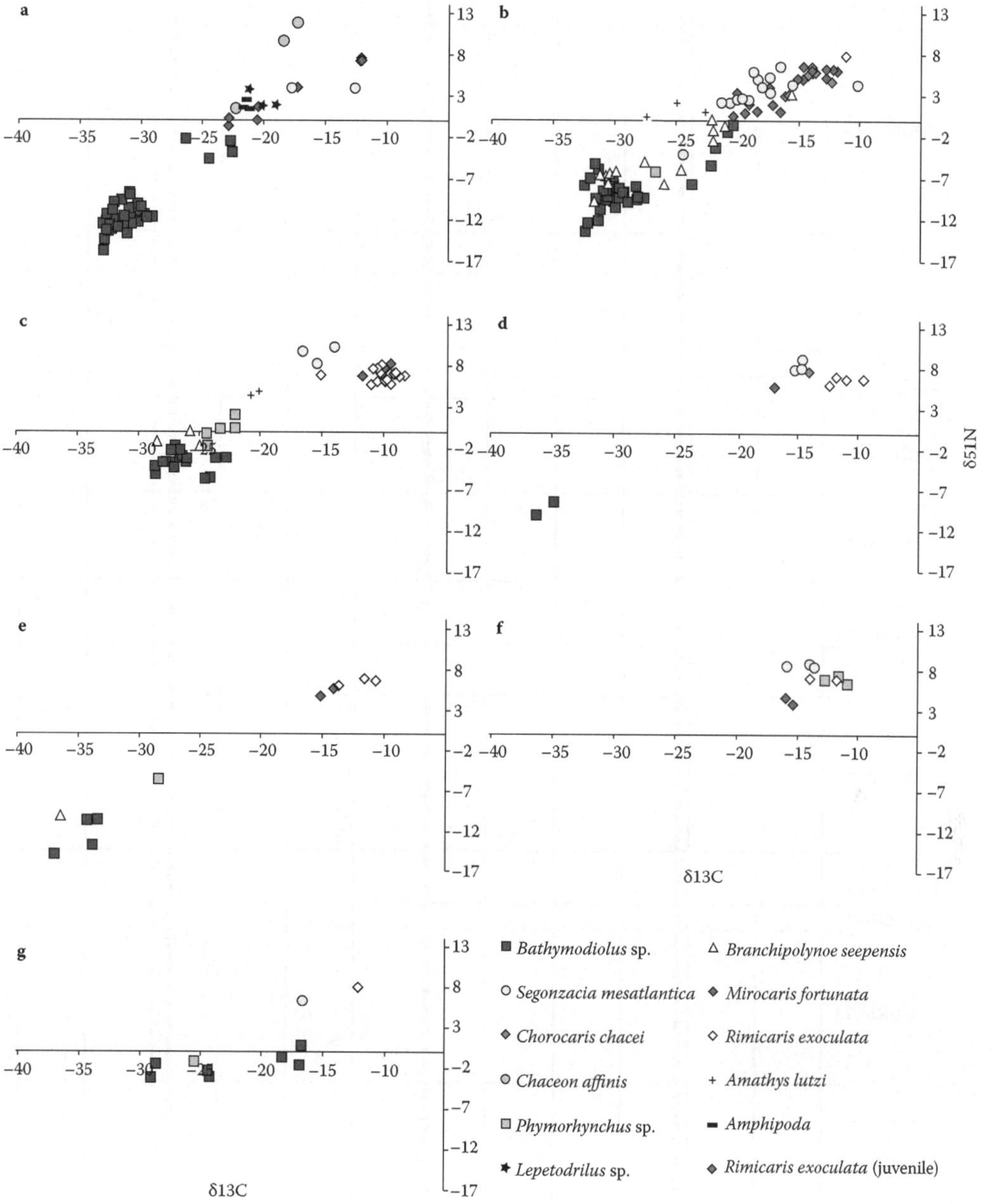

FIGURE 42.1 ∂ ^{13}C and ∂ ^{15}N values identifying the different main groups of organisms sampled at Mid-Atlantic Ridge deep-sea hydrothermal vent fields: (a) Menez Gwen; (b) LuckyStrike; (c) Rainbow; (d) Broken Spur; (e) Snake Pit; (f) TAG; (g) Logatchev. (Data from Colaço, A., F. Dehairs, and D. Desbruyeres. 2002. *Deep-Sea Research Part I* 49: 395–412.)

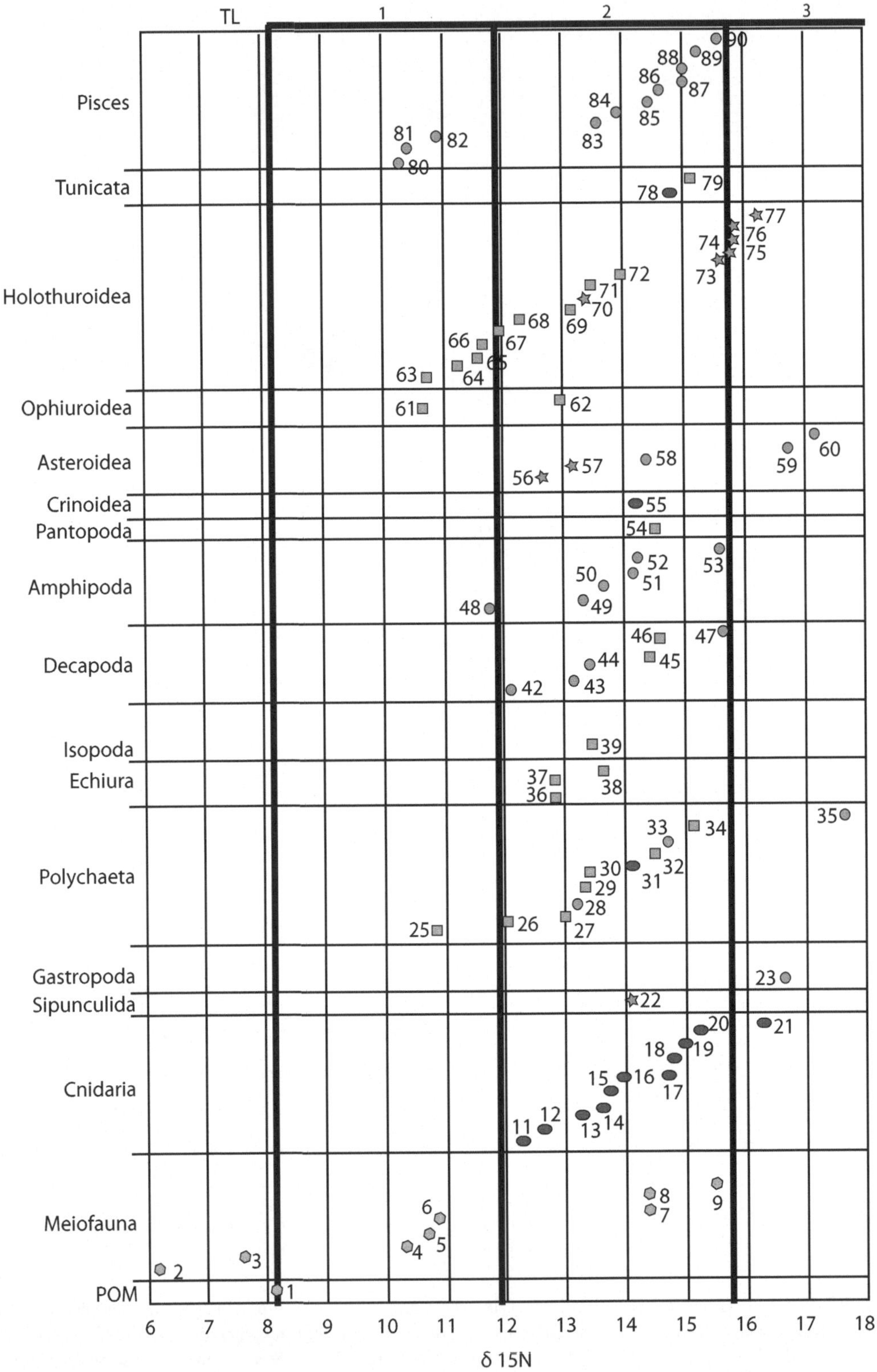

FIGURE 42.2 ∂ ¹³C Relative trophic position of benthic species/groups in deep-sea sediments based on ∂ ¹⁵N. TL: relative trophic level. Different feeding types are designated with squares (suspension feeders), crosses (surface deposit feeders), filled circles (subsurface deposit feeders), triangles (predator/scavenger), and open circles (no feeding type associated). (Data from Iken, K., T. Brey, U. Wand, J. Voigt, and P. Junghans. 2001. *Progress in Oceanography* 50: 383–405.)

24 Meiofaunal Diversity

CONTENTS

24.1 MEIOFAUNAL TAXON RICHNESS

24.1.1 Introduction

Meiofaunal diversity can be estimated as the total number of taxa identified in a sample. The identification of taxa is fundamental to defining the community structure of meiofaunal assemblages. Several studies performed at different deep-sea sites suggest that the metazoan meiofaunal community structure is comparable worldwide and is generally dominated by nematodes (>80%–90%), followed by copepods and polychaetes (Danovaro et al. 1999a; Danovaro et al. 2000; Tselepides and Lampadariou 2004; Vanhove et al. 2004; Netto et al. 2005; Gambi and Danovaro 2006; Danovaro et al. 2008). However, differences in the metazoan meiofaunal community structure were reported comparing bathyal and hadal depths in the Atacama slope and trench (Danovaro et al. 2002). Nematodes were always largely dominant, accounting for more than 80% of total abundance, but harpacticoid copepods ranged from 2%–7% along the slope and reached about 17% in the trench sediments. Surprisingly, gastrotrichs displayed similar abundance at all depths, but they represented the third most important taxon along the slope. By contrast, in the trench sediments, polychaetes were the third most abundant taxon, and kinorhynchs notably increased in their relative importance, likely as a result of the reduced grain size at hadal depths. Here we report the methods used to identify the metazoan meiofaunal taxa.

24.1.2 Field Procedure

In deep-sea soft sediments, a multiple corer is used to collect simultaneously several meiofaunal samples (e.g., 4–12 corers; Elmgren and Radziejewska 1989). The instrument allows the collection of undisturbed sediment samples of a given surface area to analyze the horizontal and vertical distribution of benthic organisms. Another common sampling instrument for soft sediments is the box corer (Fleeger et al. 1983). The weight of a box corer after sampling can range from 150 to >750 kg and collect a surface sediment of 0.02–1.0 m^2 down to a depth of 50 cm. From the box corer, meiofaunal samples could be collected using manual Plexiglas® corers (internal diameter 3.6 cm), one replicate from each different deployment. However, during sampling recovery, most box corers can allow the mixing of the sediment surface, which might alter the quality of the sample. Bett et al. (1994), Shirayama and Fukushima (1995), and Bett (2000), comparing sampling efficiency of multiple corers and box corers in deep-sea environments, pointed out that meiofaunal abundance collected using box corers was significantly lower than that determined in samples collected using multiple corers. However, for the studies, comparing NIOZ-type box corers and multiple corers reported identical results for all benthic variables investigated, suggesting that both sampling devices are adequate for the sampling of deep-sea meiofauna (Danovaro et al. 1999b).

24.1.3 Laboratory Procedures

The procedures to collect, store, extract, and sort meiofaunal samples for the determination of meiofaunal diversity are the same reported in the Chapter 17, "Abundance of Metazoan Meiofauna."

Once the extraction of organisms from the sediment is completed, it is possible to proceed with the identification of the meiofaunal taxa. The identification of meiofaunal taxa is performed by analyzing organisms fixed in formalin (4%) and stained with Rose Bengal to facilitate their identification under the stereomicroscope (Higgins and Thiel 1988; Giere 1993, 2009). This procedure can be performed at the same time of the counting of the organisms since the procedure to extract them from sediments is the same. The analysis and counting of individuals is performed under the stereomicroscope with a minimum of magnification factor ×25. Higher magnification factors ×40 and ×50 may be necessary for the smallest meiobenthic organisms and larvae. The pictorial keys reported in the most common books dedicated to the taxonomy of marine organisms facilitate the identification of the meiofaunal taxa under the stereomicroscope (Riedl, 1983; Higgins and Thiel 1988; Giere 1993).

Metazoan meiofaunal taxa (Figure 24.1) include Cnidaria, Sipuncula, Mollusca, Oligochaeta, Polychaeta, Tardigrada, Amphipoda, Cladocera, Cumacea, Harpacticoida, Isopoda, Ostracoda, Mystacocarida, Syncarida, Tanaidacea, Thermosbaenacea, Halacarida, Pycnogonida, Entoprocta, Turbellaria, Nemertea, Rotifera, Gnatostomulida, Gastrotricha, Nematoda, Priapulida, Kinorhyncha, Loricifera, Brachiopoda, and Echinodermata.

Some authors include also the taxon of Foraminifera within meiofaunal assemblages due to the high abundance and diversity of this group in deep-sea sediments (Gooday et al. 1998). However, here we have only considered metazoan meiofaunal taxa.

More sophisticated techniques have to be applied when we want to identify the specimens to genus or species level. The identification at species level of various taxa requires the collection of individuals and preparation of permanent slides for observation under the microscope. For nematode identification, detail procedures are reported in this chapter. Sample examination and counting are eased by using cuvettes with internal subdivisions (e.g., 200 cells as in the Delfuss cuvette). Alternatively, Petri dishes with retinated bottoms may be used, but the lack of separating septa makes the counting much more problematic compared with the use of Delfuss cuvettes.

24.1.3.1 Equipment and Instruments for Sorting and Identification of Meiofaunal Taxa

Stereomicroscope
Microscope

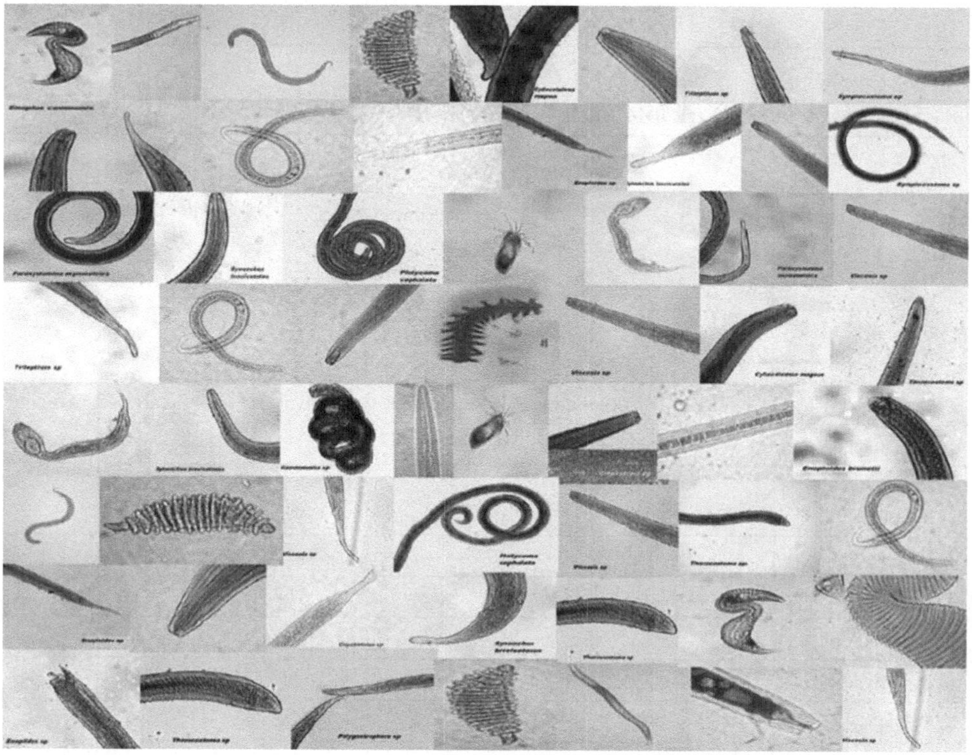

FIGURE 24.1 A miscellany of different meiofaunal taxa.

Sieve, 20/30 μm
Delfuss cuvette
Wash bottle
Handling needle
Slides
Cover glass
Pasteur pipette
Handbooks
Data sheet

24.1.3.2 Solutions

Fresh water or 0.45 μm prefiltered seawater
4% buffered formalin
Rose Bengal (0.5 g L^{-1})

A sample (extracted, fixed, and stained with Rose Bengal) is transferred in the 20–30 μm mesh sieve. The sample, washed with fresh water to eliminate the residuals of formalin, is transferred into the Delfuss cuvette with a wash bottle. Before starting the identification of meiofaunal taxa, wait a few minutes to allow the sedimentation of all organisms at the bottom of the cuvette. Start the analyses of the sample from the first cell on the left side of the first line of the Delfuss cuvette. At the end of the sorting, the sample is transferred again into the 20/30 μm sieve and transferred again into the 50 mL tube with a wash bottle full of 4% buffered formalin. Finally, the sample is stained again with Rose Bengal. Table 24.1 is an example of data sheet to annotate the sorting and counting of all meiofaunal taxa, together with abiotic environmental parameters.

TABLE 24.1
Data Sheet for Counting Meiofauna

Program:

Station: Station depth:

Coordinates: Depth of sediment core (cm):

N° Replicates: Layer (cm):

TAXA	Replicate 1	Replicate 2	Replicate 3	Replicate 4	Replicate 5	Average
1. Nematoda						
2. Copepoda Nauplii						
3. Polychaeta						
4. Bivalvia						
5. Ostracoda						
6. Kinorhyncha						
7. Turbellaria						
8. Oligochaeta						
9. Tardigrada						
10. Gastrotricha						
11. Cumacea						
12. Amphipoda						
13. Isopoda						
14. Tanaidacea						
15. Acarina						
16. Nemertina						
Incertae sedis						
Others *						
Number of other taxa						
Total number of taxa						

* List of all the taxa encountered and their abundance

Notes

24.1.3.3 Procedure

1. Transfer the samples fixed with 4% buffered formalin on the 20–30 μm sieve [NOTES 1–3].
2. Wash the sample to eliminate the 4% buffered formalin.
3. Transfer the sample into a Delfuss cuvette with water using a wash bottle.
4. Count and identify meiofaunal organisms under a stereomicroscope (minimum magnification ×25) [NOTES 4–11].
5. At the end of the sorting, transfer the sample onto a 20–30 μm sieve.
6. Transfer the samples into a 50 mL tube using 4% buffered formalin by means of a wash bottle [NOTE 12].
7. Add a few drops of 0.5 g L^{-1} Rose Bengal solution with a Pasteur pipette.
8. Samples can be stored at room temperature.

24.1.4 Notes

1. Wait few minutes before initiating the sorting and identification of the meiofaunal taxa.
2. Use the formalin carefully, dressing in all the safety devices (laboratory coat, gloves, glass, respiratory protection), and work using an extractor fan.
3. The solutions containing formalin must be collected and eliminated using the standard safety procedures.
4. Use the handling needle to move and turn the specimens to allow a complete analysis of all the body sides and symmetry (radial or bilateral). A complete vision of the body shape is fundamental to identifying the taxon.
5. Some taxa such as copepods, cumaceans, polychaetes, gastrotrichs, and amphipods displayed characteristic features that also make the identification of the taxon easier using a stereomicroscope.
6. Other taxa, in particular soft-body organisms such as oligochaetes, gnatostomulids, and turbellarians, are difficult to identify under the stereomicroscope and require the use of a microscope for a better visualization of the peculiar features of each taxon. Samples should be mounted on a slide to use under the microscope. For permanent slides, check the detail procedure in the Section 24.2, "Identification of Nematodes." For a rapid preparation of the slide, transfer the specimen into a drop of deionized water located on a slide. Cover the drop and the specimen with a cover glass. Analyze the slide using a microscope with higher magnification (×40 or ×100 with immersion oil). This type of slide can be analyzed for a few minutes before complete evaporation of the water. Do not use large drops of water because excessive water can wash out the specimen, and it may be lost when the cover glass is applied. After analysis, the specimen can be transferred again into the sample. To ease the separation of the cover glass from the slide, add a few microdrops of water between the two surfaces using a Pasteur pipette.
7. Specimens can be collected in an Eppendorf tube (containing 4% buffered formalin). In this way, organisms can be divided and maintained separately in different tubes for further analyses.
8. A general drawing of the taxa can be useful for rapid identification without having to be mounted on the slide.
9. Keep in mind that Rose Bengal stains all structures covered by a membrane (it acts on the proteins of the cellular membrane). Large protozoans or foraminifera can be present in the samples and get stained. Check with attention all these structures using the handling needle. Do not include these groups in the total metazoan meiofaunal abundance and in the composition of the community structure.
10. Do not knock against the cuvette, so as to prevent restarting the sorting from the beginning.

11. At the end of the sorting, check the surface water to count and classify the specimens that can float.
12. Pay attention to collecting all specimens from the tubes, sieve, and cuvette. Clean the sieve very well at each step during the preparation of the sample for sorting and after the counting. Each step is crucial, and can cause the loss of several specimens.

24.2 IDENTIFICATION OF NEMATODES

24.2.1 INTRODUCTION

Benthic faunal diversity provides an ideal tool for exploring the relationships between biodiversity and ecosystem functioning (Snelgrove 1999; Danovaro et al. 2008a), and among benthic faunal taxa, nematodes are ideal model organisms to estimate benthic biodiversity. Nematodes are indeed the most abundant metazoans on Earth; in terrestrial and marine ecosystems they reach comparable abundance (accounting for 80% of the abundance of multicellular animals; Lambshead 2004), and in the deep sea they can represent >90% of meiofaunal assemblages (Danovaro et al. 2000, 2008b). This phylum is also characterized by (1) very high species richness (i.e., among the most diverse marine phyla), (2) distinct and easily recognizable feeding types, and (3) life strategies that make it also possible to identify functional diversity traits (Danovaro et al. 2001; Pusceddu et al. 2007). Moreover, although comparative studies are rare, deep-sea nematode diversity appears to be related to that of other benthic components, including Foraminifera (Gooday et al. 1998), macrofauna (Levin et al. 2001), and the richness of higher meiofaunal taxa (a group which includes 22 of the 35 modern animal phyla). At present, the most practical method to describe nematode genera and species is based on the use of the light microscope and pictorial keys proposed by Platt and Warwick (1983, 1988) and Warwick et al. (1998), and recent literature on the description of new genera and species. Recently published online at the NeMys Web site is a generic online species information system that stores data on morphology, biogeography, taxonomy, literature, pictures, collections, and molecular aspects to create a large group of possible end users and to allow data analyses and comparison for a specific taxonomic group from a variety of approaches (Deprez et al. 2005). Also recently, the molecular approach has been used to identify nematode species collected in coastal sediments (Meldal, 2004; Bhadury et al. 2006; Meldal et al. 2006). Although nematode taxonomy is relatively laborious and time consuming (Coomans 2002; Floyd et al. 2002; Cook et al. 2005), there has been limited application of molecular techniques for the rapid assessment of marine nematode diversity from estuarine and marine sediments. *Caenorhabditis elegans* was the first metazoan organism to have its complete genome sequenced (the *C. elegans* Sequencing Consortium 1998), and currently, over 30 nematode genome sequencing projects are ongoing (Mitreva et al. 2005). No sequencing projects are underway for marine nematodes (i.e., Araeolaimida, Chromadorida, Desmodorida, Desmoscolecida, Enoplida, and Monhysterida), largely because it is difficult to collect enough high-quality, species-specific material. These difficulties increase when deep-sea nematodes are considered. Moreover, to our knowledge, since most of the specimens are fixed using formalin, no methodology has been exclusively developed and used to extract and amplify DNA to allow a molecular approach from long-time formalin-preserved specimens, archived marine meiofauna (such as marine nematodes) that have been stored for a long period of time (i.e., museum collection). Only Bhadury et al. (2005) has proposed a method to extract DNA from nematodes stored with formalin within few days from the collection.

24.2.2 FIELD PROCEDURE

In deep-sea soft sediments, the multiple corer is used to collect simultaneously several meiofaunal samples (e.g., 4–12 corers; Elmgren and Radziejewska 1989). The instrument allows the collection of undisturbed sediment samples of a given surface area to analyze the horizontal and vertical distribution of benthic organisms. Another common sampling instrument for soft sediments is the box corer (Fleeger et al. 1983). The weight of a box corer after sampling can range from 150 to >750 kg and collect a surface sediment of 0.02–1.0 m^2 down to a depth of 50 cm. From the box corer, meiofaunal samples could be collected using manual Plexiglas corers (e.g., internal diameter 3.6 cm), one replicate from each different deployment. However, during sampling recovery, most box corers allow surface sediment mixing, which might alter the quality of the sample. Bett et al. (1994), Shirayama and Fukushima (1995), and Bett (2000), comparing sampling efficiency of multiple corers and box corers in deep-sea environments, pointed out that meiofaunal abundance collected using box corers was significantly lower. However, some studies conducted in the Eastern Mediterranean demonstrated that the use of two different samplers (USNEL-type box corer and multicorer) provide identical results in terms of benthic parameters, suggesting an adequate sampling efficiency of both sampling devices (Danovaro et al. 1999b).

24.2.3 Laboratory Procedure to Prepare Permanent Slides

The procedures to collect, store, extract, and sort meiofaunal samples for the determination of biomass are the same as reported in Chapter 17, "Abundance of Metazoan Meiofauna."

For the determination of nematode biomass and identification of nematode genera and species, nematodes have to be transferred on slides to be analyzed under the light microscope. In order to obtain a permanent preparation, a particular procedure should be followed.

The procedure involves the preparation of permanent slides that can be used for the determination of the nematode biomass and identification of nematode genera and species. The microscope slides should be prepared with a drop of glycerin and a ring of paraffin. A quantity of 10–15 nematodes for each sample can be fixed on a slide. This method is based on the clearing, and, when combined with alcohols, dehydrating properties of glycerin. Dehydration must be gradually reached; otherwise, soft-bodied animals might be damaged and rendered irremediably useless for observation (Vincx 1996).

Nematodes should be transferred using a handling needle from the cuvette to a staining block containing a solution of formalin (4%) and glycerin (100 mL solution: 99 mL formalin 4% and 1 mL glycerin). After 12 h, nematodes are transferred to an oven at 36°C and kept overnight. During the permanence in the oven, some drops of a solution containing ethanol (95%) and glycerin are added in order to prevent the desiccation of the nematodes. Glycerin makes the nematodes transparent, and ethanol gives the best penetration. After the treatment in the oven, nematodes can be transferred onto slides.

24.2.3.1 Equipment and Instruments

Light microscope, equipped with a micrometer scale
Stereomicroscope
Sieve with a 20 μm mesh net
Delfuss cuvette
Wash bottle
Fresh water
Staining block
Slides
Cover glass
Paraffin in Petri dish
Lighter or hotplate
Oven
Handling needle
Glass tube

24.2.3.2 Solutions

4% buffered formalin
Rose Bengal
Glycerin

24.2.3.3 Methods for Preparation of Nematodes for Permanent Mounts

Transfer the samples into a cuvette using the procedures reported for meiofaunal counting and sorting [NOTES 1, 2].
Pick out the nematodes using a handling needle from the cuvette to a staining block containing a few drops of the solution containing formalin 4% and glycerin 1% [NOTE 3].
Add a few drops of Rose Bengal.
Place the staining block in a beaker containing a thick layer of ethanol (95%) overnight [NOTE 4].
The day after, place the staining block containing the nematodes (washed in the solution containing formalin 4% and glycerin 1%) in the oven at 36°C.
Add a few drops of the solution containing 95% ethanol and 5% glycerin every 2 h.
Leave the staining block in the oven overnight to let the ethanol evaporate until the nematodes are in pure glycerin.

At this step, the nematodes are ready to be transferred onto permanent slides.

24.2.3.4 Methods for Preparing the Slides

Prepare a few slides with a paraffin ring [NOTES 5–7].
Heat a glass Pirex tube (internal diameter 1.5–2 cm) and insert it in the Petri dish containing the paraffin.
Place the glass tube at the center of the slide to produce a smooth and thin paraffin ring.

24.2.3.5 Methods for Permanent Mounts

Place a small drop of glycerin at the center of the paraffin ring.
Transfer a few nematodes (10–15 specimens) from the staining block to the glycerin drop on the slide using a stereomicroscope [NOTES 8, 9].
Arrange the nematodes, using a handling needle to avoid any overlap.
Gently place a cover glass on the paraffin ring and the glycerin drop.
Heat the slide on a hotplate or using a lighter for few seconds, keeping the slide horizontal.
Wait for the paraffin to melt and, after 1–2 min, seal the slide [NOTE 10].
Write down all informations relating to the sample (project, station, depth, replicate) with a permanent waterproof ink.
Place the slides in cardboard or plastic slide trays.

24.2.3.6 Notes

This procedure is a mix of two techniques suggested by Seinhorst (1959) and De Grisse (1969). The first procedure is based on 96% ethanol and glycerin. It is relatively rapid, labor saving (2–3 d), and gives very good results. De Grisse modified the procedure described by Seinhorst, combing the killing and fixing of fresh samples. This procedure is based on the use of three solutions: (1) 4% formalin and glycerin (99:1), (2) 96% ethanol and glycerin (95:5), and (3) 96% ethanol and glycerin (50:50). The first solution is made hot to fix the sample. This method requires 3 d.

1. The procedure described here is a combination of these techniques, but results are obtained relatively rapidly and the quality of the permanent slides is high.

2. It is recommended that nematodes be sorted and picked out at the same time as the counting and sorting of meiofauna to avoid several transfers of the sample from the tube to the sieve and cuvette. All these steps are rather critical in that there is the danger of losing some specimens.
3. Care should be taken in the transfer of nematodes using the handling needle; check to see if some nematodes remain on the needle. A quantity of 100–120 nematodes for each replicate can be transferred onto permanent slides. When abundance is rather low, try to pick out all the specimens present in the sample.
4. Close the beaker containing the thick layer of ethanol and the staining block using Parafilm to avoid the total evaporation of the ethanol. The presence of Parafilm allows the accumulation of the ethanol from the bottom of the beaker to the staining block to avoid desiccation of the nematodes.
5. Exercise care in producing the paraffin rings. Avoid large paraffin blocks that can wash out the drop of glycerin containing the nematodes during the heating.
6. Avoid large drops of glycerin that can wash out the melting paraffin and prevent the closing of the slide during the heating.
7. Other seals to close the slides are "Zut" cement; nail varnish; brown cement; silicone rubber; araldite resin, and a local putty. All sealing agents are affected by immersion oil. At the end of analysis, clean your slides after use by means of soft paper.
8. If, after the treatment of the staining block into the oven, the solution containing the nematodes is too dense, add a few drops of pure glycerin to facilitate the transfer of the nematodes onto the slides using the handling needle.
9. Alternative to the glass slides, the use of aluminum double-cover glass slides (Cobb 1917) is recommended, which offer the following advantages: (1) specimens can be viewed from either side, (2) the slides stack horizontally without one touching its neighbors, and (3) the slides can withstand being dropped on a hard floor.
10. A more rapid method for preparing microscopy slides implies the use of Hoyer's medium. First, wash the specimen with a drop of distilled water on a slide; then, add a drop of Hoyer's medium (30 g extremely fine gum arabic dissolved in 50 mL distilled water with 20 mL glycerin and 125 g chloral hydrate, filtered through cotton wool) that acts as a dehydrating, clearing, and sealing agent. The organism is transferred onto this drop and a cover glass is carefully lowered onto them to prevent the formation of air bubbles. This method, despite having the advantage of being rapid, might generate an excessive clearing of the specimen, making the observation of internal organs quite difficult (Danovaro et al. 2004).

24.2.4 Nematode Identification at the Genus and Species Level

Nematodes can be identified to the genus and species level using a light microscope with the magnification ×100 and immersion oil.

Before starting the identification procedure, a map with the position of the nematodes in the permanent slide should be built (see the scheme in Tables 24.2 and 24.3). Next to the circles are lines in which the genera or species of the nematodes can be written. For each specimen, body length and width as well as sex (different stages of juveniles; female; gravid or not; and male) can also be annotated.

To arrange the nematodes in the map and in the identification list, move from the left to the right side and vice versa, and from the top to the bottom, as suggested in Table 24.3.

In order to identify the nematodes to the genus and species level, the following aspects should be considered:

1. Type of cuticle (smooth, annulated with punctuations, or dots)

TABLE 24.2
Worksheet for the Identification of Nematodes

Program: Depth (m):
Station:
Coordinates: Layer (cm):
N° Replicates:

n°	Map		Body		Juvenile	Female Gravid	Male
			Length	Width			
		1					
		2					
		3					
		4					
		5					
		6					
		7					
		8					
		9					
		10					

TABLE 24.3
Map of Nematodes

Program: Depth (m):
Station:
Coordinates: Layer (cm):
N° Replicates:

n°	Map		Body Length	Body Width	Juvenile	Female Gravid	Male
	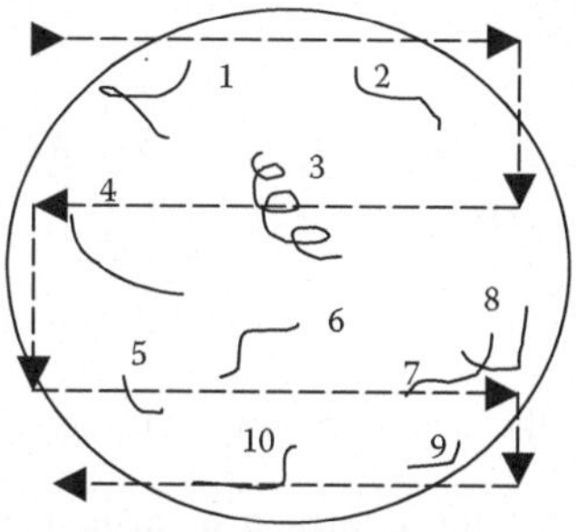	1					
		2					
		3					
		4					
		5					
		6					
		7					
		8					
		9					
		10					
	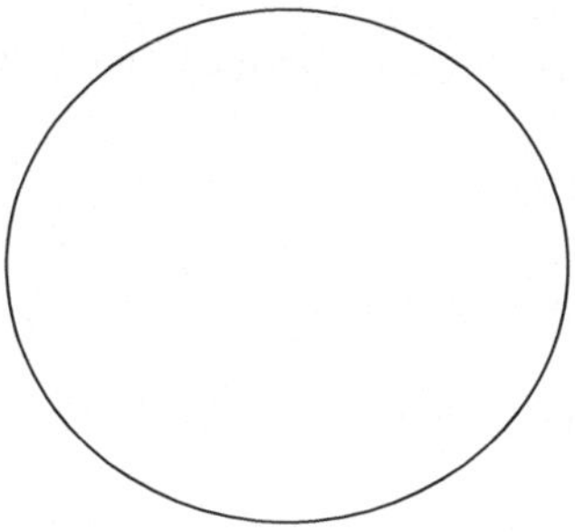						
	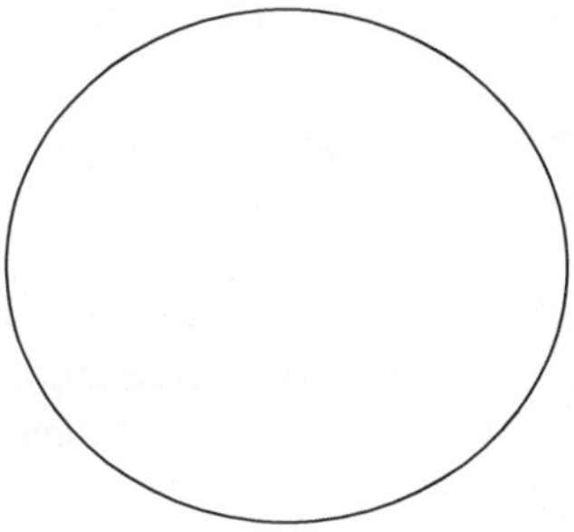						
	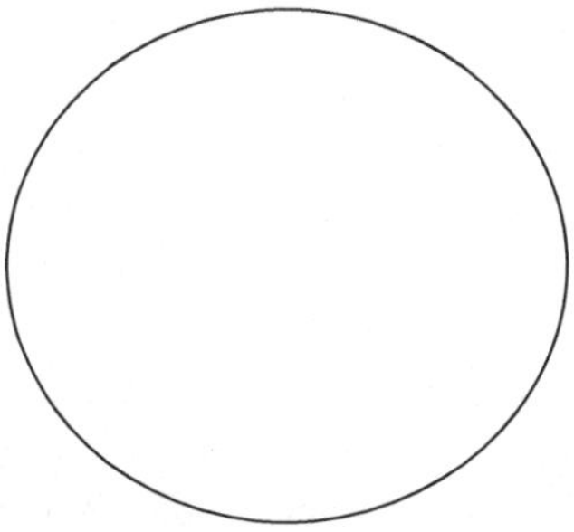						

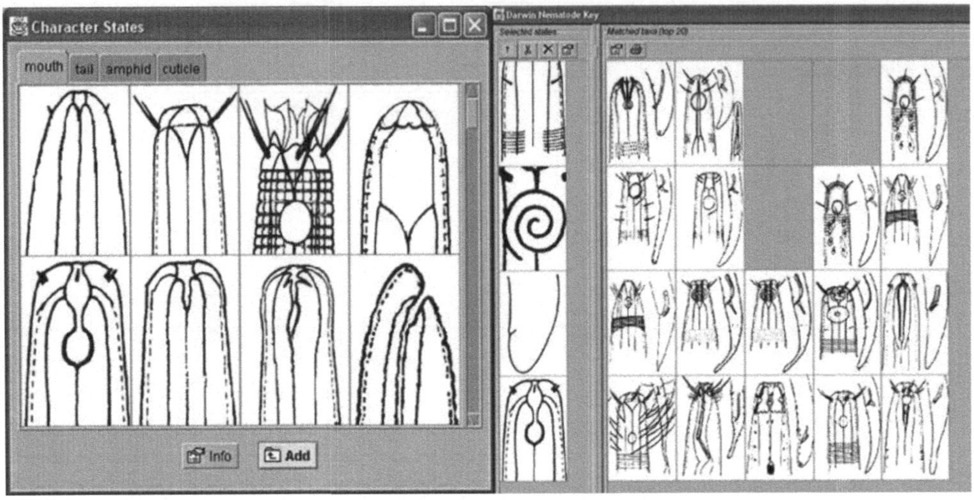

FIGURE 24.2 Preview of the program (Darwin Nematode keys) for the identification of nematodes.

2. Shape and characteristics of the buccal cavity (large, minute, unarmed or armed with tooth, or moveable mandibles)
3. Number and position of cephalic sensilla (Ghent school) or seta (German and British school) on the head
4. Shape of the amphids (pocket-like, loop-shaped, round, spiral, etc.)
5. Spicular apparatus for the male
6. Tail shape (short and round, conical, clavate, elongated, or filiform)
7. Presence or absence of the precloacal supplements and their shapes
8. De Man ratios a, b, and c (a = body length/body width; b = body length/esophagus length; c = body length/tail length)

The more detailed the descriptions, the easier and more precise the identification of the nematode species.

The identification of different morpho-species within a genus is a more rapid method to make an inventory of species presented in a sample (De Mesel et al. 2006). Species can be indicated with the genus, followed by Sp1, Sp2, etc.

The description of the morpho-species is based on simple line drawings of the most distinct characteristics and measurements of length, width, body length and greatest body width ratio, body length and distance from anterior end to junction of pharynx and intestine, and body length and body width at anus or cloaca.

The pictorial keys proposed by Platt and Warwick (1983, 1988), and Warwick et al. (1998), report the drawings of the male nematode genera. These drawings are simplified, and it is important to check the descriptions of the species reported within the cited handbooks (based on the nematodes collected along the British coasts) and in the NeMys Web site since species belonging to one genus can be rather different from the drawing reported in the handbook's pictorial key (Figure 24.2). In fact, species belonging to the same genus can be characterized by very different features such as the number of sensilla, shape and measurements of the amphid, and morphology of the buccal cavity and tail. It is important to keep in mind all these elements to distinguish different species or morpho-species.

The description of new species is time consuming and requires detailed analyses (De Coninck 1965; Gerlach and Riemann 1973; Lorenzen 1981, 1994; Gourbault and Vincx 1985; Soetaert and Vincx 1997; Muthumbi and Vincx 1997). The scientists of Ghent University suggest the analyses of body size and shape, more advanced line drawings using the drawing tube, detailed description

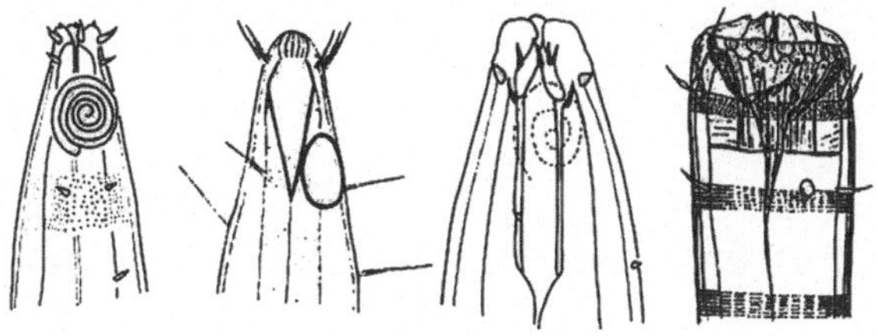

FIGURE 24.3 The four trophic groups (1A, 1B, 2A, 2B) as originally defined by Wieser.

of the male/female habitus, head-end of the male/female (inner and outer surface view), tail ends, variability of measurements (5 males/5 females, if available), illustration of juvenile series, pictures and illustration (SEM) for the description of new or little known species.

24.2.5 Nematode Trophic Diversity

The different features of the buccal cavity allow definition of the trophic composition of the nematode assemblages. According to Wieser (1953), nematodes were divided into four groups as follows: (1A) no buccal cavity or a fine tubular one—selective (bacterial) feeders; (1B) large but unarmed buccal cavity—nonselective deposit feeders; (2A) buccal cavity with scraping tooth or teeth-epistrate or epigrowth (diatom) feeders; (2B) buccal cavity with large jaws—predators/omnivores (Figure 24.3).

Jensen (1987) revised Wieser's classification and proposed the following subdivision in which Wieser's group 2B (omnivore-predators) included, first, predators that caught their living prey by protrusible claws or mandibles (i.e., Thoracostomopsidae, Enoplidae, Selachinematidae), and those that caught the prey with widely open mouth (i.e., Sphaerolaimidae). Jensen properly called the nematodes belonging to the families Oncholaimidae and Enchelidiidae the "garbage collectors."

Moens and Vincx (1997) and Moens et al. (1999) proposed a modified feeding-type classification based on (1) microvores, (2) ciliate feeders, (3) deposit feeders *sensu strictu*, (4) epigrowth feeders, (5) facultative predators, and (6) predators. However, in general, Wieser's classification is preferred because it is still widely used and because no information is available for most species encountered in deep-sea systems.

Based on the composition of the nematode trophic structure, the Index of Trophic Diversity (ITD) can be calculated as ITD= $g_1^2+g_2^2+g_3^2 \ldots +g_n^2$, where g is the relative contribution of each trophic group to the total number of individuals, and n is the number of trophic groups (Gambi et al. 2003). For n = 4 (when Wiser's classification is used), ITD ranges from 0.25 (high diversity, all trophic groups are present) to 1.00 (low diversity, only one group is present), and values are strictly related to the variability of food sources.

In deep-sea sediments, deposit feeders (as selective—1A—plus not selective—1B—deposit feeders) usually account together for more than 50% of the nematode assemblage. Their relevance is also consistent with the presence of large amounts of "fresh" organic detritus. Predator/scavengers account for a minor fraction of total nematode abundance (ca. 1%) in deep-sea sediments (Jensen 1987; Gambi et al. 2003; Tietjen 1984, 1989). Tietjen (1989) suggested that the negligible contribution of this feeding type could indicate the absence of freshly dead organisms. However, recently, Danovaro et al. (2008b) reported high contribution of predators (up to 31%) in nematode assemblages located in the eastern basin of the Mediterranean Sea.

Take the meiofauna previously
extracted and stored with 4% buffered formalin

↓

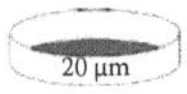

Transfer samples to the sieve 20–30 µm

↓

Wash sample

↓

Transfer sample in a Delfuss cuvette with
water using a wash bottle

↓

Count and identify meiofaunal organisms
under stereomicroscope (minimum magnification 25×)

↓

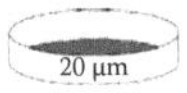

Transfer samples to the sieve, 20–30 µm

↓

Transfer sample in a 50 mL tube
using 4% buffered formalin

↓

Add some drops of Rose Bengal solution (0.5 g/L)

↓

Samples can be stored at room T

PROTOCOL 24.1 Meiofaunal counting.

Using a handling needle pick out the nematodes from the cuvette to a cavity block containing some drops of the solution formalin 4% and glycerin 1%

↓

Put the cavity block in a backer containing a thick layer of ethanol (95%) for one night

↓

The day after, put the cavity block containing the nematodes (washed in the solution formalin 4% and glycerin 1%) into the oven at 36°C

↓

Add some drops of the solution 95% ethanol and 5% glycerin every two h

↓

Transfer each organism to a slide with a drop of glycerin in the center of a paraffin ring

↓

Gently place a cover glass on the paraffin ring and the glycerin drop

↓

Heat the slide on a hotplate or using a lighter for few seconds, keeping the slide horizontal

PROTOCOL 24.2 Preparation of permanent slides for nematode identification.

REFERENCES

Bett, B. J. 2000. Comparative benthic ecology of the Rockall Trough and Faeroe-Shetland Channel. In *Atlantic margin environmental surveys of the seafloor of the UK Atlantic margin (Atlantic Frontier Environmental Network)* [CDROM], ed. Geotek Ltd, Section 4.3.2. Daventry, Northauts, U.K.

Bett, B., A. Vanreusel, M Vincx et al. 1994. Sampler bias in the quantitative study of deep-sea meiobenthos. *Marine Ecology Progress Series* 104: 197–203.

Bhadury, P., M. C. Austen, D. T. Bilton, P. J. D. Lambshead, A. D. Rogers, and G. R. Smerdon. 2005. Combined morphological and molecular analysis of individual nematodes through short-term preservation in formalin. *Molecular Ecology Notes* 5: 965–968.

Bhadury, P., M. C. Austen, D. T. Bilton, P. J. D. Lambshead, A. D. Rogers, and G. R. Smerdon. 2006. Molecular detection of marine nematodes from environmental samples: Overcoming eukaryotic interference. *Aquatic Microbiology Ecology* 44: 97–103.

Bongers, T. and H. Ferris. 1999. Nematode community structure as a bioindicator in environmental monitoring. *Trends in Ecology and Evolution* 14: 224–228.

Cobb, N. A. 1917. Notes on nemas. *Contribution to a Science Nematology* 5: 117–128.

Cook, A. A., P. Bhadury, N. J. Debenham, and B. H. M. Meldal. 2005. Denaturing gradient gel electrophoresis as a tool for the identification of marine nematodes. *Marine Ecology Progress Series* 291: 103–113.

Coomans, A. 2002. Present status and future of nematode systematics. *Nematology* 5: 573–582.

Danovaro, R., A. Dinet, G. Duineveld, and A. Tselepides. 1999a. Benthic response to particulate fluxes in different trophic environments: A comparison between the Gulf of Lions–Catalan Sea (Western Mediterranean) and the Cretan sea (Eastern Mediterranean). *Progress in Oceanography*, 44: 287–312.

Danovaro, R., D. Marrale, N. Della Croce, A. Dell'Anno, and M. Fabiano. 1999b. Heterotrophic nanoflagellates, bacteria, and labile organic compounds in continental shelf and deep-sea sediments of the eastern Mediterranean. *Microbial Ecology*, 35(3): 244–255.

Danovaro, R., A. Tselepides, A. Otegui, and N. Della-Croce. 2000. Dynamics of meiofaunal assemblages on the continental shelf and deep-sea sediments of the Cretan Sea (NE Mediterranean): Relationships with seasonal changes in food supply. *Progress in Oceanography* 46: 367–400.

Danovaro, R., A. Dell'Anno, M. Fabiano, A. Pusceddu, and A. Tselepides. 2001. Deep-sea ecosystem response to climate changes: The Eastern Mediterranean case study. *Trends in Ecology and Evolution* 16: 505–510.

Danovaro, R., C. Gambi, and N. Della Croce. 2002. Meiofauna hotspot in the Atacama Trench (Southern Pacific Ocean). *Deep-Sea Research Part I* 49: 843–857.

Danovaro, R., C. Gambi, S. Mirto, R. Sandulli, and V. U. Ceccherelli. 2004. Meiofauna. In *Mediterranean marine benthos: A manual of methods for its sampling and study* ed. M. C. Gambi, M. Dappiano, Chap. 3: 55–97. *Biologia Marina Mediterranea* 11(1).

Danovaro R., C. Gambi, A. Dell'Anno et al. 2008a. Exponential decline of deep-sea ecosystem functioning linked to benthic biodiversity loss. *Current Biology* 18: 1–8.

Danovaro, R., C. Gambi, N. Lampadariou, and A. Tselepides. 2008b. Deep-sea biodiversity in the Mediterranean Basin: Testing for longitudinal, bathymetric, and energetic gradients. *Ecography* 31: 231–244

De Coninck, L. A. P. 1965 Systématique des nématodes. In *Traité de Zoologie: Anatomie, Systématique, Biologie*, ed. P. P. Grasse, 586–731. Paris: Masson et Cie.

De Grisse, A. T. 1969. Redescription ou modifications de quelques techniques utilisées dans l'étude des nématodes phytoparasitaires. *Mededelingen Rijks Faculteit Landbouww Et Gent* 34: 351–369.

De Mesel, I., H. J. Lee, S. Vanhove, M. Vincx, and A. Vanreusel. 2006. Species diversity and distribution within the deep-sea nematode genus acantholaimus on the continental shelf and slope in Antarctica. *Polar Biology* 29: 860–871.

Deprez, T. et al. (2005) NeMys. World Wide Web electronic publication. www.nemys.ugent.be version (3/2009).

Elmgren, R. and T. Radziejewska. 1989. Recommendations for quantitative benthic meiofauna studies in the Baltic. *Baltic Marine Biologists Publication* 12: 1–23.

Fleeger, J. W., W. B. Sikora, and J. P. Sikora. 1983. Spatial and long-term temporal variation of meiobenthic–hyperbenthic copepods in lake Pontchartrain, Louisiana. *Estuarine, Coastal, and Shelf Science* 16: 441–453.

Floyd, R., E. Abebe, A. Papert, and M. L. Blaxter. 2002. Molecular barcodes for soil nematode identification. *Molecular Ecology* 11: 839–850.

Gambi, C., A. Vanreusel, and R. Danovaro. 2003. Biodiversity of nematode assemblages from deep-sea sediments of the Atacama Slope and Trench (Southern Pacific Ocean). *Deep-Sea Research Part I* 50: 103–117.

Gambi, C. and R. Danovaro. 2006. A multiple-scale analysis of metazoan meiofaunal distribution in the deep Mediterranean Sea. *Deep-Sea Research* 53: 1117–1134.

Gerlach, S. A. and F. Riemann. 1973. The Bremerhaven checklist of aquatic nematodes: A catalogue of Nematoda Adenophorea excluding the Dorylaimida. *Veröffentlichungen des Instituts für Meeresforschung in Bremerhaven.*

Giere, Olav. 1993. *Meiobenthology: The microscopic fauna in aquatic sediments.* Berlin: Springer-Verlag.

Giere, Olav. 2009. *Meiobenthology. The microscopic motile fauna of aquatic* sediments, 2nd edition. Berlin: Springer-Verlag.

Gooday, A. J., B. B. J. Bett, R. Shires, and P. J. D. Lambshead. 1998. Deep-sea benthic foraminiferal diversity in the NE Atlantic and NW Arabian sea: A synthesis. *Deep-Sea Research Part II* 45: 165–201.

Gourbault, N. and M. Vincx. 1985. Nèmatodes abyssaux (Campagne Walda du N/O "Jean Charcot") VI. Le genre Acantholaimus: Espèces nouvelles et étude de l'appareil reproducteur à spermatozoïdes géants. *Bulletin du Musum national d'histoire naturelle, Paris* 4e sér 7, section A, 3: 611–632.

Higgins, R. P. and H. Thiel. 1988. *Introduction to the study of meiofauna.* Washington, DC: Smithsonian Institution Press.

Jensen, P. 1987. Feeding ecology of free-living aquatic nematodes. *Marine Ecology Progress Series* 35: 187–196.

Lambshead, P. J. D. 2004. Marine nematode biodiversity. In *Nematology: Advances and perspectives Vol 1: Nematode morphology, physiology, and ecology*, ed. Z. X. Chen, S. Y. Chen, and D. W. Dickson, 436–467. CABI Publishing.

Levin, L. A., R. J. Etter, M. A. Rex et al. 2001. Environmental influences on regional deep sea species diversity. *Annual Review of Ecological Systems* 32: 51–93.

Lorenzen, S. 1981. Entwurf eines phylogenetischen Systems der freilebenden Nematoden. *Veröffentlichungen des Instituts für Meeresforschung in Bremerhaven* 7: 1–472.

Lorenzen, S. 1994. *The phylogenetic systematics of freeliving nematodes.* London: The Ray Society.

Meldal, B. H. M. 2004. Phylogenetic systematics of the phylum Nematoda—evidence from molecules and morphology. Ph.D. thesis, University of Southampton, U.K.

Meldal B. H. M., N. J. Debenham, P. De Ley et al. 2006. An improved molecular phylogeny of the Nematoda with special emphasis on marine taxa. *Molecular Phylogenetics and Evolution* 42: 622–636.

Mitreva, M., M. L. Blaxter, D. M. Bird, and J. P. McCarter. 2005. Comparative genomics of nematodes. *Trends in Genetics* 21(10): 573–581.

Moens, T., L. Verbeeck, and M. Vincx. 1999. Feeding biology of predatory and facultatively predatory nematode (*Enoploides longispiculosus* and *Adoncholaimus fucus*). *Marine Biology* 134: 585–593.

Moens, T. and Vincx M. 1997. Observations on the feeding ecology of estuarine nematodes. *Journal of the Marine Biological Association of the United Kingdom* 77: 211–227.

Muthumbi, A. and M. Vincx. 1997. Acantholaimus (chromadoridae: nematoda) from the Indian Ocean: Description of seven species. *Hydrobiologia* 346: 59–76.

Netto, S. A., F. Gallucci, and G. F. C. Fonseca. 2005. Meiofaunal communities of the continental slope and deep-sea sites off SE Brazil. *Deep-Sea Research Part I* 52: 845–859.

Platt, H. M. and R. M. Warwick. 1983. *A synopsis of the free-living marine nematodes. Part I: British Enoplids.* Cambridge, U.K.: Cambridge University Press.

Platt, H. M. and R. M. Warwick. 1988. *A synopsis of the free-living marine nematodes. Part II: British Chromadorids*, Cambridge, U.K.: Cambridge University Press.

Pusceddu, A., C. Gambi, E. Manini, and R. Danovaro. 2007. Trophic state, ecosystem efficiency, and biodiversity of transitional aquatic ecosystems: Analysis of environmental quality based on different benthic indicators. *Chemistry and Ecology* 23(6): 505–515.

Riedl, R. 1983. *Flora und Fauna des Mittelmeeres.* Hamburg/Berlin: Paul Parey.

Seinhorst, J. W. 1959. A rapid method for the transfer of nematodes from fixative to unhydrous glycerine. *Nematologica* 4: 67–69.

Shirayama, Y. and T. Fukushima. 1995. Comparison of deep-sea sediments and overlying water collected using multiple corer and box corer. *Journal of Oceanography* 51: 75–82.

Snelgrove, P. V. R. 1999. Getting to the bottom of marine biodiversity: Sedimentary habitats. *BioScience* 49: 129–138.

Soetaert, K. and M. Vincx. 1987. Six new Richtersia species (Nematoda, Selachinematidae) from the Mediterranean Sea. *Zoologica Scripta* 16: 125–142.

Tietjen, J. H. 1984. Distribution and species diversity of deep-sea nematodes in the Venezuela basin. *Deep-Sea Research* 31: 119–132.

Tietjen, J. H. 1989. Ecology of deep-sea nematodes from the Puerto Rico Trench area and Hatteras Abyssal Plain. *Deep-Sea Research Part I* 36: 1579–1594.

Tselepides, A. and Lampadariou. 2004. Deep-sea meiofaunal community structure in the eastern Mediterranean: Are trenches benthic hot spot?. *Deep-Sea Research Part I* 51: 833–847.

Vanhove, S., H. Vermeeren, and A. Vanreusel. 2004. Meiofauna towards the South Sandwich Trench (750–6300 m), focus on nematodes. *Deep-Sea Research Part II* 51(14–16): 1665–1687.

Vincx, M. 1996. Meiofauna in marine and freshwater sediments. In *Methods for the examination of organismal diversity in soils and sediments*, ed. G. S. Hall, 187–195. CABI Publishing.

Warwick R. M., H. M. Howard, and P. J. Somerfield. 1998. *A synopsis of the freeliving marine nematodes. Part III: Monhysterids.* Field Studies Council, Shrewsbury, U.K.

Wieser, W. 1953. Die Beziehung zwischen Mundhöhlengestalt, Ernährungsweise und Vorkommen bei freilebenden marinen Nematoden. *Arkiv für Zoolgie* 2(4): 439–484.

25 Macrofaunal and Megafaunal Diversity

CONTENTS

25.1 INTRODUCTION

The extreme environmental conditions of deep-sea habitats have important implications not only for the size structure of deep-sea communities but also for the analysis of factors controlling the pattern of biodiversity. Abyssal macrofaunal and megafaunal species diversities are depressed (Rex 1981). For one important macrofaunal taxon—mollusks—the abyssal fauna is composed largely of sparsely occupied range extensions for a subset of bathyal species with larval dispersal (Rex et al. 2005). There is little evidence of abyssal endemism (McClain et al. in press) except at chemosynthetically-driven ecosystems such as hydrothermal vents (Desbruyères et al. 2006), cold seeps (Sibuet and Olu 1998), or whale carcasses (Smith and Baco 2003). Rex et al. (2005) hypothesized that the bathyal and abyssal zones may function as a source-sink system in which abyssal populations of many macrofaunal species are not reproductively self sustaining but represent a balance between chronic local extinction from inverse density dependence and immigration from bathyal sources. Extremely low abundance and biomass may also constrain the potential for evolutionary diversification in larger size groups (Etter et al. 2005). Even if speciation is successful in terms of producing reproductively isolated populations, food levels in the abyss may be insufficient to support subsequent coexistence. Patterns of standing stock suggest that there is considerably more ecological and evolutionary opportunity for adaptive radiation in the bathyal zone (200 to 4000 m) than in the abyss (>4000 m). The dramatic decrease in total community standing stock and the ascendancy of smaller organisms with depth has important implications for deep-sea biodiversity. The bathyal zone (200 to 4000 m) affords more ecological and evolutionary opportunity in the form of energy availability for larger organisms, and consequently supports higher macrofaunal and megafaunal species diversity than the abyss (>4000 m) (Figure 25.1).

25.2 FIELD PROCEDURES

Sampling in deep-sea soft sediments for collection of macrofauna and megafauna is described in Chapter 18, "Macrofaunal Abundance" and Chapter 19, "Megafaunal Abundance."

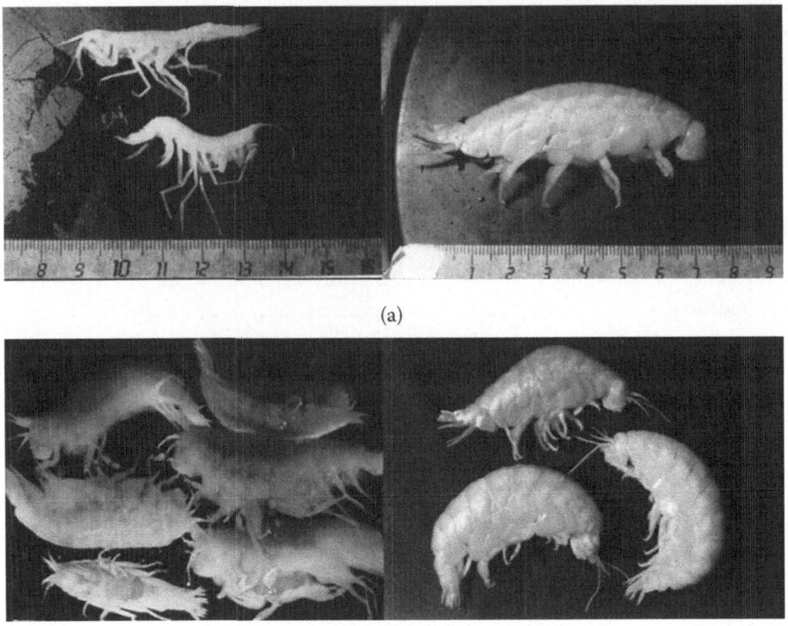

(a)

(b)

FIGURE 25.1 Reported are very large specimens of *Eurythenes gryllus* and *Hyrondellea* sp. (a) On the left, two very large specimens of *Hyrondellea* sp.; on the right, a very large specimen of *E. gryllus*. (b) Different specimens of *E. gryllus* sampled in Atacama trench.

25.3 LABORATORY PROCEDURES

The procedures to collect, store, extract, and sort macrofaunal and megafaunal samples for the subsequent determination of their diversity are those reported in Chapter 18, "Macrofaunal Abundance" and Chapter 19, "Megafaunal Abundance."

For a fine taxonomic identification of macrofaunal specimens, the use of an optical microscope is needed (high quality with ocular ×10–20 and objectives ×40–100). To observe anatomical details, it is recommended that a few drops of water or of a glycerine/ethanol solution (1:1) be placed on a microscope slide, the animal or the part of the body of interest be dipped into the solution and the slide covered with a cover slip. To avoid the formation of bubbles, the cover slip should be placed at a slanting angle on one side and then gently lowered onto the slide. The organisms can be stained at the time of fixation prior to observation in order to increase contrast in the structures to be examined.

25.3.1 Equipment and Instruments

Stereomicroscope
Light microscope
Delfuss cuvette
Handling needle
Slides
Cover glass
Petri dish
Pasteur pipette
Data sheet
Oil with low reflectivity

25.3.2 Solutions

Prefiltered seawater
4% buffered formalin
Rose Bengal (0.5 gL^{-1})
80% alcohol

A sample (extracted, fixed, and stained with Rose Bengal or other suitable stain) is transferred into the 200 μm mesh sieve (smaller than the one used for the initial extraction to reduce possible losses). The sieve to be used must be as small as possible (typically 5 cm in diameter to fit with the size of the sample) in order to reduce the risk of losing specimens during storage.

In some cases it is necessary to use an optical microscope to identify megafauna species. The specimen is placed in a Petri dish with seawater or 80% alcohol and the distinctive characteristics observed. The total weight of the sample for each species is calculated for (biomass). Depending on the studies to be conducted, each specimen is also weighed and measured individually.

25.3.3 Procedure

1. Pick up the specimen to be identified.
2. Proceed with identification.
3. At the end of the sorting, transfer the sample into tube or container of appropriate size and fix it with either 4% buffered formalin or alcohol.
4. Add a few drops of 0.5 g L^{-1} Rose Bengal solution with a Pasteur pipette.
5. Samples can be stored in the dark at room temperature.

25.4 NOTES

After staining with Rose Bengal, wait 4–8 h before starting the sorting and the identification of the macrofaunal taxa.

At the end of the sorting, check the surface water to count and classify the specimens that can float (Table 25.1).

Keep in mind that Rose Bengal stains all structures covered by a membrane (it acts on the proteins of the cellular membrane).

Exercise care in collecting the macrofaunal specimens from the sieve and jars. Clean the sieve well at each step during the preparation of the sample for sorting and after the end of counting. Each step is crucial, and can cause the loss of specimens.

Use the formalin carefully, dressing in all the safety devices (laboratory coat, gloves, glass, respiratory protection), and work using an extractor fan.

The solutions containing formalin must be collected and eliminated using the standard safety procedures.

TABLE 25.1
Example of Data Sheet Used for the Sorting of Macrofauna

Cruise/Project	
Date of sampling	Station depth:
Date of sorting	Check data quality: YES/NO
Station	Sorting efficiency: %

Taxa		Replicate 1	Replicate 2	Replicate 3
Nematodes				
Anellids	Polychaetes			
	Oligochaetes			
Mollusks	Bivalves			
	Gastropods			
	Polyplacophors			
Crustaceans	Decapods			
	Amphipods			
	Isopods			
	Tanaidacea			
Echinoderms	Echinoids			
	Ophiuroids			
Tunicates	Ascidiaceans			
Sipunculids				
Platyhelminthes				
Picnogonids				
Cephalochordates				
Others *				

Notes
Notes for preservation:
* List of other taxa:

REFERENCES

Desbruyères, D., M. Segonzac, and M. Bright. 2006. *Handbook of Deep-Sea Hydrothermal Vent Fauna.* Linz: Denisia.

Etter, R. J., M. A. Rex, M. R. Chase, and J. M. Quattro. 2005. Population differentiation decreases with depth in deep-sea bivalves. *Evolution* 59: 1479–1491.

McClain, C. R., M. A. Rex, and R. J. Etter. in press. Patterns in deep-sea macroecology. In *Marine macroecology*, ed. J. Witman and K. Roy. Chicago: University of Chicago Press.

Rex, M. A. 1981. Community structure in the deep-sea benthos. *Annual Review of Ecological Systems* 12: 331–353.

Rex, M. A., C. R. McClain, N. A. Johnson et al. 2005. A source-sink hypothesis for abyssal biodiversity. *The American Naturalist* 165: 163–178.

Sibuet, M. and K. Olu. 1998. Biogeography, biodiversity, and fluid dependence of deep-sea cold-seep communities at active and passive margins. *Deep Sea Research Part II* 45: 517–567.

Smith, C. R. and A. R. Baco. 2003. Ecology of whale falls at the deep-sea floor. *Oceanography and Marine Biology: An Annual Review* 41: 311–354.

26 Parameters for the Measurement of Marine Benthic Structural and Functional Diversity

CONTENTS

26.1 INTRODUCTION

Biodiversity encompasses a range of different levels of organization from the genetic variation between individuals and populations to species diversity, assemblages, habitats, landscapes, and biogeographical provinces (Gray 2000). Whittaker (1960) suggested that there was a range in scales of species richness and partitioned diversity into alpha, beta, gamma, and epsilon components to characterize different aspects of diversity in relation to spatial scale. Alpha diversity is the diversity within a single habitat, beta diversity measures the degree of change in the composition of samples along gradient or the extent of the difference in samples from the opposite end of a gradient (Gray 2000), and gamma diversity is the diversity within homogeneous sampling units (i.e., region, including different habitats), and epsilon diversity is the total diversity within a region (Gray 2000). The knowledge of the processes that determine patterns of species diversity in space and time is a major focus of ecological research (Ricklefs and Schluter 1993). Several factors can be invoked to explain the spatial patterns of biodiversity: historical, phylogenetic, and climate processes are important determinants of regional pools of species (i.e., gamma diversity), whereas biotic interactions play a key role in determining local diversity (i.e., alpha diversity). Changes in the physical characteristics of the habitat can strongly affect variation in composition and abundance of species among sites, thus influencing beta diversity (Danovaro et al. 2004, 2008a; Balata et al. 2007; Pusceddu et al. 2008). Biodiversity can also be measured in terms of functional traits as well as the functional role and identity of the species reported in the investigated areas (Cardinale et al. 2006). Here we report the most common diversity indices to measure benthic structural and functional diversity.

26.2 STRUCTURAL DIVERSITY

Species Richness (SR): This is a count of the number of species in the sample/study site (Figure 26.1a).

Simpson Diversity Index (D): It is a diversity index that takes into account the number of species present in a sample as well as the relative abundance of each species (Simpson 1949). This index calculates the probability that two individuals, randomly selected in the habitat, belong to the same species. The Simpson Diversity Index is calculated as follows:

$$D = \frac{\sum_{i=1}^{S} n_i (n_i - 1)}{N(N-1)} \tag{26.1}$$

where

S = the number of species.

N = the total number of organisms.

n_i = the number of organisms within a species.

Shannon Index (H'): It is used to measure diversity on categorical data (Shannon 1948; Krebs 1989) (Figure 26.1b). It takes into account the number of species and the evenness of the species. The Shannon index is calculated as follows:

$$H' = -\sum_{1=1}^{S} p_i \ln p_i \tag{26.2}$$

where

S = the total number of species.

$p_i = (n_i/N)$ where n_i is the number of individuals of the *i*th species.

N = total number of individuals (*N* is the total number of all individuals).

Logarithm can be with base *E* or 2.

Margalef Index (D):

$$D = (S-1)/\ln N \tag{26.3}$$

where

S = the number of species.

N = the number of individuals in the sample (Margalef 1958).

Evenness Index (J; Pielou's Index): It is the relative abundance of individuals amongst species (Pielou 1975). The Evenness index is calculated as follows:

$$E = \frac{H'}{\ln(S)} \tag{26.4}$$

where

S = the number of species.

Expected Number of Species ES(X): The species-abundance data are converted into rarefaction diversity indices (Sanders 1968, as modified by Hurlbert 1971) (Figure 26.1c). The expected number of species ES(X) is particularly suitable for the analysis of deep-sea biodiversity. Since species richness is strongly affected by the sample size, ES(X) therefore allows the comparison between

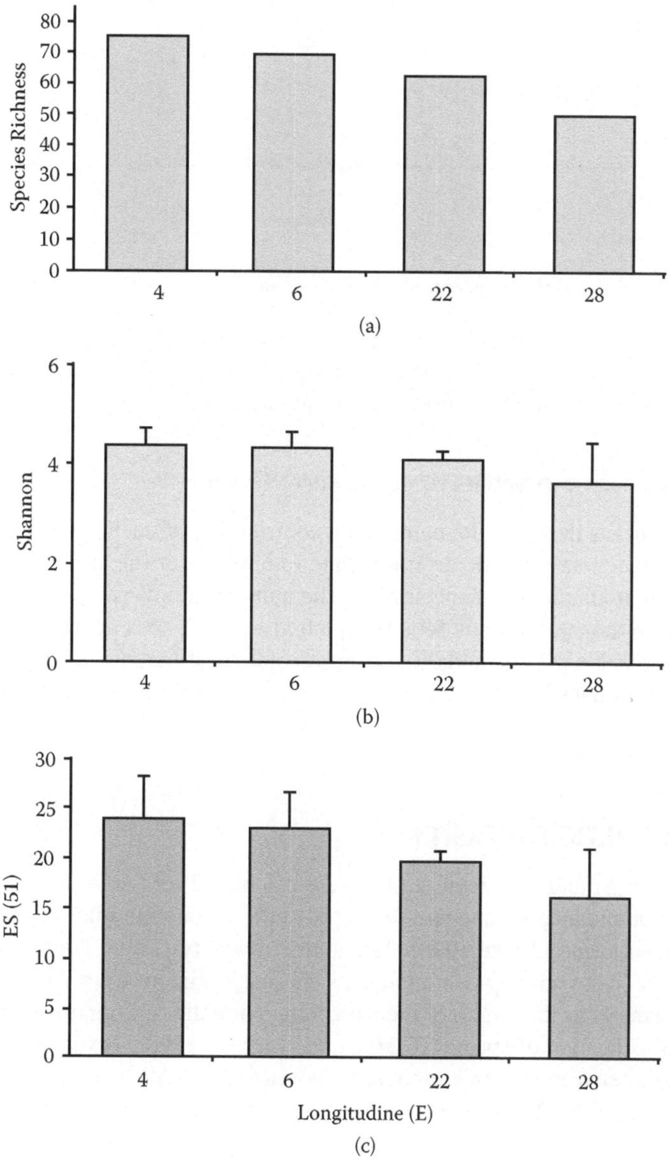

FIGURE 26.1 Structural diversity. Reported are (a) Species Richness, (b) Shannon diversity index (H′), and (c) expected nematode species number estimated for 51 individuals ES(51) along a longitudinal gradient in the deep Mediterranean Sea.

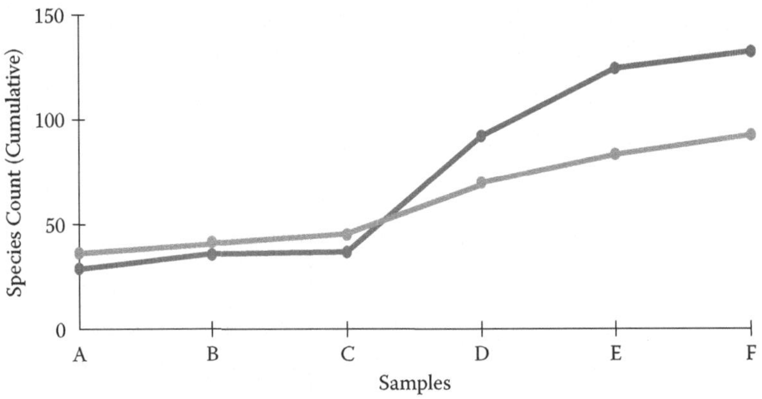

FIGURE 26.2 Two examples of the species accumulation curve in deep-sea sediment.

diversity values measured using different sampling sizes. The expected number of species for a theoretical sample of X specimens (ES(51) for nematodes and ES(100) for macrofauna, respectively) can be selected to compare samples with different sampling size.

26.2.1 K-Dominance and Species Accumulation Curves

The k-dominance curves describe the cumulative contribution of each species to total abundance (cumulative dominance) and provide an immediate impression of the structure of benthic assemblages. The accumulated number of species (i.e., the number of new species found in each successive sample added to the total already found) is plotted on the Y-axis against the quadrants (in the order tallied) on the X-axis. This modality of representation of benthic diversity can be useful for the identification of samples in which there is a significant contribution of new species. In this case, the curve will show an abrupt increase, followed by a stabilization. An example of the species accumulation curve is shown in Figure 26.2.

26.3 PHYLOGENETIC DIVERSITY

Taxonomic diversity (Δ) and distinctness (Δ*) were calculated to define the structure not only of the distribution of abundance among species but also the taxonomic relatedness of the species in each sample (Warwick and Clarke 1998). Taxonomic diversity can be considered as the average taxonomic "distance" between any two organisms chosen at random from the sample; this distance can be visualized simply as the length of the path connecting the two organisms through a Linnean or phylogenetic classification of species (Clarke and Warwick 1998). Taxonomic distinctness is the average path length between any two randomly chosen individuals belonging to different species (Clarke and Warwick 1998). For practical comparisons, both taxonomic diversity and distinctness are not dependent on the degree of sampling effort involved in the data collection.

26.4 TURNOVER DIVERSITY

Turnover diversity (i.e., beta diversity) can be estimated using different methods. Whittaker (1960) proposed the measure of turnover diversity as follows:

$$\beta = \gamma / \alpha \qquad (26.5)$$

where

γ = the number of species resulting from merging a number of individual samples.
α = the number of species in a sample.

Wilson and Shmida (1984) proposed several measures of beta diversity and concluded that the two best ones were Whittaker's and a new suggested index:

$$\beta_T = [g(H) + l(H)] / 2(SR_s) \tag{26.6}$$

where
g = number of species gained along the gradient.
l = number of species lost along the gradient.

Harrison et al. (1992) modified Whittaker's formula to allow comparisons of two transects of unequal size:

$$\beta - 1 = \{(S / SR_S - 1) / (N - 1)\} \cdot 100 \tag{26.7}$$

where
S = the total number of species in the transects.
N = the number of sites.
$\beta - 1$ ranges from 0, complete similarity to 100, complete dissimilarity.

The authors proposed a further index that also considers species dropout along the gradient but are not replaced by new species. This index is named $\beta - 2$, but it has not been widely tested.

Benthic ecologists widely use multivariate statistical analyses. The PRIMER (Plymouth Routines in Marine Environmental Research; Clarke and Warwick 1994) is one of the most widely used packages. PRIMER and BiodiversityPro (McAleece 1997) allow measuring the similarity between sites in terms of biodiversity or assemblage composition. SIMPER (PRIMER package) allows the estimation of the coefficient of dissimilarity (percentage, based on the Bray–Curtis similarity index) in species composition among sites. Generally, data of species abundance presence/absence transformed prior to the analysis.

Most of the foregoing common diversity indices can be easily calculated starting from a simple Excel matrix containing the list of species and the relative abundance of each species for each sample/replicate.

26.5 FUNCTIONAL DIVERSITY

Functional diversity is the range of functions that are performed by organisms in a system (Petchey and Gaston 2006). One possible approach is based on the assumption that the different morphologies, buccal sizes, and other traits reflect a diverse ecological role (e.g., selection of food items within the same feeding guild).

26.5.1 Trophic Diversity

The functional diversity of nematodes can be measured using the number of different functional (trophic) traits based on the analysis of the feeding types according to the classical literature (Heip et al. 1985) and updated to the most recent approaches (Moens and Vincx 1997). As an example, the trophic composition of nematode assemblages can be defined according to Wieser (1953), Jensen

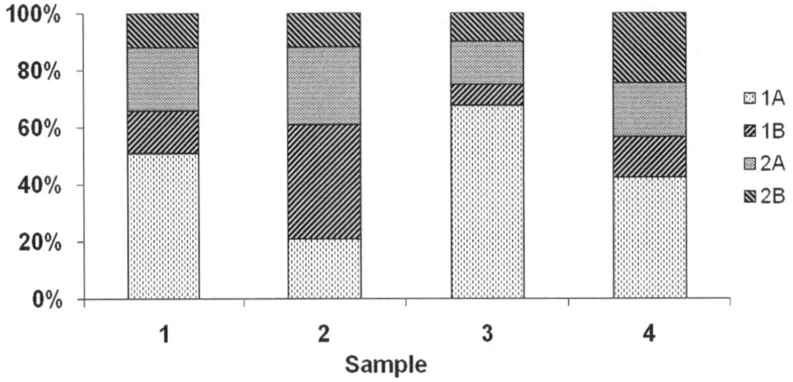

FIGURE 26.3 Nematode trophic structure according to Wieser's classification (1953). Reported are 1A (selective deposit feeders), 1B (nonselective deposit feeders), 2A (epigrowth feeders), and 2B (predators), along a longitudinal gradient in the deep Mediterranean Sea.

(1987), and Moens and Vincx (1997). According to Wieser (1953), nematodes are divided into four original groups as follows: (1A) no buccal cavity, or a fine tubular one—selective (bacterial) feeders, (1B) large but unarmed buccal cavity—nonselective deposit feeders, (2A) buccal cavity with scraping tooth or teeth—epistrate or epigrowth (diatom) feeders, and (2B) buccal cavity with large jaws—predators/omnivores (Figure 26.3). Jensen (1987) modified the previous classification and suggested identifying four feeding groups based on oral feeding. Deposit and epistrate feeders are distinguished as two feeding groups by the absence or presence of a tooth in the buccal cavity, respectively. Consequently, the two groups include different feeding types but exploit the same food source. A third group of species is recognized as scavengers, united by the presence of an onchium (tooth-like structure) in the buccal cavity penetrated by a duct through which the contents of a salivary gland cell empty. Scavengers lack mandibles, claws, and articulating plates in the buccal cavity, which characterized the fourth group, predators. Moens and Vincx (1997) proposed a modified feeding-type classification based on (1) microvores, (2) ciliate feeders, (3) deposit feeders *sensu strictu*, (4) epigrowth feeders, (5) facultative predators, and (6) predators. However, in most studies, Wieser's classification is still widely used since few or no information was available for most species encountered in deep-sea systems. A similar approach can be used for other benthic components, including macrofauna and megafauna.

On the basis of the trophic composition of benthic assemblages, the *Index of Trophic Diversity* (ITD) can be measured as follows:

$$ITD = g_1^2 + g_2^2 + g_3^2 + ... + g_n^2 \tag{26.8}$$

where

g = the relative contribution of each trophic group to the total number of individuals.
n = the number of trophic groups (Gambi et al. 2003).

Where four trophic guilds are considered, the ITD ranges from 0.25 (highest trophic diversity; i.e., the four trophic guilds account for 25% each) to 1.0 (lowest diversity; i.e., one trophic guild accounts for 100% of nematode density; Heip et al. 1985). As the highest trophic diversity scores the lowest values, Danovaro et al. (2004) proposed a modified version of this index as 1-ITD (ranging from 0 to 0.75) to better visualize the changes in functional diversity.

The *Maturity Index* (MI) can be used to describe the life strategies (r-k) characterizing an ecosystem on the basis of a taxon for which life strategies are known (Bongers, 1990). As an example, in order to identify colonization strategies of nematodes, Bongers et al. (1991) proposed a classification of nematodes into "colonizers" (comparable to r-strategists, characterized by short life cycle, high colonization ability, and tolerance to disturbance, eutrophication, and anoxybiosis) and "persisters" (with the opposite characteristics, K-strategists with low reproduction rate, long life cycle, and low colonization ability and tolerance to disturbance).

MI was calculated according to the weighted mean of the individual genus scores:

$$MI = \sum v_{(i)} \cdot f_{(i)} \tag{26.9}$$

where

$v_{(i)}$ = the c-p value (colonizers-persisters) of genus i.
$f_{(i)}$ = the frequency of that genus.

The c-p scale ranges from 1 for colonizers to 5 for persisters as given in the Appendix of Bongers et al. (1991).

26.6 PREDATOR SPECIES RICHNESS

Functional diversity can also be estimated using the number of predator species (Danovaro et al. 2008b), which depended upon the assumption that the number of species at the top of the benthic food web reflects a higher functional diversity of the entire benthic assemblage (Ngai and Srivastava 2006).

Functional diversity estimated as number of predators is based on the assumption that predatory species influence the structural and functional diversity of meio-, macro-, and megafaunal assemblages by preying selectively on the organisms (or the larvae), especially those displaying lower mobility (Giere 1993; Danovaro et al. 1995, 2008). This variable might have important consequences on the structural and functional diversity of higher trophic levels (Danovaro et al. 1995).

REFERENCES

Balata, D., L. Piazzi, and L. Benedetti-Checchi. 2007. Sediment disturbance and loss of beta diversity on subtidal rocky reefs. *Ecology* 88: 2455–2461.

Bongers, T. 1990. The maturity index: An ecological measure of environmental disturbance based on nematode species composition. *Oecologia* 83: 14–19.

Bongers, T., R. Alkemade, and G. W. Yeates. 1991. Interpretation of disturbance induced maturity decrease in marine nematode assemblages by means of the Maturity Index. *Marine Ecology Progress Series* 76: 135–1472.

Cardinale, B. J., D. S. Srivastava, J. E. Duffy et al. 2006. Effects of biodiversity on the functioning of trophic groups and ecosystems. *Nature* 443: 989–992.

Clarke, K. R. and R. M. Warwick. 1994. *Change in marine communities: An approach to statistical analysis and interpretation*. Cambridge, U.K.: Natural Environment Research Council.

Clarke, K. R. and R. M. Warwick. 1998. A taxonomic distinctness index and its statistical properties. *Journal of Applied Ecology* 35: 523–531.

Danovaro, R., S. Fraschetti, A. Belgrano et al. 1995. The potential impact of meiofauna on the recruitment of macrobenthos in a subtidal coastal benthic community of the Ligurian Sea: A field result. In *Biology and ecology of shallow coastal waters*, ed. A. Eleftheriou, A. D. Ansell, and C. J. Smith, 115–122. Fredensborg, Denmark: Olsen and Olsen.

Danovaro, R., A. Dell'Anno, and A. Pusceddu. 2004. Biodiversity response to climate change in a warm deep sea. *Ecology Letters* 7: 821–828.

Danovaro, R., C. Gambi, A. Dell'Anno et al. 2008a. Exponential decline of deep-sea ecosystem functioning linked to benthic biodiversity loss. *Current Biology* 18: 1–8.

Danovaro, R., C. Gambi, N. Lampadariou, and A. Tselepides. 2008b. Deep-sea biodiversity in the Mediterranean Basin: Testing for longitudinal, bathymetric, and energetic gradients. *Ecography* 31: 231–244.

Gambi, C., A. Vanreusel, and R. Danovaro. 2003. Biodiversity of nematode assemblages from deep-sea sediments of the Atacama Slope and Trench (Southern Pacific Ocean). *Deep-Sea Research Part I* 50: 103–117.

Giere, Olav. 1993. *Meiobenthology: The microscopic fauna in aquatic sediments*. Berlin: Springer-Verlag.

Gray, J. S. 2000. The measurement of marine species diversity, with an application to the benthic fauna of the Norwegian continental shelf. *Journal of Experimental Marine Biology and Ecology* 250: 23–49.

Harrison, S., S. J. Ross, and J. H. Lawton. 1992. Beta diversity on geographic gradients in Britain. *Journal of Animal Ecology* 61: 151–158.

Heip, C., M. Vincx, and G. Vranken. 1985. The ecology of marine nematodes. *Oceanography and Marine Biology: An Annual Review* 23: 399–489.

Hurlbert, S. H. 1971. The non-concept of species diversity: A critique and alternative parameters. *Ecology* 52: 577–586.

Jensen, P. 1987. Feeding ecology of free-living aquatic nematodes. *Marine Ecology Progress Series* 35: 187–196.

Krebs, C. 1989. *Ecological Methodology*. New York: Harper & Collins.

Margalef, R. 1958. Information theory in ecology. *General Systems* 3: 36–71.

McAleece, N. 1997. BioDiversityPro: http://www.nrmc.demon.co.ukibdpro/.

Moens, T. and M. Vincx. 1997. Observations on the feeding ecology of estuarine nematodes. *Journal of the Marine Biological Association of the United Kingdom* 77: 211–227.

Ngai J. T. and Srivastava D. S. 2006. Predators accelerate nutrient cycling in a bromeliad ecosystem. *Science* 314: 963.

Petchey, O. L. and Gaston K. J. 2006. Functional diversity: Back to basics and looking forward. *Ecology Letters* 9: 741–758.

Pielou, E. C. 1975. *Ecological diversity*. New York: John Wiley & Sons.

Pusceddu, A., C. Gambi, D. Zeppilli, S. Bianchelli, and R. Danovaro. 2008. Organic matter composition, meiofauna and nematode biodiversity in deep-sea sediments surrounding two seamounts. *Deep-Sea Research Part II* doi: 10.1016/j.dsr2.2008.10.012.

Ricklefs, R. E. and D. Schluter. 1993. *Species diversity in ecological communities: Historical and geographical perspectives*. Illinois, Chicago: University of Chicago Press.

Sanders, H. L. 1968. Marine benthic diversity: A comparative study. *American Naturalist* 102: 243–282.

Shannon, C. E. 1948. A mathematical theory of communication. *Bell System Technical Journal* 27: 379–423 and 623–656.

Simpson, E. H. 1949. Measurement of Diversity. *Nature* 163: 688.

Warwick, R. M. and Clarke, K. R. 1998. Taxonomic distinctness and environmental assessment. *Journal of Applied Ecology* 35(4): 532–543.

Whittaker, R. H. 1960. Vegetation of the Siskiyou mountains: Oregon and California. *Ecological Monographs* 30: 279–338.

Wieser, W. 1953. Die Beziehung zwischen Mundhöhlengestalt, Ernährungsweise und Vorkommen bei freilebenden marinen Nematoden. *Arkiv für Zoolgie* 2(4): 439–484.

Wilson, M. V. and A. Shmida. 1984. Measuring beta diversity with presence and absence data. *Journal of Ecology* 72: 1055–1064.

Section V

Deep-Sea Benthic Functioning

27 Deep-Sea Experiments and Manipulations

CONTENTS

27.1 INTRODUCTION

Experimental design and sample manipulation in the deep sea are extremely difficult, time consuming, and expensive. This is due to several factors, including the ship time and high technology required for carrying on experiments under high pressures and in the open ocean. So far, a very limited number of experiments has been conducted at depths >1000 m, and most of them have been carried out only in the last 30 years following the interest in the hydrothermal vents and cold seeps ecosystems. Some experiments have also been conducted on common soft deep-sea sediments and other habitats, but the experimental replication and overall quality of the information gathered from these studies is very low when compared with experiments carried out in shallow soft-bottom habitats or in the intertidal communities. Given the relevance of deep-sea ecosystems to global biogeochemical cycles, their key role in the overall functioning of the biosphere, and their fundamental contribution to the total global biodiversity, experimental manipulations are clearly needed in the future. It can also be anticipated that future investigations in the deep sea will follow the trend of development pointed out in other research fields, with a progressive increase of the importance of manipulative and experimental research when compared with descriptive studies and with an increase in the level of replication of the experiments, scale of investigation, and complexity of the phenomena investigated.

27.2 *IN SITU* DEEP-SEA MANIPULATIVE EXPERIMENTS

In this chapter we have summarized the available information on different typologies of experiments and manipulations carried out in the deep sea. In particular we have focused our attention on seven major categories of experiments: (1) predator exclusion experiments, (2) *in situ* incubation/manipulation experiments, (3) feeding-preference experiments, (4) *in situ* measurement of metabolism of deep-sea organisms, (5) food fall, carcass deposition, and food input experiments, (6) *in situ* colonization experiments, (7) laboratory experiment on deep-sea organisms.

27.2.1 Predator Exclusion Experiments

To determine whether predators influence the establishment and structure of natural assemblages of benthic invertebrates in deep-sea ecosystems, it is possible to conduct mid- (1-month-long) to long-term (12-month-long) predator exclusion experiments. These experiments are typically conducted using a deep-sea submersible. The effects of predation on abundance, number of taxa, assemblage structure, and production can be quantified comparing caged with one uncaged (control) plots. For exclusion experiments on hard bottoms, the experiments typically rely on the deployment of clean recruitment substrates (basalt cubic blocks, roughly 10 cm on each side) within each of the faunal zones to be investigated. In the case of experiments to be conducted on hard bottoms (e.g., canyons, hydrothermal vents, etc.), blocks with rough, unpolished surfaces similar to the texture of natural basalt rocks at ridge crests can be used. To determine the impact of predation, the recruitment blocks are placed inside or outside cubic mesh cages 20 × 20 × 20 cm, constructed of plastic mesh (6 mm mesh size), reinforced by a frame of 1.1 cm diameter PVC pipes, designed to exclude the larger (>6 mm in smallest dimension) local predators (e.g., fish, crabs, whelks, octopus, etc.). Cages of larger size (e.g., length 70/150 cm, width 70/150 cm, height 20/50 cm) would be preferred, but the complexity of their deployment can make their use extremely difficult. To assess whether cages caused experimental artifacts, for example, by modifying water flows around blocks and/or providing additional structure for animal recruitment, an additional set of blocks (a third block treatment) within cage controls is placed.

Cage controls are identical to full cages except for one missing side. At the end of the experimental period on the seafloor, recruitment blocks are recovered and placed in separate, solid-walled recovery boxes on the submersible. Recovery boxes with lids can be closed after each block was placed in the box to prevent the recruits from escaping or being lost during ascent. All animals on blocks, and those that fell from blocks into the recovery boxes during the ascent to the surface and are subsequently retained by a 1 mm sieve, are counted and identified to the lowest possible taxonomic level. For exclusion experiments on soft sediments, the cages can be made of zinc-iron (length 70 cm, width 70 cm, height 12 cm). When no specific hypothesis on which predator species among those visually encountered *in situ* could be responsible for possible differences of benthic assemblages between caged and uncaged plots, a typical cage (with top and lateral sides covered by a plastic net of 1 mm mesh to allow the immigration and emigration of meiofauna) can be utilized to exclude larger epibenthic predators. Before cage deployment, the substrates are accurately checked to make sure that large epibenthic predators are not included. The lateral panels of the cages are inserted into soft sediments penetrating by ca. 20 cm and fixed into the substratum. Considering that cages can alter the hydrodynamic regime at the sediment–water interface, modifying sediment transport with possible consequences on benthic fauna, four vertical metal panels were inserted into the sediment (down to ca. 20 cm depth) at each side of the cage. These panels can be successful in avoiding sediment alteration (e.g., by erosion) around and within the cage.

The experimental design in deep-sea systems are typically characterized by a low number of replicated, that is, with only one cage or a limited number of cages versus one uncaged plots. This limits data interpretation as differences among treatments are difficult to discriminate in the absence of a rigorous quantification of spatial variability of prey and predators. A second aspect is the lack of procedural controls, with the possibility that the effects of predator exclusion could be partially confounded with caging effects (e.g., reduced water movement, presence of the structure). As a consequence, results must be discussed with caution, in the light of a suite of factors that could have influenced the outcome of the experiment. Sampling of the benthic fauna can be carried out by means of a submersible using either push corers or other devices suitable for collecting organisms of larger sizes (e.g., arms, meshes, scoops, etc.). At each sampling date, samples are collected in 3–5 replicates in caged and uncaged plots. The same extraction and preservation procedures are generally used for soft- and hard-bottom samples. After the completion of the experiment, all animals are counted and classified under a stereomicroscope.

27.2.2 *In Situ* Enrichment Experiments

Enrichment experiments can be carried out using three identical benthic chamber landers, each equipped with three chambers (0.2 × 0.2 m). Each chamber of one lander carried two O_2 optodes that continuously record the O_2 concentration in the chamber water. To simulate a sedimentation event, a suspension of 0.2 g freeze-dried ^{13}C-labeled phytodetritus (e.g., *Thalassiosira rotula*; 98 ± 1% ^{13}C; C:N = 13), equivalent to 1 g C org, has been injected into each chamber (Witte and Pfannkuche 2000). During the incubations, a pump system exchanged part of the water in the chamber. From each chamber, seven water samples (50 mL) were taken during the incubation, one before and after each pumping cycle. Sediment community oxygen consumption (SCOC) can be determined by Winkler titration of syringe water samples and by optode readings and comparing these values with data from the continuous O_2 record. Short (1–7 d), medium (7–14 d), and long (>24 d) incubations can be carried out. In each lander, one chamber serves as control without particulate organic matter (POM) enrichment. Additional sediment cores are taken for the determination of background ^{13}C signatures of the organisms. After lander recovery, the biotic components are taken, sorted, identified, and the relative variables measured.

27.2.3 Feeding-Preference Experiments

Feeding-preference experiments are conducted by deploying an array of common invertebrates and/or vertebrate remains onto the seafloor of the investigated deep-sea area. The identities, activities, and locations of all consumers attracted to the array are then recorded using an automated time-lapse still-camera system. A large array of invertebrates potentially eaten by local predators can be constructed by securing dead individuals of the deep-sea assemblage as well as other organisms potentially reaching the seafloor by falling through the water column. A control lacking toxic constituents is attached to an 85 × 71 cm piece of plastic mesh, reinforced by a flexible frame constructed with 5 cm wide PVC strips. Organisms are immediately frozen upon recovery from the submersible and later attached to the frame just prior to its deployment in the field. To control for a possible effect of food availability on consumer choice, an equal number of each food type is offered. Before deployment, the food array, with food already attached, is rolled up and placed in a plastic cylinder fitted to the still-camera system. On the seafloor, the submersible maneuvers the camera system into position on a section of the horizontal surface, removes the food array from the sheath on the camera system, unrolls it, and places it beneath the camera within its field of view. The camera system takes pictures of the positions of consumers on the array at 15 s/1 min intervals for a total duration of up to 24 h. To standardize results across the experimental trials, only selected images are used. The images produced (generally >200) during each trial are analyzed by recording the identity and location of all consumers on the food array. Because invertebrates offered to consumers in the feeding preference experiment are dead, this assay tests for consumer scavenging preferences rather than predatory choices.

27.2.4 *In Situ* Metabolism of Large Deep-Sea Organisms

The *in situ* metabolism of deep-sea organisms can be investigated by means of incubations in controlled condition. The Benthic Incubation Chamber System 2 (BICS2) is a respirometer system used to measure the respiration rate of deep-sea megafauna. BICS2 is comprised of two watertight respirometry chambers (~14 L each) housed within an external protective frame. One oxygen probe is present inside both chambers and measures the oxygen concentration and temperature of the enclosed water. The oxygen optodes are rated to work up to 6000 m. The O_2 measurements are recorded in text file format. The chambers' frame is composed of glass-reinforced plastic with dimensions 1000 mm × 580 mm × 642 mm. The system is deployed by a remotely operated vehicle (ROV) elevator and remains attached to the elevator throughout the deployment. When at the seafloor, the ROV searches for suitable animals for use in BICS2, which are picked up by collection scoop and deposited into the chambers. The lids are closed by the ROV turning the t-handles in an anticlockwise direction. The chamber is activated when the push switch is pressed and is confirmed as activated by an LED flashing.

27.2.5 Food Fall, Carcass Deposition, and Food Input Experiments

Whale falls, first recognized in the 1980s, are whale carcasses that fall to the deep-ocean floor where, like an oasis in the desert, they attract a specialized group of clams, crabs, and worms that feed for up to decades on the oil-rich bones and tissues. Some scientists think these random, deep-ocean oases are stepping stones for organisms moving from one ocean floor environment to another—whether a hot vent, a cold seep, or a whale carcass—in search of sustenance from energy-rich chemicals (Smith and Baco 2003).

The short-term fate of cetacean carcasses in the deep sea can be investigated using autonomous deep-sea lander vehicles incorporating time-lapse camera systems, and fish and amphipod traps. Lander deployments can place cetacean carcasses at depths up to 4800 m for periods of 36–276 h before being recovered. Photographic sequences are used to reveal that carcasses are rapidly consumed by fish and invertebrate scavengers with removal rates ranging from 0.05 to 0.4 kg h^{-1}. In the longest experiment, the carcass was skeletonized within 5 d. In each deployment, approximately 1 h after emplacement, the grenadier *Coryphaenoides* (*Nematonurus*) *armatus* and large numbers of lysianassid amphipods arrive at the food fall. The initially high numbers of grenadiers decline once the majority of the bait has been consumed and a variety of other fish and invertebrates are observed, some taking up residence at the site. Funnel traps indicate a succession in the species composition of amphipods, with the specialist necrophages such as *Paralicella* spp. being replaced by more generalist feeders of the *Orchomene* species complex. In addition, the skeletons and bones of the cetaceans can be subsequently colonized by other organisms and new species. The distribution of the experiments is, so far, highly localized, and there is a clear need for an implementation of these experiments in different areas of the world (Smith and Baco 2003).

27.2.6 *In Situ* Colonization Experiments

Most of the deep-sea floor is highly oligotrophic and therefore nutrient limited. Exceptions are vent and seep ecosystems fueled by focused sources of chemical energy such as methane and sulfide, which attract high biomasses of chemosynthetic organisms. Wood and kelp falls can also lead to the development of chemosynthetic environments in the deep sea that might act as stepping stones in the evolution and distribution of chemoautotrophic communities. Sunken wood (Figure 27.1) provides a concentrated, locally restricted input of organic material to the deep sea, but so far it has not been well understood how and when its degradation could lead to the production of high sulfide concentrations attracting chemosynthetic organisms. Wood colonization experiments are performed by deploying wood on the seafloor and waiting for their colonization. These experiments simulate the

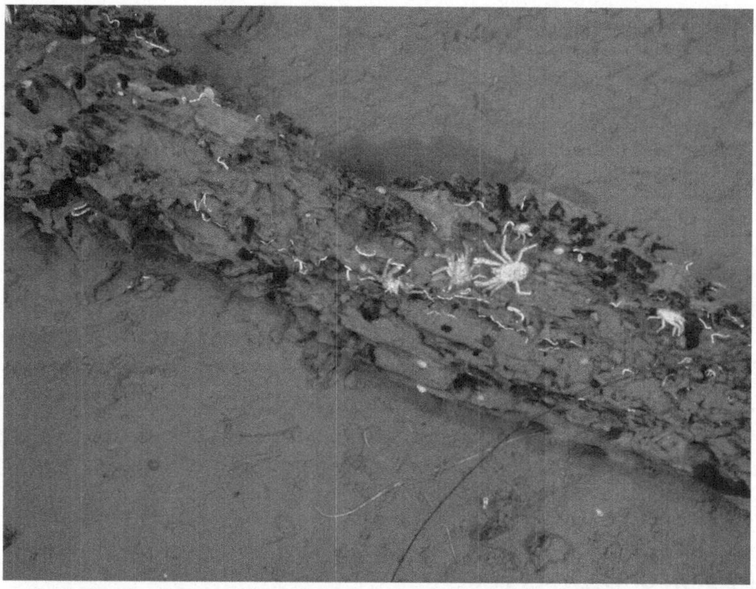

FIGURE 27.1 Sunken woods provide unique opportunities for colonization of deep-sea organisms. (Image courtesy of the Monterey Bay Aquarium Research Institute. © 2009 MBARI. Used with permission.)

fall of large wood in deep-sea ecosystems. Most specialized herbivores of the deep sea can bore into wood and other plant remains, and among them are the species *Xylophaga* and *Xyloredo*. There is typically a dramatic response to wood deposition as colonizing organisms show typical opportunistic behaviors. When the sunken wood is retrieved on board, macrofaunal, meiofaunal, and microbial colonizations of wood are quantified by means of classical and molecular techniques.

27.3 LABORATORY EXPERIMENTS ON DEEP-SEA ORGANISMS

Pressurized aquaria can be constructed with stainless steel. On the basis of effective experiments, such a kind of aquaria is equipped with three windows having outside diameter of 8 cm and inside diameter of 11 cm and a 5-cm-thick wedge-shaped acrylic resin. The side window lids have 6 cm outside and 9 cm inside diameters and a 5-cm-thick acrylic resin wedge. The windows are glued with a silicon bonding agent to the aquarium. The aquarium is 36 cm in diameter, is formed from a 1-cm-thick stainless steel ball, weighs 45 kg, and holds a volume of 20 L.

The deep-sea organisms can be collected by a specially designed suction servomotor through aquarium openings 3 cm (lower) and 6 cm (upper) in diameter. The maximum suction power is approximately 250 L min^{-1}. After capture of an organism, the openings in the inner lids are closed by a spring mechanism that removes the pins. The calculated minimum resistance pressure limit of the aquarium is 30 MPa, and the inner seawater can escape through the pressure control valve when the inner pressure increases to more than 20 MPa. Therefore, the aquarium system maintains an inner pressure of 20 MPa (corresponding to the depth of ca. 2000 m) even when the submersible surfaces from the deepest sea bottom. A gas-filled balloon is placed in the aquarium to reduce the pressure loss arising from metal expansion. Insulation is placed around the aquarium to protect the fish from the warm surface seawater (about 18°C).

Delivery of seawater to the aquarium is achieved through a high-pressure pump. To enhance seawater supply, the pressure aquarium system is designed to allow incremental increases in the high-pressure pumps. Compressed seawater supply can be controlled at a rate of up to 40 mL min^{-1}. A pressure controller, consisting of an electrically powered motor valve, modulates the draining rate and decompression speed through a pressure-sensor monitor. When the pressure value and amount of compressed seawater supply are set up, a feedback circuit of the pressure controller and pressure sensor adjusts the amount of seawater drainage and maintains the pressure in the aquarium (autotuning). After autotuning, pressure is maintained at 17±0.1 0.2 MPa during a 2-month period when the compressed seawater supply is set at the rate of 40 mL min^{-1}. To avoid fluctuation in pressure due to autotuning, compression and decompression speeds can also be controlled up to approximately 1.0 MPa h^{-1} under manual control. Narrow stainless steel tubing with an inside diameter of 2 mm and outside diameter of 3 mm provides oxygen to, and removes waste products from, the seawater surrounding the deep-sea multicellular organisms. Thick 316 stainless steel tubing with an inside diameter of 10 mm and outside diameter of 12 mm is used to circulate the seawater. Circulation of seawater is controlled up to a rate of 2 mL min^{-1}, and large waste particles such as uneaten bait worms are removed by the filter in the feed box. Small waste products from deep-sea fish are eliminated with the exchange of seawater. The temperature inside the pressure aquarium is monitored and controlled by an electrode sensor and cooler that cools to 5°C. Dissolved oxygen is monitored for both injection and drainage seawater with a portable dissolved oxygen sensor. In oxygen consumption experiments, coral gravel on the bottom net in the aquarium and in the inner filter of the feed box is removed.

27.3.1 Collection of Deep-Sea Fishes

A specimen of the deep-sea fish *Zoarcidae* sp. (Bond 1996; Figure 27.2) has been collected with the pressure aquarium system at depths >1100 m (*Shinkai 2000*, 35°00.07′N, 139°13.50′E) and has been maintained at 10 MPa and 5°C. Salinity at the collection site was 34.3, and the specimen was placed in 10 L of salinity 34.3 artificial seawater, and 20 L of the collected deep-sea water with the fish in the aquarium.

A total of 21 specimens of the deep-sea fish *Ebinania brephocephala* (13–15 cm in total length; Masuda et al. 1984) were collected at depths of 500 m in the trawl net of a fishing boat in Suruga Bay, Japan (138°28′E, 34°45′N). One specimen was transferred through the side window lid into the pressure aquarium filled with 40 gL^{-1} and 4°C fresh artificial seawater. All conditions (salinity, temperature, food, oxygen concentration, removal of waste products) were the same as in the pressure-stat aquarium except hydrostatic pressure. Despite the release of gases in the intestines of *Ebinania brephocephala* with a syringe, 20 of the specimens only remained alive for a few days at atmospheric pressure. Therefore, only one of the specimens was kept in the pressure-stat aquarium system.

After these procedures, fresh artificial seawater was circulated, compressed to 5 MPa at the rate of 0.5 MPa min^{-1}, and exchanged at the rate of 30 mL min^{-1} by the pressure aquarium system. Fresh bait worms, purchased from a fishing-tackle store, were fed to the deep-sea fish 2–3 times weekly.

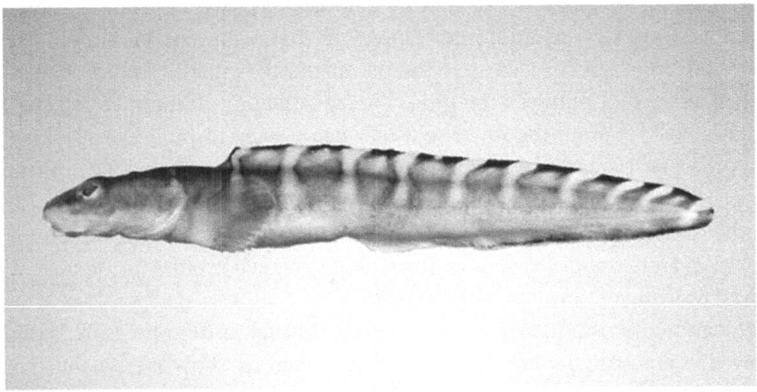

FIGURE 27.2 An image of deep-sea fish *Zoarcidae* sp. (Photo by Tashiyuki Suzuki © Copyright Kanagawa Prefectural Museum of Natural History. Used with permission.)

The potential to acclimate deep-sea fish to atmospheric pressure is particularly exciting. If successful, deep-sea fish could be not only better investigated for their feeding habits and metabolism, but also displayed in the aquarium's exhibit. In the late 1990s, more than 60 different species were displayed in the Monterey Aquarium, but none of them came from water deeper than 1000 m. Trials have been done with some species relatively common in deep-sea habitats that have the potential to survive in aquaria (Figure 27.3).

Fish such as the rattail also named the "real ambassador" of the deep sea, can survive at depths up to 6000 m (Figure 27.4) and have large heads and mouths with long, tapering tails. Some species weigh up to 8 kg and can grow up to 1.8 m long. They are very sensitive to pressure, and the development of new pressurized traps might be able to slowly acclimate these specimens to one atmosphere pressure.

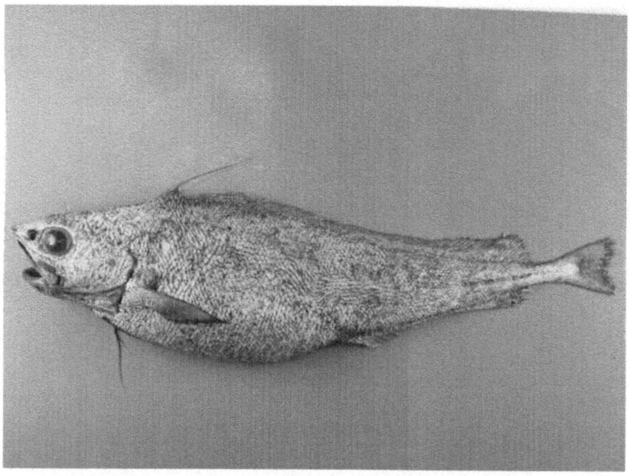

(a)

(b)

FIGURE 27.3 Images of common deep-sea fish species on which there are attempts to collect and maintain living specimens in aquaria. Reported are (a) *Antimora rostrata*, (b) *Bathypterois dubius*, (c) *Halosauropsis macrochir*, (d) *Lepidion eques*, and (e) *Nezumia aequalis*. (Photos by Alexei Orlov, VNIRO-Russia. Used with permission.)

Continued

(c)

(d)

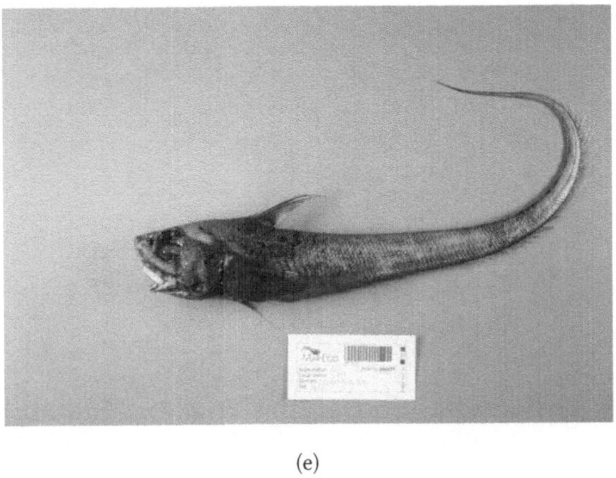

(e)

FIGURE 27.3 *Continued.*

(a) (b)

FIGURE 27.4 Image of the common deep-sea fish *Coryphaenoides* sp. shown are an image captured by ROV (a) and a specimen curious for the sampling activities conducted in the deep-sea sediments of the Portuguese Margin (b). (Courtesy of Doug Masson, NOCS, Southampton. Used with permission.)

27.3.2 Collection of Deep-Sea Invertebrates

27.3.2.1 Predatory Tunicate (*Megalodicopia hians*)

So far, a very limited number of deep-sea invertebrates have been maintained alive in aquaria. The predatory tunicate (*Megalodicopia hians;* Figure 27.5) has been found at depths of 200–5325 m. Within this depth range, dissolved oxygen levels can be 0.5–2.5 mL L^{-1}. It feeds mainly on copepods, ostracods, amphipods, isopods, and mysids (Monniot and Monniot 1978). Deep-sea predatory tunicates have been collected using an ROV equipped with a robotic arm, which collects the tunicate by breaking the substrate, with the animal attached. On the deck of the ship, the tunicates

FIGURE 27.5 Deep-sea predatory tunicates. (© Copyright Monterey Bay Aquarium Foundation. Used with permission.)

are quickly transferred to fresh 6°C seawater that has had nitrogen bubbled through it to strip out the oxygen to 1.5 mL/L. The animals are transported within 5 h from the ship to the aquarium in temperature-controlled conditions. Predatory tunicates survived 6 to 8 months.

Tunicates are fed three times per week with *Artemia,* enriched with an amino acid complex. Food may take up to 5 d to pass through the digestive tract, and then small flake-like fecal pellets are excreted from the siphon. Uneaten food can be dislodged with a gentle current or removed using a small-volume eyedropper. In summary, proper collection with the ROV, low oxygen conditions, specialized feeding techniques, and delicate handling are the key elements in displaying these unique deep-sea tunicates.

27.3.2.2 Mushroom Soft Coral

The mushroom soft coral is a large, spectacular, temperate deep-sea alcyonacean living on hard substrates down to >2000 m depth (Figure 27.6). There are 22 species of *Anthomastus* distributed worldwide. Little information exists on soft corals from temperate regions and deep waters. Mushroom soft coral colonies are composed of a proximal stalk attached to the substratum and a distal capitulum containing zooids. The capitulum may be red or white in color and may grow to a diameter of 20 cm. The autozoids are feeding polyps, each consisting of an anthocodium (8 pinnate tentacles and the pharynx) protruding from the surface of the capitulum and the gastrovascular cavity extending into the stalk tissue. Tentacles harboring sticky nematocysts opportunistically trap and grab food items from the prevailing currents. Siphonozooids contain the reproductive structures and are responsible for controlling water flow within the colony. Stalk tissue is sterile, containing rod-like spicules throughout and double capstan-type spicules at the base. Stalk length is variable and can reach up to 30 cm. Mushroom corals internally brood and release planula larvae; they have internal fertilization, and the sexes are separate (Cordes 2001). Female gonads are yellowish in color, and the male gonads (spermaries) are white. Details regarding the gonadal development are described in Cordes (2001).

Mushroom soft corals are collected using a remotely operated vehicle. Collecting sites are selected based on rock substrate of suitable size and shape for the robotic arm. Once on deck, the

FIGURE 27.6 Deep-sea mushroom soft coral. (Courtesy of Government of Canada. Used with permission.)

mushroom corals attached to rock are removed from the drawer and placed in ice chests containing chilled seawater and transported to the aquarium. Mushroom corals from the Pacific Ocean have been maintained more than 5 years in captivity. Water temperature is held at a constant 6°C. Mushroom corals are tolerant to different oxygen levels. Water motion is important for maintaining the corals and stimulating polyp deployment.

Colonies are fed frozen small krill *(Euphausia pacifica)* and adult brine shrimp three to four times per week. Small colonies are fed diced small krill, and large colonies are given whole small krill. Krill and/or enriched brines are suspended in a container with chilled seawater and fed to individual polyps using a turkey baster. An interesting and convenient feature of the mushroom coral is the transparency of each polyp, which gives an opportunity to observe ingestion of food into the colony. Lacerations on the stalk, particularly at the base, often result in complete detachment from the rock. Ciliates are the most common pathogen associated with this problem. The consumption of tissue by ciliates is rapid and difficult to control.

Large planula larvae (3.3 mm long and 0.5 mm wide) are released through the oral opening of the siphonozoids at the capitulurn surface. Planulae crawl by ciliary action or are passively transported until settlement. The larvae settle within 2 to 3 d, but nonsettled planulae can survive up to 4 months. Settlement and metamorphosis appear to be induced by the presence of an organic film on the substratum. Larvae enter the Edwardsia stage with the formation of 8 primary mesenteries approximately 2 d after settlement. After 2–3 d, larvae metamorphose, forming 8 tentacle buds and permanent adhesion to the substrate. Tentacles become pinnate within a week, marking the onset of feeding. Additional autozoids form 6–8 weeks after feeding begins but can be slow and variable. Capitulum development and size increases at 8 to 10 months, as does stalk diameter. Colony growth is initially slow increasing during intermediate stages and reaching an asymptote at approximately 25–30 years (Cordes 2001). Mushroom corals can survive for more than 5 years. They are characterized by slow growth and long life span.

27.3.2.3 Pom-Pom Anemone (*Liponema brevicornis*)

The pom-pom anemones are recorded at depths of 100–1000 m and lives unattached on sand, mud, and gravel bottoms, and occasionally on rocks (Figure 27.7). The firm pedal disc is circular in outline, slightly wider than the column, somewhat convex, and nonadhesive. The shape of this

FIGURE 27.7 Deep-sea "pom-pom" anemone (*Liponema brevicornis*). (Courtesy of John Butler, NMFS-NOAA. Used with permission.)

anemone is extremely plastic. They can form 20-cm-high domes or flatten into 5-cm-high discs. When distressed, they can reduce up to 500% in volume. In order to move, this anemone draws itself into a barrel or tumbleweed shape and is swept about by benthic currents.

Deep-sea anemones are collected using an ROV, exercising care in handling as it can easily cause loss of tentacles or damage to the organism. The ultra sticky tentacles detach easily upon contact with most surfaces. Moreover, anemones are only slightly negatively buoyant, so currents readily move them. Gloves are recommended when handling the anemones to prevent tentacle detachment and potential stinging by nematocysts. Pom-pom anemones have been maintained up to 3 years in captivity at 6.0°C on a fine-grained sandy substrate. A gentle-to-medium current is important to keep the anemones expanded and aids in the removal of ejected fecal material. As with other deep-sea cnidarians, these anemones are light sensitive and can tolerate only low levels of light. Small krill is directed at the oral surface of the pom-pom. After several days of digestion, the anemone literally evacuates all the fluid from the body cavity, including the tentacles, and purges the gastrovascular cavity. A gentle current directed over the oral surface removes uneaten or regurgitated food and organic material. These anemones rarely recover from massive loss of tentacles. Spawning (eggs are orange, diameter ca. 1 mm) has been observed on several occasions, but no further development was observed.

27.3.2.4 Deep-Water Corals (*Lophelia pertusa* L. 1758)

Lophelia is a stony coral (Scleractinia) belonging to the family Caryophylliidae. It is distributed throughout the world oceans except in the polar regions (Zibrowius 1980; Cairns, 1994). The preferred temperature range seems to be 6–8°C (Frederiksen et al. 1992; Freiwald 1998), but it has also been found in the Mediterranean Sea at temperatures >14°C. The bathymetric distribution of *Lophelia pertusa* is typically from 200 to 1000 m (Zibrowius 1980; Freiwald 1998). The shallowest record of a living *Lophelia* reef is at 39 m in Norway (Rapp and Sneli 1999), while the deepest records extend down to 3000 m in the Atlantic (Squires 1959). *Lophelia pertusa* and *Madrepora oculata* form colonies (*Madrepora* is generally less abundant than *Lophelia* and does not build reefs; Dons 1944; Frederiksen et al. 1992; Figure 27.8).

Lophelia has been known for centuries to scientists and fishermen, especially those using passive gear such as gillnets and long-lines in deep water. These species can be collected easily and maintained in aquaria.

Lophelia is a passive suspension feeder or passive carnivore that has been observed to take zooplankton such as calanoid copepods and cumaceans in natural conditions, and live zooplankton such as chaetognaths and small crustaceans (ca. 1 mm e.g., copepods) and larger species such as krill (ca. 2 cm in length) in aquaria (Mortensen 2001; Mortensen et al. 2001). Mortensen (2001) demonstrated that *Lophelia* could also take a variety of foods, including dead food particles of

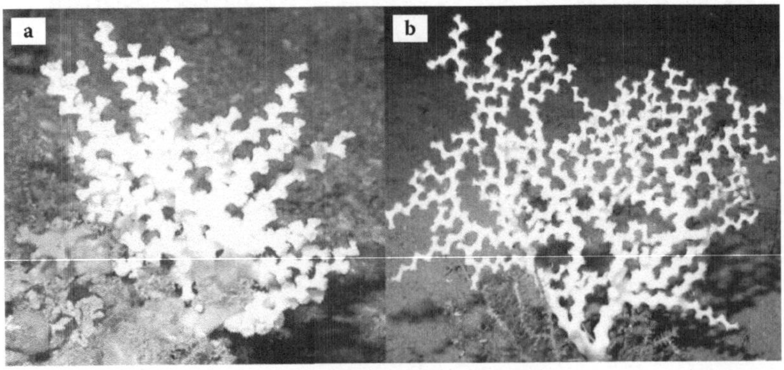

FIGURE 27.8 *Lophelia pertusa* (a) and *Madrepora oculata* (zig-zag coral, b)

FIGURE 27.9 *In situ* manipulation activities on a deep coral colony. (Courtesy of Karine Olu. Used with permission.)

krill, shrimp, herring, and squid, and was able to reject unsuitable material, including sediment. Mortensen (2001) concluded that *Lophelia* could use small organic particulates as food, as well as live zooplankton. It probably also feeds on small invertebrates crawling over the coral surface. Once collected using an ROV (Figure 27.9) or accidentally by means of trawls, the samples/branches are carefully transferred into large buckets filled with fresh and clean seawater (Figure 27.10). After a period of acclimatization the living *Lophelia* fragments and their associated fauna can be transferred from the buckets into large transportation tanks prepared in the wet laboratory.

Large tanks (750 L) are equipped with a powerful dive pump for strong water current, an airstone supported with fresh air through an air pump, and a fiberglass lattice on the bottom for sample fixation. The water in the transportation tanks is fresh seawater from the previously filled aquaria tanks. Due to biological processes and warming up, the water is renewed at regular intervals. In aquaria and under controlled conditions, deep-water corals can survive for several months.

FIGURE 27.10 Transportation of deep corals in tanks within the wet laboratory.

(a)

(b)

FIGURE 27.11 Images of specimens of deep-sea megafauna from the Portuguese Margin. Reported are a starfish (Brisingid) and crinoids (a) and pycnogonid (b). (Courtesy of Doug Masson, NOCS, Southampton. Used with permission.)

27.3.2.5 Other Invertebrates

Occasionally, crabs, pycnogonids, and other invertebrates (Figure 27.11) have been collected and tentatively kept in aquaria. Their average survival is from hours to weeks, depending on the sampling depth, modality of collection, and feeding.

27.4 COLLECTION AND EXPERIMENTS WITH UNICELLULAR AND MICROBIAL ORGANISMS

The collection of deep-sea sediment samples and their transfer in the laboratory in pressurized conditions is the prerequisite for performing experiments on deep-sea microbial organisms, especially psychropiezophiles. These samples can be transferred into pressurized aquaria without

FIGURE 27.12 An image of the deployment of the deep-sea laboratory GEOSTAR used for sampling and *in situ* experiments within the frame of the BIODEEP program. (Courtesy of Cesare Corselli—University of Bicocca, Milan, and the Biodeep team. Used with permission.)

decompression using systems recently developed for the transfer and manipulation of deep-sea microbiota (Figure 27.12). The behavior of deep-sea benthic foraminifera in response to seasonal inputs of food material can be also examined through feeding experiments. The experiments using deep-sea sediments containing living foraminifera are carried out in microcosms or aquaria with thin-walled glass sides. Three different nutrient levels are tested in order to observe how benthic foraminifera reacted to deposited organic matter. Observations of both behavior and microhabitat segregation with respect to the sediment-water interface were conducted using a sideways-mounted binocular microscope.

Benthic foraminifera are vertically distributed according to three types of microhabitat segregation patterns, comparable to those of natural populations in the natural environment: surface infaunal, intermediate infaunal, and deep infaunal. After addition of food into the aquaria, many foraminifera migrate upwards, and some of them ingest the food. Results from the Sagami Bay (1425 m water depth) indicate that surface and intermediate infaunal species react faster to newly deposited food materials than deep infaunal species. Among deep infaunal species, *Globobulimina affinis* react very slowly, whereas *Chilostomella ovoidea* did not respond at all. Reactions to newly deposited organic matter observed in this experiment vary from one microhabitat to another. These different food preferences are an important factor for understanding foraminiferal microhabitats and the degradation of phytodetritus.

The size distribution of foraminiferal tests after the experiment was completed suggests that some species reproduced during the experimental run, although there was no significant difference in total numbers of foraminifera between fed and unfed aquaria. These results indicate that some benthic foraminifera may reproduce in response to organic matter arriving before the spring bloom. Similar experiments can be performed on protozoa, and the measured variable and equipment used are described in other chapters of this book.

27.5 ICHTHYOFAUNAL ASSESSMENT USING BAITED CAMERA LANDERS IN THE ROBUST BIODIVERSITY (ROBIO) LANDER

The RoBio lander is a free-fall lander equipped with a digital still camera (Kongsberg Maritime, OE14-208), flash unit (Kongsberg Maritime, OE11-242), Aquadop current meter, rechargeable battery pack, and twin acoustic ballast release system (MORS AR and RT). The current meter is programmed to measure depth, temperature, and current velocity and direction at 30 s intervals throughout the deployment. The camera is programmed to take digital photographs at 60 s intervals, with an average of 700 photos per deployment. Dominant species attending bait are highly specific, depending on the depth. The abyssal grenadier, *Coryphaenoides (Nematonurus) armatus*, and the deepwater arrow tooth eel, *Histiobranchus bathybius* (Günther, 1877), are the most common species at depths > 3000 m. Invertebrates can also be observed at shallower depths, such as decapod shrimps, asteroidea, gastropoda, amphipoda, and paguridae. Invertebrates are less common at the deeper sites; however, specimens of decapod shrimps, ophiuroidea, amphipoda, and scyphozoa can be observed.

REFERENCES

Bond, C. E. 1996. *Biology of fishes*. Tokyo: Saunders College Publishing.

Cairns, S. D. 1994. Scleractinia of the temperate north Pacific. *Smithsonian Contributors to Zoology* 557: 1–150.

Cordes, E. E., J. W. Nybakken, and G. Van Dykhuizen. 2001. Reproduction and growth of *Anthomastus ritteri* (Octocorallia: Alcyonacea) from Monterey Bay, California, USA. *Marine Biology* 138: 491–501.

Freiwald, A. 1998. Geobiology of *Lophelia pertusa* (Scleractinia) reefs in the north Atlantic. Habilitation thesis, University of Bremen.

Gooday, A. J. and Rathburn A. E. 1999. Temporal variability in living deep-sea benthic foraminifera: A review. *Earth-Science Reviews* 46: 187–212.

Masuda, H., K. Amaoka, C. Araga et al. 1984. *The fishes of the Japanese archipelago*. Tokyo: Tokai University Press.

Micheli, F., C. H. Peterson, L. S. Mullineaux et al. 2002. Predation structures communities at deep-sea hydrothermal vents. *Ecological Monographs* 72: 365–382.

Monniot, C. and F. Monniot 1978. Recent work on the deep-sea tunicates. *Oceanography and Marine Biology: An Annual Review* 16: 181–228.

Mortensen, P. B. 2001. Aquarium observation on the deep water coral *Lophelia pertusa* (L., 1758) (Scleractinaria) and selected associated invertebrates. *Ophelia* 54: 83–104.

Mortensen, P. B., M. Hovland, J. H. Fosså, and D. M. Furevik. 2001. Distribution, abundance and size of *Lophelia pertusa* coral reefs in mid-Norway in relation to seabed characteristics. *Journal of the Marine Biological Association of the United Kingdom* 781: 581–597.

Smith, C. R. and A. R. Baco. 2003. Ecology of whale falls at the deep-sea floor. *Oceanography and Marine Biology: An Annual Review* 41: 311–354.

Turley, C. M., A. J. Gooday, and J. C. Green. 1993. Maintenance of abyssal benthic foraminifera under high pressure and low temperature: Some preliminary results. *Deep-Sea Research Part I* 40: 643–652.

Witte, U. and O. Pfannkuche. 2000. High rates of benthic carbon remineralisation in the abyssal Arabian Sea. *Deep-Sea Research Part II* 47: 2785–2804.

28 Effect of Pressure on Enzymatic Activities, Viral Production, and Prokaryotic Heterotrophic Production

CONTENTS

28.1 INTRODUCTION

With an average depth of approximately 3850 m of the deep-sea biosphere, the organisms living in the benthic domain are subjected to an average pressure of ca. 390 atmospheres. It is therefore important to test the effect of pressure on some specific variables. Microbial communities in the deep-sea sediments may include both autochthonous species adapted to *in situ* temperature and pressure, and allochthonous taxa attached to settling particles. The activity of allochthonous microbes can also change with depth (Turley et al. 1995; Tamburini et al. 2006) as a result of high-pressure conditions. Thus, microbial activity measured at both atmospheric and *in situ* deep-sea conditions may reflect different components of the microbial physiological status (Eardly et al. 2001; Tamburini 2006; Deming 2007).

Research into the effects of high-pressure conditions on the physiology of deep-sea bacteria have been developed over the last century (see ZoBell 1970). The first assays to measure bacterial activity without decompression were performed by Seki and Robinson (1969). They compared glucose uptake (250 μg L^{-1} final concentration) measured on samples decompressed during retrieval and then incubated at atmospheric pressure, or recompressed in the laboratory. Additional samples were incubated *in situ* for 4 h. They found that samples collected at 200 m displayed similar activity, but decompression due to sample recovery from 400 m depth decreased the measured activities up to 60%. At that time, however, the effects of pressure on microbial activity were investigated by means of a double-pressure shock, the first due to sample decompression during retrieval, and the second due to sample recompression of the *in situ* conditions. This double stress was likely to affect the measurements of microbial activities. The need for developing new technologies to collect deep-sea samples avoiding decompression led to the creation of new systems capable of recovering deep-sea samples without decompression and allowing their manipulation on board without changing the pressure conditions.

Jannasch and coworkers performed a series of incubations at the water-sediment interface at 1830 m depth using the manned submersible Alvin (Jannasch and Wirsen 1973). Sterile incubation bottles containing nutrient solution were deposited on the seafloor and filled with bottom seawater. Incubation was *in situ* until recovery by way of the submersible, that is, over a 51-week period. Added substrates were starch, agar, or gelatine at 1g L^{-1}. Other experiments were made with ^{14}C-labeled mannitol, acetate, glutamate, or other amino acids at 30, 10, 5, and 2 μg L^{-1} (final concentrations), and these were incubated *in situ* for 14 weeks. Samples were also decompressed during the filling of the sampling bottles (Jannasch and Taylor 1984). These experimental artifacts possibly explain the low response of deep-sea microbial populations under ambient pressure conditions relative to their decompressed controls.

Jannasch et al. (1973) designed a pressure-retaining sampler/incubation vessel to compare microbial activity at *in situ* pressure and in decompressed deep-sea samples (Jannasch et al. 1976). Jannasch and Wirsen (1977) designed a gear to retrieve deep water at ambient pressure. Jannasch and Wirsen (1982) retrieved water samples from depths of between 1800 and 6000 m in the Atlantic Ocean, reporting a single "barophilic" response in samples collected from a depth of 4500 m. In all the other assays, the rates of ^{14}C incorporation and respiration measured were lower at *in situ* pressure than in the decompressed samples. These observations led the authors to conclude that high pressures had a negative effect on microbial activities. Deming et al. (1980) measured the microbial incorporation of labeled substrates in the Puerto Rico trench at depths from 3450 to 7730 m and found that the percentage of substrate utilized was surprisingly low relative to previous data. Tabor et al. (1981) found the rates at *in situ* pressure were 38 times higher than those measured in decompressed samples in the deep Atlantic Ocean.

Alongi (1990) conducted similar studies on benthic bacteria by comparing decompressed samples with decompressed and recompressed (to *in situ* pressure conditions) sediment samples. Using a high-pressure sampler and a pressure-retaining transfer bench, contrasting responses of bacteria to pressure were observed depending on water-column stratification. Bianchi and Garcin (1993, 1994) measured rates at *in situ* and in decompressed samples during stratified water-column conditions, and found that bacteria exhibited barophilic behavior with an activity nearly double that found in decompressed conditions, likely due to the bacteria having adapted to the deep-sea pressure. However, during water-column mixing, bacteria responded differently, as decompressed samples displayed a significantly (ca. 10 times) higher value than at *in situ* pressure. These results indicate that bacterial activities can vary depending on the water-column stratification.

Tholosan et al. (1999) designed a 5-L high-pressure sampler to incubate water samples at *in situ* pressure and decompressed subsample withdrawn from the high-pressure sampler. Over 39 incubations, 34 showed a decrease of potential metabolic activity when the samples were analyzed in decompressed conditions; one sample showed a negative effect of *in situ* pressure, and four samples were unaffected by pressure. Bianchi et al. (1999) also developed a high-pressure serial sampler (HPSS) based on a commercially available multisampling device (Figure 28.1) and found that bacterial heterotrophic production on decompressed samples was underestimated. The effects of sample decompression might be relevant also on some functional variables (Bartlett 2002; Bartlett et al. 2007; Gardner et al. 1983; Kato et al. 2008; Yayanos 1995). For instance, it is know that ectoenzymes and their activities play a key role in organic matter degradation (Chrost 1991). Poremba (1994, 1995) observed that microbial degradation of phytodetritus in deep-sea sediments was positively influenced by elevated pressure, and Tamburini et al. (2002) used the HPSS to measure aminopeptidase and phosphatase activities, and bacterial production rates, on decompressed and at *in situ* pressure, and found that sample decompression caused a slowdown of hydrolytic activity by ca. 50%. The effect of pressure appears evident at 1000 m depth in the water column but only at more than 4000 m depth in the benthic compartment.

Overall, it can be concluded that microbial activity measurements published so far from deep-water samples have been possibly underestimated. A review by Bianchi and collaborators based on 77 measurements indicates that, on average, the pressure effect is 3.2 ± 4.2 (± SE) with a median value of 2.2, but the poor statistical significance of these calculations make the use of a correction factor difficult for measurements reconducted on decompressed samples.

Here we provide the details to test the effects of pressure on different benthic microbial variables (i.e., viral production, prokaryotic heterotrophic production, and enzymatic activities). Samples collected at the benthic boundary layer are incubated at *in situ* pressure and temperature conditions.

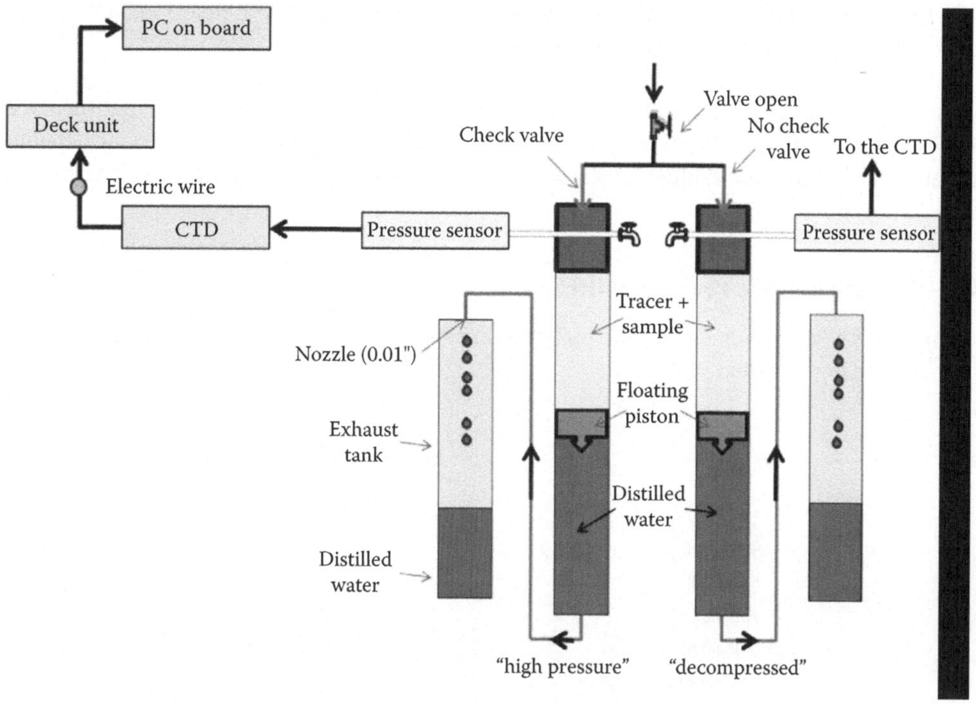

FIGURE 28.1 Schematic representation of the high-pressure bottles (HPBs). (Modified from Tamburini et al. 2003. *Aquatic Microbial Ecology* 32: 209–218.)

28.2 FIELD PROCEDURES

There are four different experimental strategies to investigate the effects of pressure on organisms and functional variables:

a. *In situ* incubation of the sediment sample using a benthic lander. Such processing implies that the research vessel is immobilized for the entire incubation period, or that the lander is recovered after the incubation period.
b. *In situ* incubation of samples collected at the water sediment interface or from the surface sediment by means of an ROV or a human-operated submersible.
c. Sampling using classical equipment and incubation on board using pressure-controlled equipment.
d. Sampling and incubation using pressure-retaining samplers (high-pressure bottles [HPB]). The advantage of this equipment is that time course experiments may be run as subsamples can be taken off without decompressing the main sample.

This chapter refers to the latter two approaches, as the use of landers or ROVs to manipulate deep-sea samples is described in other chapters of this book.

28.2.1 Sampling and Sample Storage

Deep-sea sediment samples can be collected by means of a multiple corer, retrieved on board, immediately transferred in sterilized syringes, and repressurized in HPB in the dark and under *in situ* temperature conditions for subsequent viral and prokaryotic production measurements.

Alternatively, the sample is preferably collected at *in situ* pressure, maintained in pressurized conditions, and the performance of each incubation compared by comparing pairs of samples (*in situ* pressure versus decompressed). The availability of several pairs of samplers allows time independent series experiments to be performed.

All variables are typically determined by time-course experiments and, at each sampling time, aliquots of the incubated samples were withdrawn directly from the pressurized experimental system. For all experiments, additional samples are collected using the same sampling equipment without pressurization for evaluating the effect of incubation at 0.1 MPa on all of the measured microbial variables.

28.3 LABORATORY PROCEDURES

After incubation, measurements of the different variables are conducted as described elsewhere in this book (Figures 28.2 and 28.3).

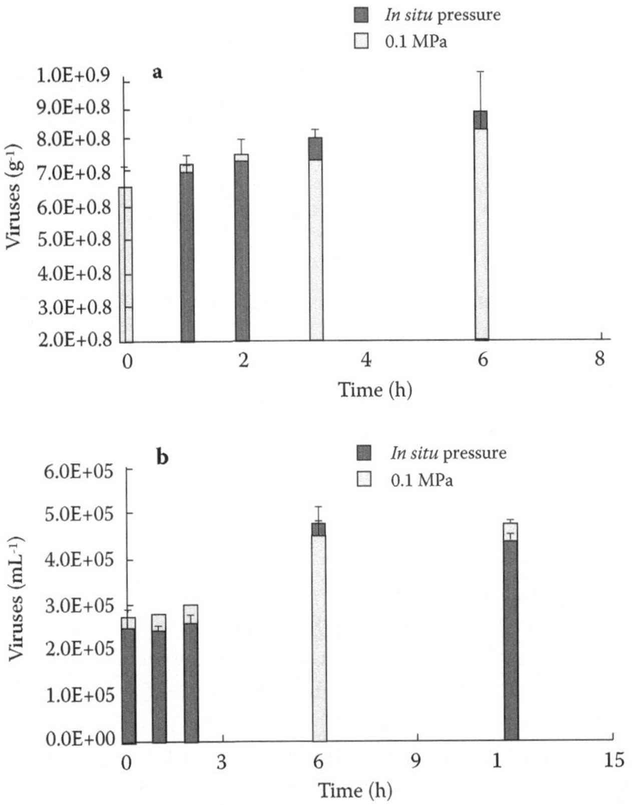

FIGURE 28.2 Effect of pressure on deep-sea viral production. Reported are (a) time-course experiments of the increase of viral abundance in sediment samples maintained at *in situ* pressure and at 0.1 MPa; (b) time-course experiments of the increase of viral abundance at the sediment–water interface from samples maintained at *in situ* pressure and at 0.1 MPa. (Data from Danovaro et al. 2008. *Nature* 454: 1084–1087.)

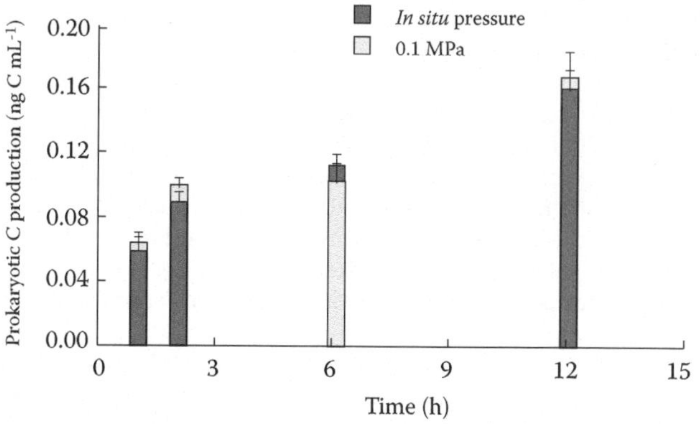

FIGURE 28.3 Effect of pressure on prokaryotic C production in samples maintained at *in situ* pressure and at 0.1 MPa. Reported are time-course experiments of prokaryotic C production in sediment samples maintained at *in situ* pressure and at 0.1 MPa. (Data from Danovaro et al. 2008. *Nature* 454: 1084–1087.)

REFERENCES

Alongi, D. M. 1990. Bacterial growth rates, production, and estimates of detrital carbon utilization in deep-sea sediments of the Solomon and Coral Sea. *Deep-Sea Research Part I* 37: 731–746.

Bartlett, D. H. 2002. Pressure effects on in vivo microbial processes. *Biochimica et Biophysica Acta* 1595: 367–381.

Bartlett, D. H., F. M. Lauro, and E. A. Eloe. 2007. Microbial adaptation to high pressure In *Physiology and biochemistry of extremophiles*, ed. C. Gerday and N. Glandsdorf, 333–348. Washington, DC: American Society for Microbiology Press.

Bianchi A. and Garcin J. 1993. In stratified waters the metabolic rate of deep-sea bacteria decreases with decompression. *Deep-Sea Research Part I* 40: 1703–1710.

Bianchi A. and Garcin J. 1994. Bacterial response to hydrostatic pressure in seawater samples collected in mixed-water and stratified-water conditions. *Marine Ecology Progress Series* 111: 137–141.

Bianchi A., Garcin J., and Tholosan O. 1999. A high-pressure serial sampler to measure microbial activity in the deep sea. *Deep-Sea Research Part I* 46: 2129–2142.

Chrost, R. J. 1991. Environmental control of the synthesis and activity of aquatic microbial ectoenzymes. In *Microbial enzymes in aquatic environments*, ed. R. J. Chrost, 29–59. New York: Springer-Verlag.

Danovaro, R., A. Dell'Anno, C., Corinaldesi et al. 2008. Major viral impact on benthic deep-sea ecosystem functioning. *Nature* 454: 1084–1087.

Deming, J. W., P. S. Tabor, and R. R. Colwell. 1980. Deep ocean microbiology. In *Advanced concepts in ocean measurements for marine biology*, ed. F. Diemer, J. Vernberg, and M. Mirkes, 285–305. University of South Carolina Press.

Deming, J.W. 2007. Extreme high-pressure marine environments. In *ASM manual of environmental microbiology*, ed. C. J. Hurst, R. L. Crawford, J. L. Garland, A. L. Mills, and L. D. Stetzenbach, 575–590. Washington, DC: ASM Press.

Eardly, D. F., M. W. Carton, J. M. Gallagher, and J. W. Patching. 2001. Bacterial abundance and activity in deep-sea sediments from the eastern North Atlantic. *Progress in Oceanography* 50: 245–259.

Gardner, W. D., K. R. Hinga, and J. Marra. 1983. Observations on the degradation of biogenic material in the deep ocean with implications on accuracy of sediment trap fluxes. *Journal of Marine Research* 41: 195–214.

Jannasch, H. W. and C. O. Wirsen. 1982. Microbial activities in undecompressed and decompressed deep-sea water samples. *Applied and Environmental Microbiology* 43: 1116–1124.

Jannasch, H. W. and C. D. Taylor. 1984. Deep-sea microbiology. *Annual Review of Microbiology* 38: 487–514.

Jannasch, H. W. and C. O. Wirsen. 1973. Deep-sea microorganisms: *In situ* response to nutrient enrichment. *Science* 180: 641–643.

Jannasch, H. W., C. O. Wirsen, and C. L. Winget. 1973. A bacteriological pressure-retaining deep-sea sampler and culture vessel. *Deep-Sea Research* 20: 661–664.

Jannasch, H. W., C. O. Wirsen, and C. D. Taylor. 1976. Undecompressed microbial populations from the deep sea. *Applied and Environmental Microbiology* 32: 360–367.

Jannasch, H. W. and C. O Wirsen. 1977. Retrieval of concentrated and undecompressed microbial populations from the deep sea. *Applied and Environmental Microbiology* 33: 642–646.

Kato, C., T. Sato, F. Abe et al. 2008: Protein adaptation to high-pressure environments. In *Protein adaptation in extremophiles, series*: *Protein adaptation in extremophiles (molecular anatomy and physiology of proteins)*, ed. T. Thomas and K. S. Siddiqui, 167–191, Nova Science Publisher.

Patching, J. W. and D. Eardly. 1997. Bacterial biomass and activity in the deep waters of the eastern Atlantic—evidence of a barophilic community. *Deep-Sea Research Part I* 44: 1655–1670.

Poremba, K. 1994. Simulated degradation of phytodetritus in deep-sea sediments of the NE Atlantic (47ON, 19"W). *Marine Ecology Progress Series* 105: 291–299.

Poremba, K. 1995. Hydrolytic enzymatic activity in deep-sea. *FEMS Microbiology Ecology* 16(3): 213–222.

Seki, H. and D. G. Robinson. 1969. Effect of decompression on activity of microorganisms in seawater. *International Review of Hydrobiology* 54: 201–205.

Tabor, P. S., J. W. Deming, K. Ohwada, H. Davis, M. Waxman, and R. R. Colwell. 1981. A pressure-retaining deep ocean sampler and transfer system for measurement of microbial activity in the deep sea. *Microbial Ecology* 7: 51–65.

Tamburini, C., J. Garcin, G. Grégori, K. Leblanc, P. Rimmelin, and D. L. Kirchman. 2006. Pressure effects on surface Mediterranean prokaryotes and biogenic silica dissolution during a diatom sinking experiment. *Aquatic Microbial Ecology* 43: 267–276.

Tamburini, C., J. Garcin, M. Ragot, and A. Bianchi. 2002. Biopolymer hydrolysis and bacterial production under ambient hydrostatic pressure through a 2000 m water column in the NW Mediterranean. *Deep-Sea Research Part II* 49: 2109–2123.

Tamburini, C., J. Garcin, and A. Bianchi. 2003. Role of deep-sea bacteria in organic matter mineralization and adaptation to hydrostatic pressure conditions in the NW Mediterranean Sea. *Aquatic Microbial Ecology* 32: 209–218.

Tamburini, C. 2006. Life under pressure: Deep-sea microbial ecology. In *Life as we know it.* Series: *Cellular Origin and Life in Extreme Habitats and Astrobiology,* Vol. 10, ed. J. Seckbach, 125–143. Dordrecht, The Netherlands: Springer.

Tholosan, O., J. Garcin, and A. Bianchi. 1999. Effects of hydrostatic pressure on microbial activity through a 2000 m deep water column in the NW Mediterranean Sea. *Marine Ecology Progress Series* 183: 49–57.

Turley, C. M., K. Lochte, and R. S. Lampitt. 1995. Transformations of biogenic particles during sedimentation in the northeastern Atlantic. *Philosophical Transactions of the Royal Society of London B: Biological Sciences* 348: 179–189.

Yayanos, A. A. 1995. Microbiology to 10,500 meters in the deep sea. *Annual Review of Microbiology* 49: 777–805.

ZoBell C. E. 1970. Pressure effects on morphology and life processes of bacteria. In *High pressure effects on cellular processes*, ed. H. M. Zimmerman, 85–130. New York: Academic Press.

29 Organic Carbon Remineralization Rates in Marine Surface Sediments

CONTENTS

29.1 INTRODUCTION

The remineralization of organic carbon in marine sediments occurs through different oxidative reactions in which the ultimate electron acceptors are O_2, NO_3^-, Mn-oxides, Fe-oxides, SO_4^{2-}, and CO_2 (Froelich et al. 1979). The relative importance of the various processes changes with sediment type and sedimentation rate, but generally, oxic mineralization becomes more important with increasing water depth (Canfield et al. 1993).

The majority of the reduced species produced by anaerobic degradation are ultimately reoxidized by an equivalent amount of O_2. The O_2 uptake of sediments has been widely used as a proxy of the remineralization rate of organic C and benthic community metabolism. The benthic O_2 uptake is either measured directly by incubation of enclosed sediment core (total benthic O_2 uptake, TOU) or calculated indirectly from the O_2 concentration gradients measured across the benthic interface (diffusive oxygen uptake, DOU). Here we present a simple approach to calculate the DOU of sediments in the oxic surface zone by means of microelectrode measurements of O_2 concentration profiles above and in the sediment (Reimers, et al. 1986; Revsbech and Jorgensen 1986; Revsbech 1989). The DOU rate, due to sediment bacteria capable of aerobic respiration, together with the advective oxygen uptake rate, generally governed by benthic faunal activities, represent; the total O_2 uptake of the sediment (Sediment Community Oxygen Consumption).

Oxygen microprofiles are measured shipboard with guarded Clark-type microelectrodes (Figure 29.1). Shipboard measurements are generally comparable to the in situ ones for sediments at water

FIGURE 29.1 Shipboard pore-water oxygen microprofiling on a subcore collected by means of a box corer from deep-sea sediments. Figure illustrates the subcore within the thermostated water tank and the glass oxygen microelectrode ready to be inserted in the sediment.

depths shallower than 1000 m. At greater depths, shipboard measurements of O_2 concentrations can be biased to a large extent due to decompression during sample recovery (Glud et al. 1994; Sauter et al. 2001; Epping et al. 2002).

Care should be taken in the collection of samples and in all subsequent manipulations on board in order to minimize disturbing the sediments as much as possible.

29.2 FIELD PROCEDURES

29.2.1 Equipment and Instruments

Acrylic glass liners for sediment subsampling (i.d. approximately 6 cm, h approximately 20 cm)
Thermostated water tank
Clark-type (with guard electrode) oxygen microelectrodes with outer tip diameters ranging from 10 to 15 micron
Picoammeter (instrument for low-current measurement)
PC-controlled motorized micromanipulator
Heavy and stable laboratory stand
Oxygen sensor calibration chamber
Digital thermometer
Aquarium air pump
Nitrogen tanks
Sediment resistivity probe and meter

29.2.2 Sample Collection

1. Collect two cores with acrylic glass liners from the box corer or multicorer after visual check of the sample condition (avoid turbid water and crakes zone). Usually, the central zone of the box corer is less disturbed and preserve the in situ temperature.

2. If the original bottom water is not preserved on the top of the sediment, it has to be carefully replaced with the bottom water collected using Niskin bottles mounted directly on the box corer or multicorer frame or mounted on a Rosette carousel.
3. Immediately after recovery, the cores are transferred to a water tank thermostated at the in situ temperature.

29.2.3 Shipboard Laboratory Procedures

1. Oxygen sensor preparation: If the sensor is new or has not been used for several days, it must be polarized for at least 2 h before measurements. Polarization is necessary because the sensor electrolyte could contain oxygen that should be eliminated before use.
2. Prior to and during profiling, the supernatant bottom water is gently flushed with air (by an aquarium air pump) to reach the 100% air-saturated oxygen concentration.
3. The microelectrode must be calibrated with the 100% air-saturated bottom seawater and with 100% nitrogen-saturated bottom seawater (i.e., as zero signal).
4. Start O_2 microprofiling down to at least 4–5 cm along the vertical profile of the sediment, recording data at 100–250 μm resolution, after 10 s of stabilization at each depth. Repeat profiles at least twice for each core.
5. Perform resistivity profiles to calculate the sedimentary formation factor *(f)* in the undisturbed portion of the core by means of a 4-terminal resistivity probe (Andrews and Bennet 1981). The formation factor *f* (based on resistivity data; McDuff and Ellis 1979) is related to sediment porosity according to Archie's Law (Boudreau 1996). Repeat profiles at least twice for each core.

29.3 NOTES

1. As oxygen sensors are sensitive to temperature, and oxygen solubility to temperature and salinity, it is necessary to perform calibration and subsequent measurements at the in situ temperature and salinity.
2. It is necessary to determine the oxygen concentration of the bottom seawater by Winkler titration (Grasshoff 1983). The bottom seawater could be sampled directly from the multicorer or box corer (if the box corer has a tightly closed shutter). Alternatively, bottom seawater can be sampled from a Niskin bottle mounted on the box-corer frame or from the deepest bottle of the Rosette.
3. The depth scale of the oxygen profiles has to be adjusted taking account of the position of the sediment–water interface. The position is determined from the shape of the O_2 profile characterized by a sharp decrease of concentration at the interface. Oxygen microelectrodes provide data in Volt, which have to be transformed to oxygen concentration on the basis of a two-point calibration performed prior to each profile.

29.4 FLUX CALCULATIONS

Oxygen flux J_{O_2} through the sediment–water interface is calculated according to Fick's first law (Boudreau 1996):

$$J_{O_2} = \phi D_{sed} \left(d[O_2] / dz \right) \tag{29.1}$$

where

ϕ = the porosity.

D_{sed} = the sediment diffusion coefficient.

The sediment diffusion coefficient could be estimated on the basis of the following relationship (McDuff and Ellis 1979):

$$D_{sed} = D_O / \phi f \tag{29.2}$$

where

D_0 = the molecular diffusion coefficient of oxygen at bottom-water temperature.
f = formation factor.

Formation factor *f* is calculated as follows:

$$f = R_{sed} / R_O \tag{29.3}$$

where

R_{sed} and R_0 = the electrical resistivity of sediment and overlying water, respectively.

The values of the sedimentary formation factor are calculated from the average of the measured resistivity profiles after correction of the depth scale to place the sediment-water interface on the basis of the point of inflection in the profiles.

Finally, O_2 flux is calculated as follows:

$$J_{O_2} = D_O / f\left(d\left[O_2\right] / dz\right) \tag{29.4}$$

A flux calculation could be performed for each measurement along the sediment profile and then averaged among replicate measurements carried out on independent sediment cores.

It has been suggested that O_2 fluxes may be calculated by assuming a linear gradient between the O_2 concentration at the bottom of first centimeter of the sediment and the in situ concentration of O_2 in the bottom water (measured by Winkler titration). This approach allows reduction of the artifacts generated by the modification of O_2 concentrations due to decompression during sample recovery (Reimers et al. 1984).

Diffusive oxygen flux are translated into the flux of remineralizable organic carbon by applying the Redfield ratio. The traditional P:N:C_{org} Redfield ratios (Redfield et al. 1963) are 1:16:106, based on planktonic decomposition studies (Fleming 1940); an O_2:P ratio of 138 was inferred by effectively assuming an organic matter stoichiometric composition of $(CH_2O)_{106}(NH_3)_{16}$ H_3PO_4 and balancing the equation representing the utilization of primary nutrients by marine organisms during the synthesis and subsequent remineralization of organic matter.

REFERENCES

Andrews, D. and A. Bennet. 1981. Measurements of diffusivity near the sediment water interface with a fine-scale resistivity probe. *Geochimica et Cosmochimica Acta* 45: 2169–2175.

Boudreau, B. P. 1996. A method-of-lines code for carbon and nutrient diagenesis in aquatic sediments. *Computers and Geosciences* 22: 479–496.

Canfield, D. E., B. B. Jorgensen, H. Fossing et al. 1993. Pathways of organic carbon oxidation in three continental margin sediments. *Marine Geology* 113: 27–40.

Epping, E., C. van der Zee, K. Soetaert, and W. Helder. 2002. On the oxidation and burial of organic carbon in sediments of the Iberian margin and Nazare Canyon (NE Atlantic). *Progress in Oceanography* 52: 399–431.

Fleming, R. H. 1940. Composition of plankton and units for reporting populations and production. *Proceedings of the Sixth Pacific Science Congress* 3: 535–540.

Froelich, P. N., G. P. Klinkhammer, M. L. Bender et al. 1979. Early oxidation of organic matter in pelagic sediments of the eastern equatorial Atlantic: Suboxic diagenesis. *Geochimica et Cosmochimica Acta* 43: 1075–1090.

Glud, R. N., J. K. Gundersen, B. B. Jorgensen, N. P. Revsbech, and H. D. Schulz. 1994. Diffusive and total oxygen uptake of deep-sea sediments in the eastern South Atlantic Ocean: In situ and laboratory measurements. *Deep-Sea Research Part I* 41: 1767–1788.

Grasshoff, K. 1983. Determination of oxygen. In *Methods of seawater analysis*, ed. K. Grasshoff, M. Ehrhardt, and K. Kremling, 61–72. Weinheim: Verlag Chemie GmbH.

McDuff, R. E. and R. A. Ellis. 1979. Determining diffusion coefficients in marine sediments: A laboratory study of the validity of resistivity techniques. *American Journal of Science* 279: 666–675.

Redfield, A. C., B. H. Ketchum, and F. A. Richards. 1963. The influence of organisms on the composition of sea-water. In *The Sea, 2nd* ed., M. N. Hill, 26–77. New York: Interscience.

Reimers, C. E., K. M. Fischer, R. Merewether, K. L. Smith, Jr., and R. A. Jahnke. 1986. Oxygen microprofiles measured in situ in deep ocean sediments. *Nature* 320: 741–744.

Reimers, C. E., S. Kalhorn, S. R. Emerson, and K. H. Nealson. 1984. Oxygen consumption rates in pelagic sediments from the Central Pacific: First estimates from microelectrode profiles. *Geochimica et Cosmochimica Acta* 48: 903–910.

Revsbech, N. P. and Jorgensen, B. B. 1986. Microelectrodes: Their use in microbial ecology. In *Advances in microbial ecology*, ed. K. C. Marschall, 293–352. New York: Plenum.

Revsbech, N. P. 1989. An oxygen microelectrode with a guard cathode. *Limnology and Oceanography* 34: 474–478.

Sauter, E. J., M. Schluter, and E. Suess. 2001. Organic carbon flux and remineralization in surface sediments from the northern North Atlantic derived from pore-water oxygen microprofiles. *Deep-Sea Research Part I* 48: 529–553.

30 Degradation and Turnover of Organic Matter in Marine Sediments

CONTENTS

30.1 INTRODUCTION

In marine sediments, the organic matter pool is largely dominated by organic detritus mostly composed of high-molecular-weight compounds, recycling of which is largely dependent upon prokaryotic-mediated degradation processes (Deming and Baross 1993). Degradation processes are mainly dependent upon extracellular enzymatic hydrolysis, which allows the breakdown of high-molecular-weight compounds into bioavailable monomers/oligomers suitable for prokaryotic uptake (Hoppe 1991). Degradation rates of organic matter pools in the sediment are influenced by their biochemical composition (Arnosti and Repeta 1994; Fabiano and Danovaro 1998). Organic compounds such as humic and fulvic acids and black carbon, being resistant to decomposition, tend to accumulate in the sediment (Zegouagh et al. 1999; Middelburg et al. 1999), whereas biopolymers (i.e., proteins, carbohydrates, lipids, and nucleic acids), being more labile, are efficiently degraded and recycled by extracellular enzymatic activities (Poremba 1995; Keith and Arnosti 2001). Therefore, the determination of the enzymatic hydrolytic potential of the main biopolymeric component of organic matter in the sediment represents a key functional variable for a better understanding of organic matter cycling and diagenesis and microbial loop functioning (Hoppe 1991; Meyer-Reil 1991; Meyer-Reil and Koster 1992). Enzymatic activity in the sediment can be measured by the use of fluorogenic analogs (Hendel and Marxen 1997). The cleavage of the substrates, releasing the fluorogenic components, provides a measure of the potential degradation rates. The degradation activities of l-aminopeptidase, β-d-glucosidase, and alkaline-phosphatase can be quantified fluorometrically using, respectively, l-Leucine-4-methylcoumarinyl-7-amide (Leu-MCA), 4-methylumbelliferone β-d-glucopyranoside (MUF-Glu), and 4-methylumbelliferone phosphate (MUF-P) as substrates (Hoppe 1993; Figure 30.1).

30.2 FIELD PROCEDURES

30.2.1 SAMPLING

Sediments should be collected using multiple corers or any other device that allows the recovery of undisturbed sediment samples. Immediately after recovery, sediment subsamples, collected by using sterile syringes, are transferred into sterile tubes for further processing, detailed as follows.

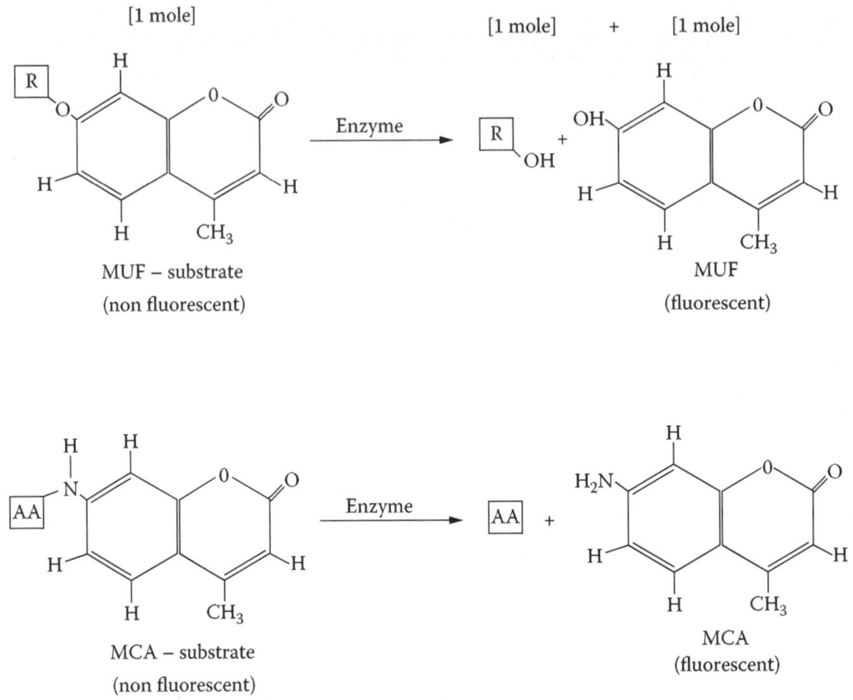

FIGURE 30.1 Molecular structure and enzymatic hydrolysis products of 4-methylumbelliferyl (MUF) substrates and 4-methylcoumarinyl-7-amide (MCA) substrates.

30.2.2 Equipment and Instruments

Centrifuge for 15 mL tube
Analytical balance
Vortex
Sterile 15 mL tubes
Sterile syringes
Sterile and prefiltered seawater [NOTE 1]
Filters of 0.2 μm pore size (Whatman Puradisc AS 25 mm ϕ)
2-20 μL, 100–1000 μL, 1–5 mL pipettes and sterile tips
Substrates
- l-Leucine-4-methylcoumarinyl-7-amide (Leu-MCA; SIGMA)
- 4-methylumbelliferone β-d-glucopyranoside (MUF-Glu; SIGMA)
- 4-methylumbelliferone phosphate (MUF-P; SIGMA)

Standards
- 7-Amino-4-methylcoumarinyl (used as standard for Leu-MCA; SIGMA)
- 4-Methylumbelliferone (used as standard for MUF-Glu and MUF-P; SIGMA)

Ethyleneglycolmonomethylether (SIGMA)
Fluorometer
Glass cuvettes
Thermostatic incubator/water bath

30.2.3 Solutions

Stock solutions of each substrate [NOTE 2] are prepared by dissolving the pure reagent in methylcellosolve (ethylenglycolmonomethylether) at a final concentration of 10 mM.

Stock solutions of standards (5 mM) are prepared in methyl cellosolve (Ethylene glycol monomethyl ether) and then diluted in prefiltered and autoclaved seawater. Standard solutions should be freshly prepared at concentrations ranging from 0.1 to 10.0 μM [NOTE 3].

30.3 LABORATORY PROCEDURES

1. Collect 0.5–1 cm^3 of sediment.
2. Add prefiltered (0.2 μm pore size) and autoclaved seawater containing the specific substrate up to a final volume of 5 mL.
3. Mix the sample gently until a homogeneous slurry is obtained.
4. Incubate in the dark for 1 h at *in situ* temperature.
5. Centrifuge at 800 × g for 10 min.
6. Collect the supernatant and measure its fluorescence at 380 nm excitation and 440 nm emission wavelengths for Leu-MCA, and at 365 nm excitation and 455 nm emission wavelengths for Glu-MUF and MUF-P.
7. Convert the fluorescence of the sample into nmol of MUF or MCA released per gram of sediment dry weight h^{-1} as follows:

$$\text{nmol (MUF or MCA) } g^{-1}h^{-1} = \{[(Fluo_c - Fluo_{blk}) \cdot C + K] \cdot v\} \cdot Ps^{-1} \cdot T_{inc}^{-1} \quad (30.1)$$

where

$Fluo_c$ is the fluorescence of the sediment sample.

$Fluo_{blk}$ is the fluorescence of the sediment blank (average of three replicates).

C is the slope of the linear regression obtained by calibration curve.

K is the constant of the linear regression obtained by calibration curve.

v is the volume of the slurry in mL.

Ps is the sediment dry weight (grams).

T_{inc} is the incubation time.

8. Aminopeptidase and ß-glucosidase activities can be transformed into equivalents of C mobilized assuming that 1 nmol of substrate hydrolyzed enzymatically corresponds to 72 ng of mobilized C.

30.4 NOTES

1. Sediment samples incubated without fluorogenic substrate are used as blanks. These samples are processed according to the procedure followed for fresh sediment samples.
2. Sterile seawater (i.e., autoclaved and filtered through 0.2 μm pore sizes filters) must be prepared before the beginning of the sample processing in order to dilute sediment samples appropriately. The substrates are stable at −20°C for several months.
3. Blanks are prepared as reported earlier and centrifuged at 4°C at 2000 × g for 5 min. After centrifugation, the supernatant is collected and analyzed fluorometrically. Kinetic parameters (i.e., Vm and Ks) can be calculated using Lineweaver–Burke plots of reaction velocity versus substrate concentrations. Linearity of enzymatic hydrolysis should be checked by time-course analysis of substrate degradation.

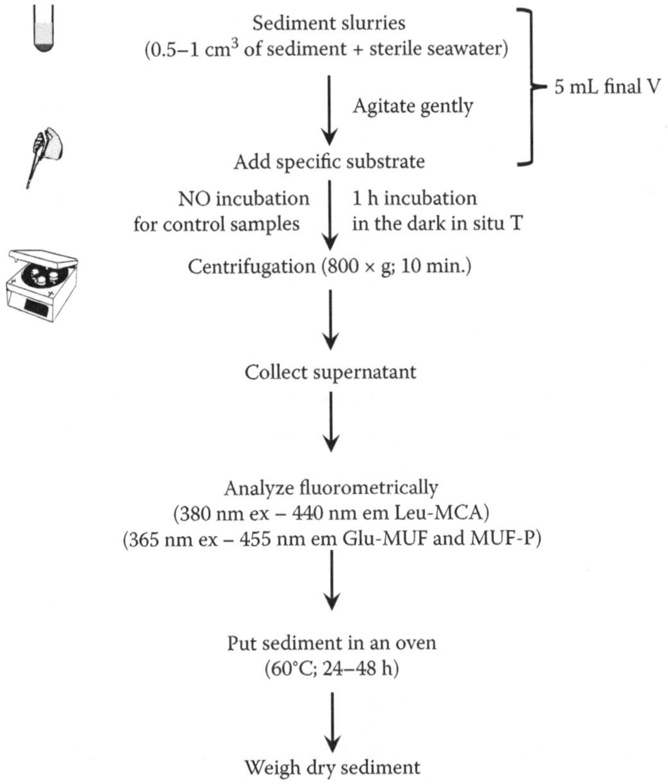

PROTOCOL 30.1 Determination of extracellular enzymatic activities in marine sediments.

REFERENCES

Arnosti, C. and D. J. Repeta. 1994. Oligosaccharide degradation by anaerobic marine bacteria: Characterisation of an experimental system to study polymer degradation in sediments. *Limnology and Oceanography* 39: 1865–1877.

Deming, J. W. and J. A. Baross. 1993. The early diagenesis of organic matter: Bacterial activity. In *Organic geochemisry: Principles and applications*, ed. M. H. Engel and S. A. Macko, 119–144. New York: Plenum Press.

Fabiano, M. and R. Danovaro. 1998. Enzymatic activity, bacterial distribution, and organic matter composition in the sediments of the Ross Sea (Antarctica). *Applied and Environmental Microbiology* 64(10): 3838–3845.

Hendel, B. and J. Marxen. 1997. Measurement of low-level extracellular enzyme activity in natural waters using fluorogenic model substrates. *Acta hydrochimica et hydrobiologica* 25: 253–258.

Hoppe, H. G. 1991. Microbial extracellular enzyme activity: A new key parameter in aquatic ecology. In *Microbial enzyme in aquatic environments*, ed. J. Chrøst, 60–79. New York: Springer-Verlag.

Hoppe, H. G. 1993. Use of fluorogenic model substrates for extracellular enzyme activity (EEA) measurement of bacteria. In *Handbook of methods in aquatic microbial ecology*, ed. P. F. Kemp, B. F. Sherr, E. B. Sherr, and J. Cole, 423–431. Boca Raton: Lewis Publishers.

Keith, S. C. and C. Arnosti. 2001. Extracellular enzyme activity in a river-bay-shelf transect: Variations in polysaccharide hydrolysis rates with substrate and size class. *Aquatic Microbial Ecology* 24: 243–253.

Meyer-Reil, L. A. 1991. Ecological aspects of enzymatic activity in marine sediments. In *Microbial enzymes in aquatic environments*, ed. R. J. Chrost, 84–95. Berlin, Germany: Springer-Verlag.

Meyer-Reil, L. A. and M. Köster. 1992. Microbial life in pelagic sediments: The impact of environmental parameters on enzymatic degradation of organic matter in deep-sea sediments. *Marine Ecology Progress Series* 81: 65–72.

Middelburg, J. J., J. Nieuwenhuize, and P. Van-Breugel. 1999. Black carbon in marine sediments. *Marine Chemistry* 65: 245–252.

Poremba, K. 1995. Hydrolytic enzymatic activity in deep-sea sediments. *FEMS Microbiol Ecology* 16: 213–222.

Zegouagh, Y., S. Derenne, C. Largeau et al. 1999. Refractory organic matter in sediments from the NorthWest African upwelling system: Abundance, chemical structure, and origin. *Organic Geochemistry* 30: 101–117.

31 Extracellular DNA Extraction from Marine Sediments

CONTENTS

31.1 INTRODUCTION

All aquatic sediments are characterized by high DNA concentrations (three to four orders of magnitude greater than those found in the water column), mostly (up to 90%) represented by extracellular DNA (Danovaro et al. 1999; Dell'Anno and Danovaro 2001; Dell'Anno et al. 1998; Dell'Anno et al. 2002; Figure 31.1). Previous studies reported that complex refractory organic molecules and/or inorganic particles are able to bind, adsorb, and stabilize free DNA in sediments (Romanowski et al. 1991), leaving less than 5% of the extracellular DNA as free fraction (Dell'Anno et al. 2002). The adsorption of extracellular DNA in the sediment might reduce its degradability, and indeed only about 50% of this DNA can be hydrolyzed by nucleases (Dell'Anno et al. 2002). Consequently, the residence time of extracellular DNA in sediments can be much longer than the residence time in the water column (Lorenz and Wackernagel 1994). The presence and persistence of large amounts of extracellular DNA in the deeper sediment layers (Dell'Anno et al. 1999) have important implications for prokaryote metabolism, providing a source of nitrogen and phosphorus and exogenous nucleotides (Danovaro et al. 1999; Dell'Anno et al. 1999), and may also contribute to horizontal gene transfer through natural transformation (Lorentz and Wackernagel 1994; Figure 31.2).

In the last decade, several protocols for the extraction of DNA from soils and sediments have been developed (Zhou et al. 1996; Hurt et al. 2001; Juniper et al. 2001; Rose-Amseleg et al. 2001). The application of culture-independent techniques involving DNA extraction and molecular analyses have allowed the detection and identification of microorganisms in natural environments (Head et al. 1998). Much effort has been devoted also to increase the DNA extraction efficiency and to minimize biases due to DNA contamination (Hurt et al. 2001; Juniper et al. 2001; Miller et al. 1999). The most commonly utilized technique for DNA extraction is based on direct *in situ* cell lysis by physical (e.g., bead mill homogenization, ultrasonication, and freeze-thawing) and/or chemical (e.g., the use of sodium dodecyl sulfate [SDS] or Sarkosyl) procedures (Hurt et al. 2001; Miller et al. 1999). In these procedures extracellular DNA is co-extracted with nucleic acids released from the lysed cells, thus prevent the discrimination of the extracellular fraction of DNA from the intracellular one (Frostegard et al. 1999). Nonetheless, the discrimination between intracellular and extra-

FIGURE 31.1 *(A color version of this figure follows page 234.)* An image showing the release of DNA from prokaryotic cells. Extracellular DNA is a source of bioavailable N and P, which sustains deep-sea heterotrophic metabolism. (Courtesy of M. Tangherlini. Used with permission.)

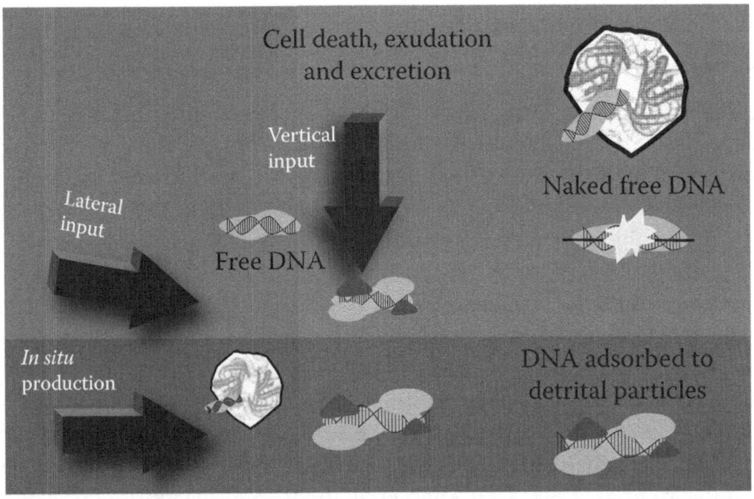

FIGURE 31.2 DNA sources and diagenesis in aquatic systems.

cellular DNA in marine sediments is crucial for carrying out simultaneous molecular studies of the two DNA pools.

Here we describe the procedure for extracting the extracellular DNA from marine sediment samples, avoiding the concurrent extraction of intracellular DNA and providing a DNA yield and purity adequate for any molecular analysis (Corinaldesi et al. 2005).

31.2 FIELD PROCEDURES

31.2.1 Sampling

Sediment samples may be collected both manually, by means of Plexiglas® cores or syringes, or through instruments such as multicorers, box corers, or piston corers, depending on the sampling depth. Sampling must be performed, limiting sediment perturbation and avoiding sediment mixing. Slicing of subsurface sediment horizons must be performed using sterile tools (i.e., tips, spatulas). All of the material used for sampling and the subsequent processing of sediment samples must be sterilized by autoclave and processed under a laminar air flow hood in order to avoid any contamination. All working solutions must be prepared using 0.02 μm filtered water and autoclaved.

It is recommended to clean the glassware utilized for nucleic acid analysis by soaking it in 1 N NaOH, 10% HCl, and reagent-grade water in order to remove organic matter contamination. Moreover, it can be useful to treat the glassware at 250°C to avoid nuclease contamination.

31.2.2 Sample Storage

In order to limit DNA degradation and damage, sediment samples must be flash frozen in liquid nitrogen and preserved at –20°C if the samples are analyzed within a few weeks after sampling. For longer preservations samples can be stored at –80°C.

31.3 LABORATORY PROCEDURES

31.3.1 Equipment and Instruments

Autoclave
Laminar and chemical air flow hoods
Horizontal shaker
Refrigerated centrifuge
Spectrophotometer with monochromator emitting at UV lambdas (<300 nm)

31.3.2 Materials

2 mL sterile tubes
50 mL sterile tubes
5 mL sterile syringes
Anodisc filters (Anotop 25, Whatman) 0.02 μm pore size, 25 mm ø
2–20 μL, 100–1000 μL, 1–5 mL pipettes
Sterile tips
Glassware for reagents

31.3.3 Solutions

0.1 M sodium phosphate buffer (pH 8.0) in reagent-grade water: prepare 500 mL of 1 M Na_2HPO_4 solution (A) and 100 mL of NaH_2PO_4 solution (B) in reagent-grade water. Combine 93.32 mL of the solution A and 6.8 mL of the solution B in a 1 L volumetric flask and add reagent-grade water to ca. 950 mL volume. Check the pH of the solution. If the pH is ca. 8, add reagent-grade water to a final volume of 1 L. Otherwise, prior to reaching the final volume of the solution, adjust the pH with HCl or NaOH 1 N.

Acid-washed polyvinylpolypyrrolidone (PVPP): 100 g of PVPP in 1 L of HCl 3 M. Stir overnight and filter through 20 μm filter. Collect the retentate in a beaker and add 500 mL of reagent-grade water. Stir for 1 h and filter through a 20 μm filter. Recover the retentate and add 500 mL of $NaPO_4$ 0.1 M (pH 7.4). Stir for 1 h, filter as described earlier and recover the retentate. Wash as above with $NaPO_4$ until pH is around 7. Bring to pH 7.0 using $NaPO_4$ 20 mM. [NOTE 1].

SDS (sodium dodecyl sulfate) solution: Final concentration 0.1% in reagent-grade water [NOTE 2].

CTAB (cetyltrimethylammonium bromide) solution: 1% CTAB in 50 mM Tris, 10 mM [NOTE 3].
Tris-EDTA buffer: 10 mM Tris-HCl, 0.1 mM EDTA, 1 M NaCl (pH 8.0).
Equilibrated phenol-chloroform-isoamyl alcohol: 25:24:1, vol/vol/vol [NOTE 4].
Chloroform-isoamyl alcohol (24:1, vol/vol).
Ethanol (EtOH) final concentration in water 70%.
NaCl (final concentration, 0.2 M) in reagent-grade water.
DNA-free reagent-grade water (autoclaved and filtered through 0.02 μm filters).

31.3.4 Procedure

1. Collect 2.5 g (wet weight) of sediment in 50 mL sterile tubes.
2. Add 7.5 mL of 0.1 M sodium phosphate buffer solution.
3. Add 0.5 g of PVPP.
4. Homogenize the sample with a horizontal shaker at a low speed (150 horizontal shakes per minute) for three 1-min cycles, with 1 min of cooling on ice between cycles.
5. Add SDS (final concentration, 0.1%) to the samples and shake again for 10 s.
6. Chill on ice the samples.
7. Centrifuge at 500 × g for 10 min at 4°C.
8. Collect and transfer the supernatant to 50 mL sterile tubes.
9. Wash the sediment pellets twice by adding 7.5 mL of 0.1 M sodium phosphate buffer (pH 8.0) and centrifuge as described above (no addition of SDS).
10. Combine supernatants and centrifuge for 20 min at 10,000 × g (at 4°C).
11. After centrifugation, filter the supernatants containing extracellular DNA through 0.02 μm-pore-size membrane filters (Anotop, Whatman) to remove any contaminating virus or prokaryotic cell [NOTE 5].
12. Add 1 volume of CTAB solution.
13. Incubate the samples at 65°C for 30 min.
14. Centrifuge at 5,000 × g for 10 min at 4°C.
15. Discharge the supernatants and resuspend the pellet in high-salt TE buffer.
16. To precipitate DNA, add 0.6 volume of cold isopropanol to each sample, and incubate for 1 h on ice.
17. Centrifuge at 10,000 × g for 15 min at 4°C.
18. Discharge the supernatants and resuspend the pellets in Tris-EDTA buffer [NOTE 6].
19. Add an equal volume of equilibrated phenol-chloroform-isoamyl alcohol solution to each sample.
20. Centrifuge at 10,000 × g for 5 min.
21. Mix each supernatant with an equal volume of chloroform-isoamyl alcohol solution (24:1, vol/vol).
22. Centrifuge at 10,000 × g for 5 min.
23. Precipitate the supernatant with cold ethanol (final concentration, 70%) and sodium chloride (final concentration, 0.2 M) and incubate at 20°C for 1 h [NOTE 7].
24. Centrifuge at 10,000 × g for 15 min.
25. Wash the pellet twice with ethanol and dry under laminar air flow [NOTE 8].
26. Resuspend in reagent-grade water.
27. Check the purity of DNA the 260/280 nm absorbance ratio.

31.3.5 Notes

1. PVPP is a water-soluble polymer used for removing humic acids and phenolic substances from the sediment sample.
2. SDS at low concentration (0.1%) is used for inhibiting electrostatic bonds among mineral particles and extracellular DNA. Higher concentrations of SDS (e.g., 10%–20%) determine cell lysis.
3. CTAB is used to remove polysaccharides and proteins from DNA.
4. Equal volumes of a phenol:chloroform mixture and the aqueous DNA sample are mixed, forming a biphasic mixture. The proteins will partition into the organic phase, while the DNA (as well as other contaminants such as salts, sugars) remain in the aqueous phase. Phenol and chloroform are both hazardous reagents, so it is recommended to work under a chemical air flow. Moreover since the phenol:chloroform is also very laborious, in alternative, it is possible to purify DNA by using purification columns (e.g., Wizard DNA purification columns, Promega). Phenol removes proteins during nucleic acid extractions. Chloroform is also a protein denaturant and it stabilizes the inter-phase, reducing the amount of aqueous phase retained in the organic phase, increasing the mixture's density, and facilitating the removal of lipids. Isoamyl alcohol is added to phenol:chloroform mixture to prevent foaming of the solution during vortexing and enhance phase separation.
5. At this step, the pellets containing prokaryote cells can be processed as described by Steffan et al. (1988) in order to extract intracellular DNA.
6. Alternatively to TE-buffer, DNA can be resuspended with reagent-grade water.
7. The ratio between the DNA absorbance at 260 nm and 280 nm has to be >1.7 to ensure a sufficient DNA purity.
8. In many cases the pellet is not naked-eye visible even if the extracellular DNA is present.

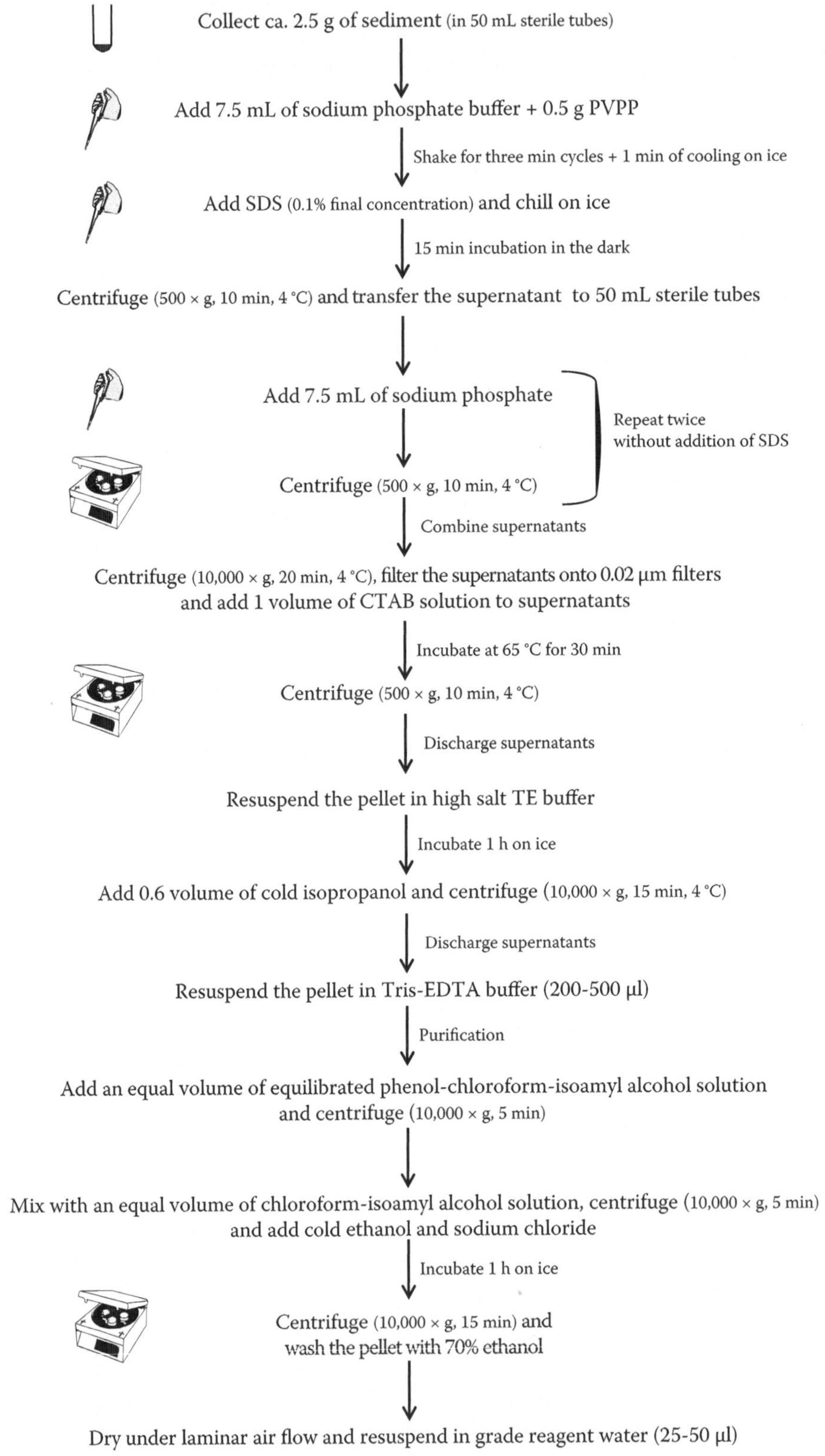

PROTOCOL 31.1 Extraction of extracellular DNA from marine sediments.

REFERENCES

Corinaldesi, C., R. Danovaro, and A. Dell'Anno. 2005. Simultaneous recovery of extracellular and intracellular DNA suitable for molecular studies from marine sediments. *Applied and Environmental Microbiology* 71: 46–50.

Danovaro, R., A. Dell'Anno, A. Pusceddu, and M. Fabiano. 1999. Nucleic acid concentrations (DNA, RNA) in the continental and deep-sea sediments of the Eastern Mediterranean: Relationships with seasonal varying organic inputs and bacterial dynamics. *Deep-Sea Research* 46: 1077–1094.

Dell'Anno, A., M. Fabiano, G. C. A. Duineveld, A. Kok, and R. Danovaro. 1998. Nucleic acid (DNA, RNA) quantification and RNA/DNA ratio determination in marine sediments: Comparison of spectrophotometric, fluorometric, and high-performance liquid chromatography methods and estimation of detrital DNA. *Applied and Environmental Microbiology* 64: 3238–3245.

Dell'Anno, A., M. Fabiano, M. L. Mei, and R. Danovaro. 1999. Pelagic benthic coupling of nucleic acids in abyssal location of the northeastern Atlantic ocean. *Applied and Environmental Microbiology* 65: 4451–4457.

Dell'Anno, A. and R. Danovaro. 2001. Nucleic acid turnover in aquatic environments. 1. Determination of total and extracellular DNA in marine sediments. In *Molecular microbial ecology manual*, ed. A. D. L. Akkermans, J. D. van Elsas, and F. J. De Bruijn, 1–9. The Netherlands, Dordrecht: Kluwer Academic Publishers.

Dell'Anno, A., S. Bompadre, and R. Danovaro. 2002. Quantification, base composition, and fate of extracellular DNA in marine sediments. *Limnology and Oceanography* 47: 899–905.

Frostegard, A., S. Courtois, V. Ramisse et al. 1999. Quantification of bias related to the extraction of DNA directly from soils. *Applied and Environmental Microbiology* 65: 5409–5420.

Head, I. M., J. R. Saunders, and R. W. Pickup. 1998. Microbial evolution, diversity, and ecology: A decade of RNA analysis of uncultivated microorganisms. *Microbial Ecology* 35: 1–21.

Hurt, R. A., X. Qiu, L. Wu et al. 2001. Simultaneous recovery of RNA and DNA from soils and sediments. *Applied and Environmental Microbiology* 67: 4495–4503.

Juniper, S. K., M. A. Cambon, F. Lesonger, and G. Barbier. 2001. Extraction and purification of DNA from organic rich subsurface sediments (ODP Leg169S). *Marine Geology* 174: 241–247.

Lorenz, M. G. and W. Wackernagel. 1994. Bacterial gene transfer by natural genetic transformation in the environment. *Microbiology Reviews* 58: 563–602.

Miller, D. N., J. E. Bryant, E. L. Madsen, and W. C. Ghiorse. 1999. Evaluation and optimization of DNA extraction and purification for soils and sediment samples. *Applied and Environmental Microbiology* 65: 4715–4724.

Romanowski, G., M. G. Lorenz, and W. Wackernagel. 1991. Adsorption of plasmid DNA to mineral surfaces and protection against DNase I. *Applied and Environmental Microbiology* 57: 1057–1061.

Rose-Amseleg, C. L., E. Garnier-Sillam, and M. Harry. 2001. Extraction and purification of microbial DNA from soil and sediment samples. *Applied Soil Ecology* 18: 47–60.

Steffan, R. J., J. Goksoyr, A. K. Bej, and R. M. Atlas. 1988. Recovery of DNA from soils and sediments. *Applied and Environmental Microbiology* 54: 2908–2915.

Zhou, J., M. A. Bruns, and J. M. Tiedje. 1996. DNA recovery from soils of diverse composition. *Applied and Environmental Microbiology* 62: 316–322.

32 Degradation of Extracellular DNA in Marine Sediments

CONTENTS

32.1 INTRODUCTION

Extracellular DNA is an ubiquitous component of dissolved and particulate pools of all deep-sea sediments (Dell'Anno et al. 1998; 2002). Although extracellular DNA is a minor component of organic matter pools in the oceans (Jiang and Paul 1995; Dell'Anno et al. 1999a) it may represent an important source of nitrogen and phosphorus for bacterial metabolism (Turk et al. 1992; Jørgensen et al. 1993; Jørgensen and Jacobsen 1996; Danovaro et al. 1999; Dell'Anno et al. 1999b), as well as a source of exogenous nucleotides that are recycled by bacteria for the synthesis of new DNA (Paul et al. 1988; Paul et al. 1989). Moreover, extracellular DNA may participate in horizontal gene transfer through the natural transformation of bacterial cells (Ochman et al. 2000).

The removal of extracellular DNA from marine environments occurs through two basic mechanisms: (1) by competent bacterial cells internalizing DNA fragments (Lorenz and Wackernagel 1994; Dubnau 1999), and (2) by cell-associated and free DNases converting DNA into nucleosides and nucleobases suitable for bacterial uptake (Paul et al. 1988; Paul and Pichard 1989; Ammerman and Azam 1991). DNases being present in all aquatic environments, this latter process is expected to be the main route for extracellular DNA cycling (Lorenz and Wackernagel 1994).

Information dealing with degradation and turnover rates of extracellular DNA in deep-sea sediments is limited (Dell'Anno and Corinaldesi 2004), but this is of great importance for understanding the biogeochemical pathways of this component and the lateral gene transfer through natural transformation (Hermansson and Linberg 1994; Davison 1999).

Degradation rates of DNA in aquatic environments have been previously determined analyzing (1) the decrease of acid-precipitable labeled DNA (Novitsky 1986; Paul et al. 1989), (2) the conversion of supercoiled into relaxed-circular or linear plasmid DNA (Phillipps et al. 1989), and (3) the loss of hybridization signals of plasmid DNA in Southern transfer or dot blots (DeFlaun and Jeffrey 1987). However, these methods have been mainly applied for understanding survival of specific DNA sequences, rather than for providing quantitative information on actual degradation rates of extracellular DNA (Lorenz and Wackernagel 1994; Davison 1999).

In order to accurately estimate extracellular DNA turnover, after the measurement of extracellular DNA concentrations (Paul et al. 1987; 1988; Phillipps et al. 1989), the subsequent step is quantifying its degradation rates. In this chapter a procedure for estimating extracellular DNA degradation rates in marine sediments is reported. On the basis of the concentrations of extracellular DNA and its degradation rates, it is possible to calculate turnover times of extracellular DNA in marine sediments.

32.2 FIELD PROCEDURE

Sediments should be collected using multiple corers or any other device that allows the recovery of undisturbed sediment samples. Immediately after recovery, sediment subsamples, collected by using sterile syringes, are transferred into sterile tubes for further processing as detailed in the following text. Sediment samples should be analyzed immediately after collection. Alternatively, sediments can be stored at –20°C and kept frozen for a few weeks before analysis.

32.3 LABORATORY PROCEDURE

The method reported here is based on fluorometric detection of nuclease activity by means of fluorescent DNA analog [poly (dεA), poly deoxyribo-1-N^6ethenoadenylic acid]. Such method has been applied to in vitro studies, and is based on the increase of fluorescence of poly (dεA) due to degradation of polynucleotides (Chabbert et al. 1987); it is highly specific for detecting exonuclease activity (Takahashi and Ling 1991).

Here, we describe this fluorometric assay for estimating DNase activity in marine sediments (Dell'Anno and Corinaldesi 2004). By this procedure it is possible, quantifying the amount of nucleotides released from the degradation of the fluorescent DNA analog. The conversion of extracellular DNA into nucleotides represents, indeed, the key step for the subsequent bacterial uptake of nucleosides and nucleobases (Ammerman and Azam 1991; Paul et al. 1988; Paul and Pichard 1989).

32.3.1 Equipment and Instruments

Autoclave
Laminar air flow
Centrifuge for 15 mL tube
Analytical balance
Vortex
Sterile 15 mL tubes
Sterile syringes
Filters of 0.2 μm pore size
Pipettes
Fluorometer
Quartz cuvettes
Thermostatic incubator/water bath

32.3.2 Solutions

Substrate stock solutions: Fluorescent DNA analog [i.e., poly (dεA), poly deoxyribo-1-N^6ethenoadenylic acid] [NOTE 1].

Standard solutions: 1,N^6-ethenoadeninedeoxyribose-5′-monophosphate (dεAMP) dissolved in prefiltered and autoclaved seawater.

32.3.3 Procedure

1. Collect 0.5–1 cm^3 of sediment [NOTE 2].
2. Add prefiltered (0.2 μm pore size) and autoclaved seawater containing poly (dεA) up to a final volume of 3 to 5 mL [NOTE 3].

3. Agitate gently the sample until a homogeneous slurry is obtained.
4. Incubate samples in the dark and at *in situ* temperature for 1 h [NOTE 3].
5. Centrifuge at 2000 × g for 5 min.
6. Collect the supernatant and read its fluorescence at 320 nm excitation and 410 nm emission wavelengths [NOTE 4].
7. Convert the fluorescence of the sample into concentrations using standard solutions of dεAMP (1,*N*^{6}-ethenoadeninedeoxyribose-5′-monophosphate).

32.3.4 Notes

1. Poly (dεA) substrate is prepared by a chemical modification of poly (dA) (fragment length of 50 bases) by means of chloroacetaldehyde (Chabbert et al. 1987). Poly (dA), at concentration between 4 and 8×10^{-4} M (expressed in mol of nucleotides) is incubated for 48 h at 37°C with 0.19 M chloroacetaldehyde in a 20 mM phosphate buffer (pH 6.8) and then is purified using G-50 spin column chromatography (Amersham-Pharmacia) to remove chloroacetaldehyde prior to use. In these conditions, almost the totality of adenine residues are converted to their ethenoderivates. Polynucleotide concentration are determined by HPLC as the concentration of dεAMP (1,*N*^{6}-ethenoadeninedeoxyribose-5′-monophosphate) residues liberated after nuclease digestion. Once prepared, the fluorescent DNA analog is stable at −20°C for several months.
2. Autoclaved sediment samples must be utilized as blanks and processed according to the procedure for fresh sediment samples.
3. Control samples are prepared as reported in the points 1–3, chilled on ice for 1 min and immediately centrifuged at 4°C at 2000 × g for 5 min. After centrifugation, supernatant is collected and analyzed fluorometrically according to point 6. Kinetic parameters (i.e., Vm and Ks) can be calculated using Lineweaver-Burke plots of reaction velocity versus substrate concentrations (generally from 5 to 100 ng poly (dεA) mL^{-1}). Linearity of extracellular DNA hydrolysis should be checked by time-course analysis of poly (dεA) degradation.
4. Fluorescence of samples obtained after 1 h of incubation must be corrected by fluorescence of the control samples.
5. Calibration curves are obtained from standard solutions of dεAMP at concentrations ranging from 2.5 to 100 ng dεAMP mL^{-1}. DNase activity is expressed as ng nucleotides mL^{-1} of wet sediment h^{-1} or as ng nucleotides g^{-1} of sediment dry weight h^{-1}.

Collect 0.5–1 cm^3 of sediment

↓

Add pre-filtered (0.2 μm pore size) and autoclaved sea water with poly dεA (final volume 3–5 mL)

↓

Agitate gently

↓

Incubate samples in the dark (room T, 1 h)

↓

Centrifuge (2000 × g; 5 min)

↓

Read fluorescence of supernatant (320 nm excitation and 410 nm wavelengths)

↓

Convert the fluorescence of the sample into concentrations using standard solutions of dεAMP (1,N^6-ethenoadeninedeoxyribose-5'-monophosphate)

PROTOCOL 32.1 Determination of the extracellular DNA degradation rates.

REFERENCES

Ammerman, J. W. and F. Azam. 1991. Bacterial 5′-nucleotidase activity in estuarine and coastal marine waters: Role in phosphorous regeneration. *Limnology and Oceanography* 36: 1437–1447.

Chabbert, M., C. Cazenave, and C. Helene. 1987. Kinetic studies of recA protein binding to a fluorescent single-stranded polynucleotide. *Biochemistry* 26: 2218–2225.

Danovaro, R., A. Dell'Anno, A. Pusceddu, and M. Fabiano. 1999. Nucleic acid concentrations (DNA, RNA) in the continental and deep-sea sediments of the eastern Mediterranean: Relationships with seasonally varying organic inputs and bacterial dynamics. *Deep-Sea Research* 46: 1077–1094.

Davison, J. 1999. Genetic exchange between bacteria in the environment. *Plasmid* 42: 73–91.

Dell'Anno, A., M. Fabiano, G. C. A. Duineveld, A. Kok, and R. Danovaro. 1998. Nucleic acid (DNA, RNA) quantification and RNA/DNA ratio determination in marine sediments: Comparison of spectrophotometric, fluorometric, and high-performance liquid chromatography methods and estimation of detrital DNA. *Applied and Environmental Microbiology* 64: 3238–3245.

Dell'Anno, A., D. Marrale, A. Pusceddu, and R. Danovaro. 1999a. Particulate nucleic acid dynamics in a highly oligotrophic system: The Cretan Sea (Eastern Mediterranean). *Marine Ecology Progress Series* 186: 19–30.

Dell'Anno, A., M. Fabiano, M. L. Mei, and R. Danovaro. 1999b. Pelagic-benthic coupling of nucleic acids in abyssal location of the North-Eastern Atlantic. *Applied and Environmental Microbiology* 65: 4451–4457.

Dell'Anno, A., S. Bompadre, and R. Danovaro. 2002. Quantification, base composition, and fate of extracellular DNA in marine sediments. *Limnology and Oceanography* 47: 899–905.

Dell'Anno, A. and C. Corinaldesi. 2004. Degradation and turnover of extracellular DNA in marine sediments: Ecological and methodological considerations. *Applied and Environmental Microbiology* 70: 4384–4386.

Dubnau, D. 1999. DNA uptake in bacteria. *Annual Review of Microbiology* 53: 217–244.

Hermansson, M. and C. Linberg. 1994. Gene transfer in the marine environment. *FEMS Microbial Ecology* 15: 47–54.

Jiang, S. C. and J. H. Paul. 1995. Viral contribution to dissolved DNA in the marine environment as determined by differential centrifugation and kingdom probing. *Applied and Environmental Microbiology* 61: 317–325.

Jørgensen, N. O. G., N. Kroer, R. B. Coffin, X. H. Yang, and C. Lee. 1993. Dissolved free amino acids, combined amino acids, and DNA as sources of carbon and nitrogen to marine bacteria. *Marine Ecology Progress Series*, 98: 135–148.

Jorgensen, N. O. G. and C. S. Jacobsen. 1996. Bacterial uptake and utilisation of dissolved DNA. *Aquatic Microbial Ecology* 11: 263–270.

Lorenz, M. G. and W. Wackernagel. 1994. Bacterial gene transfer by natural genetic transformation in the environment. *Microbiology and Molecular Biology Reviews* 58: 563–602.

Ochman, H., J. G. Lawrence, and E. A. Groisman. 2000. Lateral gene transfer and the nature of bacterial innovation. *Nature* 405: 299–304.

Paul, J. H., W. H. Jeffrey, and M. F. DeFlaun. 1987. Dynamics of extracellular DNA in the marine environment. *Applied and Environmental Microbiology* 53: 170–179.

Paul, J. H., M. F. DeFlaun, and W. H. Jeffrey. 1988. Mechanisms of DNA utilisation by estuarine microbial populations. *Applied and Environmental Microbiology* 54: 1682–1688.

Paul, J. H., W. H. Jeffrey, A. W. David, M. F. DeFlaun, and L. H. Cazares. 1989. Turnover of extracellular DNA in eutrophic and oligotrophic freshwater environments of Southwest Florida. *Applied and Environmental Microbiology* 55: 1823–1828.

Paul, J. H. and S. L. Pichard. 1989. Specificity of cellular DNA-binding sites of microbial populations in a Florida reservoir. *Applied and Environmental Microbiology* 55: 2798–2801.

Phillipps, S. J., D. S. Dalgarn, and S. K. Young. 1989. Recombinant DNA in wastewater: pBR322 degradation kinetics. *Research Journal of the Water Pollution Control Federation* 61: 1588–1595.

Takahashi, M. and C. Ling. 1991. Use of a fluorescent DNA analog for fluorometric detection of DNase activity. *Analytical Biochemistry* 198: 246–249.

Turk, V., A. S. Rehnstam, E. Lundberg, and A. Hagstrom. 1992. Release of bacterial DNA by marine nanoflagellates, an intermediate step in phosphorus regeneration. *Applied and Environmental Microbiology* 58: 3744–3750.

33 Viral Production in Marine Sediments

CONTENTS

33.1 INTRODUCTION

The measure of virus abundance is not useful for inferring the dynamics of viruses in ecosystems. Therefore, estimates of viral production in different benthic environments are needed to assess the viral impact on their hosts, and to understand the role of the viral shunt in benthic food webs and biogeochemical cycles. Values of viral production in deep-sea sediments range from 1.4 to 19.8 × 10^8 virus g^{-1} h^{-1} (Danovaro et al. 2008a; Corinaldesi et al. 2007). Recent studies carried out in deep-sea sediments suggest that viral infection may be a primary cause of prokaryote mortality, ranging from 10% to 100% and increasing with increasing water depth (Danovaro et al. 2008b; Corinaldesi et al. 2007; Figure 33.1). Viral lysis of prokaryotic cells determines the switch of microbial biomass into organic debris (dissolved and particulate) that, for its intrinsic lability, is rapidly metabolized by the microbial community. Such processes generate a trophic loop in which prokaryotic biomass, removed through viral lysis, might be utilized again by noninfected prokaryotes. At the same time, the viral loop may determine a decrease of the C-transfer efficiency to the higher trophic levels (Danovaro et al. 2008b).

Different approaches have been utilized to determine viral production in aquatic sediments, and a standardized protocol is not available yet. The protocol described here illustrates the procedures that have been optimized for determining viral production deep-sea sediment samples (Dell'Anno et al. 2009). This protocol is based on the dilution technique (i.e., the dilution of the sediment sample with virus-free seawater; Mei and Danovaro 2004) and allows quantifying directly any increase in viral abundance over time without using conversion factors.

The idea behind this approach is that by reducing viral and host densities, the occurrence of new infections is also reduced. At the same time, sediment dilution renders almost negligible the impact of protozoan predation on viruses, and reduces viral losses due to the possible enzymatic degradation (Danovaro et al. 2008a).

Previous studies reported that differences between viral production in nondiluted and diluted sediments varied within a narrow range (7.5%–11.4%, Mei and Danovaro 2004). Moreover, Danovaro et

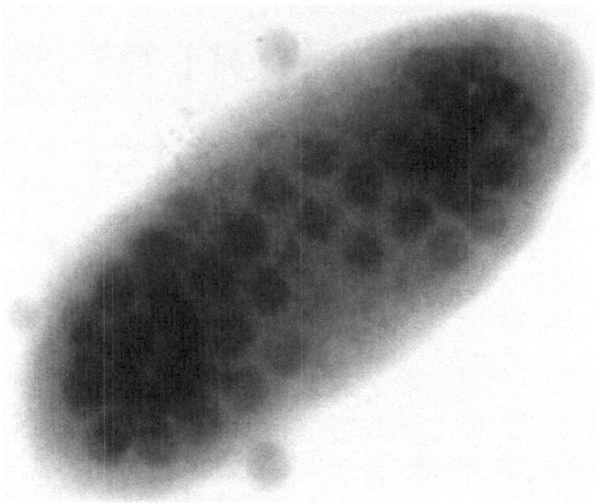

FIGURE 33.1 Transmission electron microscopy image of a prokaryotic cell containing viruses before the lysis of the cell. Some viruses are also evident around the cell wall.

al. (2008b), comparing samples with various degrees of dilution, demonstrated that viral production is not affected by increasing sediment dilution. This finding suggests that estimates of viriobenthos production are not affected by the sediment dilution, although comparative analyses based on other independent approaches (e.g., radiolabeled incorporation into viruses) are still lacking. Therefore, the comparison among different procedures represents the new challenge for a better assessment of the limits and confidence intervals of the methodologies available so far to estimate virus production in marine sediments.

33.2 FIELD PROCEDURES

33.2.1 Sampling

Surface sediment samples may be collected by means of specific equipments such as multiple corers or box corers. Since vertical distribution of viruses and viral production in sediments is expected to be heterogeneous (Corinaldesi et al. 2007), it is recommended that any manipulation be minimized during sampling and recovery, and that the sediment horizons are not mixed. Slicing of sediment horizons must be performed using sterile tools (i.e., tips, spatulas). Finally, all of material and solutions used for the subsequent processing of sediment samples must be washed using virus-free water (i.e., water prefiltered using 0.02 μm pore size filters) and sterilized by autoclave in order to maximize avoidance of any contamination.

33.2.2 Sample Storage

We recommend estimating viral production on fresh sediment samples by viral extraction, recovery, and slide preparation directly in the field. Slides, once prepared, should be stored at −20°C, and counts should be carried out as far as possible (within 30 days of storage) to avoid potential viral loss (Patel et al. 2007). Alternatively, sediment samples collected during time course experiments can be flash frozen in liquid nitrogen and stored at −20°C until laboratory analysis. Fresh samples should be filtered in the field and the filters stored at −20°C to avoid significant losses of viruses. Alternatively, filtration can be carried out in laboratory after sample freezing at −20°C.

33.3 LABORATORY PROCEDURE

33.3.1 Equipment and Instruments

Autoclave

Laminar air flow hood
Filtration apparatus: Sterile filter holder for 25 mm diameter filters with 15 mL funnel (Millipore), vacuum pump, vacuum flask (for discharge)
Sonicator
Centrifuge
Epifluorescence microscopy
Dewar for liquid nitrogen

33.3.2 Materials

Sterile 15 mL tubes
Sterile 5 mL syringes
Virus-free seawater [NOTE 1]
Tetrasodium pyrophosphate solution (SIGMA)
0.02 μm pore size, 25 mm ø, Anodisc, Al_2O_3 filters (Whatman)
0.02 μm pore size, 25 mm ø disposable syringe filters (Anotop, Whatman)
Sterile Petri dishes
2–20 μL, 100–1000 μL, 1–5 mL pipettes and sterile tips
Sterilized forceps
Glass slides and coverslips
DNase I from bovine pancreas
SYBR Gold stain (Invitrogen) [NOTE 2]
Antifade solution

33.3.3 Solutions

Tetrasodium pyrophosphate solution: Prepare 100 mM tetrasodium pyrophosphate (SIGMA) in reagent-grade water prefiltered through 0.02 μm filters (Anotop, Whatman).

DNase I working solution: Dilute the stock solution to 100 UmL^{-1} final concentration. For example, if the initial concentration of the stock solution is 10 $U\mu L^{-1}$, the working solution can be prepared diluting 100 μL of the stock solution in 10 mL virus-free reagent-grade water (working solution) [NOTE 3].

TE buffer: 10 mM Tris-HCl, 1 mM EDTA in reagent-grade water prefiltered onto 0.02 μm filters (Anotop, Whatman), pH 7.4–7.6.

SYBR Gold or Green working solution: The manufacturer's stock solution is in concentrated form (10,000×). An ideal working solution of SYBR has to be diluted till to a 2× final concentration. Generally, the stock solution (10,000×) is diluted in two further steps: Firstly, it is diluted 100 folds (100×) with 0.02 μm filtered TE buffer, than the solution obtained is diluted other 50 folds (2×). The working solution (2×) must be prepared fresh on the day of the sample filtration and slide preparation.

Antifade solution: 50% glycerol, 50% phosphate buffered saline (FLUKA, pH 7.4–7.6), and 0.5% ascorbic acid (weight/volume) [NOTE 4].

33.3.4 Procedures

1. Before sampling, prepare a set of 15 tubes as shown in Figure 33.2. The tubes should be prepared in three replicates for each time-interval incubation [NOTE 5].
2. Transfer 0.5–1 g of sediment in each tube.
3. Add 4.5 mL of virus-free seawater to each tube up to a final volume of 5 mL.
4. Shake manually the tubes in order to homogenize the sediment slurry.

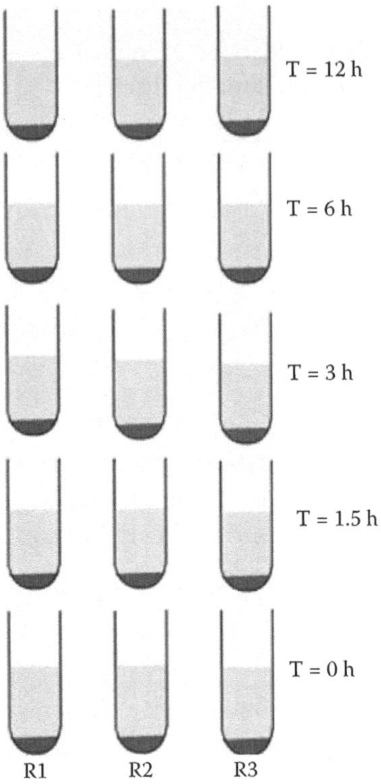

FIGURE 33.2 Scheme showing the set up of the time course for the determination of the viral production rates.

5. Stop the t_0 replicates possibly by freezing with liquid nitrogen, and then store the samples at –20°C until analysis [NOTE 6].
6. Incubate the other samples in the dark at *in situ* temperature conditions.
7. After each time interval, stop the viral production with liquid nitrogen, and then store them at –20°C until analysis.
8. Follow the protocol described in the procedures section of the paragraph dedicated to the viral abundance in marine sediments.

33.3.5 Viral Abundance Calculations

$$\text{Viral abundance (viruses g}^{-1}) = [\text{Average virus number for each optical field}] \cdot \frac{\text{Optical field coefficient} \cdot \text{extraction coefficient} \cdot \text{dilution factor}}{\text{sediment dry weight}}$$

where

Optical field coefficient = (filtration area/counting area).
Extraction coefficient = 1.66.
Sediment dry weight = sediment weight after desiccation (60°C at least for 24 h).

33.3.6 Viral Production Calculations

Viral production rates are determined on the basis of first-order regressions of viral abundance versus time. Viral production rates could be expressed as produced viruses $g^{-1}\ h^{-1}$. An example of regression is reported in Figure 33.3.

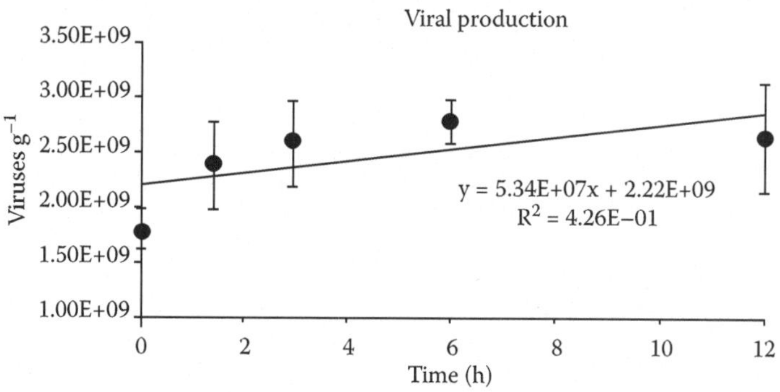

FIGURE 33.3 Viral production rates, determined from the slope of the first-order regression of viral abundance versus time for triplicate incubations. Bars represent standard deviations.

33.3.7 Problems

Viral abundance typically increases over time within 12 h, from the beginning of the experiment. In most cases, viral abundance increases linearly up to 6 h of incubation, then virus abundance begins to decay. In this case, viral production rate is determined from the slope of the first-order regression in the interval t_0–t_{6h}. However, it can also happen that viral abundance does not increase linearly with time. In this case, we have to select the incubation interval during which the highest increment of viral abundance is observed. The increment is calculated as the difference between viral abundance t_n and t_0 normalized to the number of hours of incubation.

When a decreasing pattern (when compared to values observed at t_0) is observed, the time-course experiments must be reperformed with shorter incubation times.

33.3.8 Notes

1. Virus-free seawater should be prepared before all other procedures by filtering seawater through Anotop 0.02 μm filters.
2. A detailed information on SYBR Gold is reported in the section Notes of the chapter discussing the determination of viral abundance in marine sediments.
3. Higher DNase concentrations than those recommended could damage virus particles, thus altering the values measured.
4. We recommend to add glycerol after PBS in order to avoid difficulties in the mixing of the reagents.
5. Viral production experiments typically include 5 time intervals for each set of replicates: t_0 (which corresponds to the beginning of the experiment), t_1 (after 1.5 h from the beginning of the experiment), t_2 (after 3 h), t_3 (after 6 h), and t_4 (after 12 h). We recommend selecting short time intervals in the first part of the time course experiment, as generally the highest increase of viral abundance is observed within 3–6 h from the beginning of the experiment.
6. The best way for stopping and preserving viral production is the use of liquid nitrogen. In absence of liquid nitrogen, viral production could be stopped by freezing at −20°C. As described in the previous section on the determination of viral abundance in marine sediments. The use of preservatives such as formaldehyde results in less viral abundance and in a large underestimation of viral abundances at each time interval.

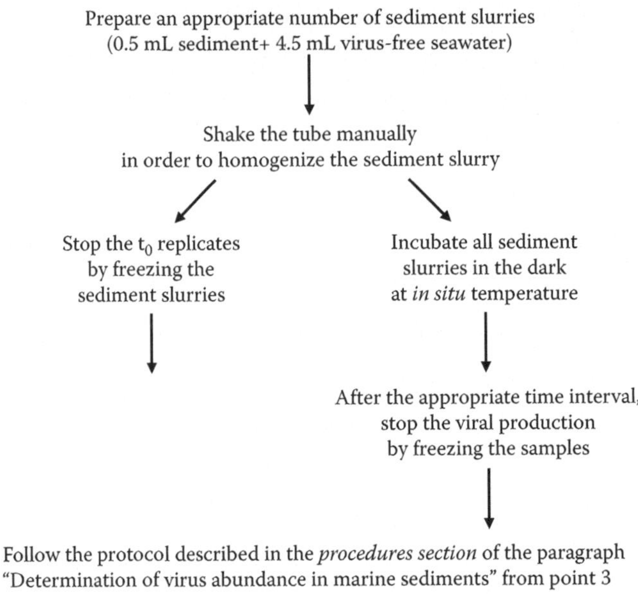

PROTOCOL 33.1 Determination of viral production in marine sediments.

REFERENCES

Corinaldesi, C., A. Dell'Anno, and R. Danovaro. 2007. Viral infection plays a key role in extracellular DNA dynamics in marine anoxic systems. *Limnology and Oceanography* 52: 508–516.

Danovaro, R., C. Corinaldesi, M. Filippini et al. 2008a. Viriobenthos in freshwater and marine sediments: A review. *Freshwater Biology* 53: 1168–1213.

Danovaro, R., A. Dell'Anno, C. Corinaldesi et al. 2008b. Major viral impact on the functioning of benthic deep-sea ecosystems. *Nature* 454: 1084–1087.

Dell'Anno, A., C. Corinaldesi, M. Magagnini, Danovaro R. 2009. Determination of viral production in aquatic sediments using the dilution-based approach. *Nature Protocols* 4: 1013–1022.

Mei, M. L. and R. Danovaro. 2004. Virus production and life strategies in aquatic sediments. *Limnology and Oceanography* 49: 459–470.

Patel, A., R. T. Noble, J. A. Steele, M. S. Schwalbach, I. Hewson, and J. A. Fuhrman. 2007. Virus and prokaryote enumeration from planktonic aquatic environments by epifluorescence microscopy with SYBR Green I. *Nature Protocols* 2: 269–276.

34 Determination of Living/Dead and Active/Dormant Bacterial Fractions in Marine Sediments

CONTENTS

34.1 INTRODUCTION

In the last years, a number of methodologies have been proposed for estimating the metabolically active fraction of marine prokaryotes. These methods include the analysis of electron transport system by the use of a redox dye (Rodriguez et al. 1992; Proctor and Souza 2001), the incorporation of radiolabeled substrates (Carman 1993), the direct viable count after addition of an antibiotic cocktail (Joux and Lebaron 1997), and the detection of nucleoid-containing cells (Zweifel and Hagström 1995; Luna et al. 2002). All of these technique have advantages and disadvantages, depending also on the kind of samples in which they are applied (e.g., water versus sediments).

The most commonly used method for discriminating between living and dead prokaryotes in environmental samples is based on the use of fluorescent stains. Among these stains, the most frequently used are propidium iodide (PI), ethidium bromide, ethidium monoazide, Calcofluor white, SITOX Green and Live/Dead Bac-Light (Lloyd and Hayes 1995; McFeters et al. 1995; Porter et al. 1996; Lebaron et al. 1998; Williams et al. 1998; Boulos et al. 1999; Sherr et al. 1999; Luna et al. 2002; Quéric et al. 2004). Due to their large molecular sizes and shapes, these can have steric hindrance effects, as they can penetrate only those prokaryotic cells in which membrane integrity is compromised (McFeters et al. 1995). The evidence of membrane damage (i.e., a lack of membrane integrity) is a widely accepted criterion for identifying dead cells, which are typically discriminated from living cells by simultaneous staining with a permeant DNA-binding dye (e.g., SYBR Green; molecular probes). However, results based on membrane integrity should be interpreted with caution. Differential permeabilities of intact membranes of different prokaryotic strains present limitations to this approach (Gregori et al. 2001). This problem is also particularly evident in prokaryotic cells that are competent for natural transformation, which can display higher membrane permeabilities under the influence of particular environmental conditions (Lorenz and Wackernagel 1994). This method is also dependent on the properties of each specific stain (e.g., molecular weight and potential for steric hindrance during membrane diffusion), which can lead to bias due to false staining (Williams et al. 1998). All these considerations highlight the need for a careful validation of any procedure in order to avoid erroneous interpretations. Moreover, recent studies have indicated that the discrimination between living and dead prokaryotes is not sufficient to understand whether the living fraction is metabolically active or not (Luna et al. 2002; Manini and Danovaro 2006).

This protocol describes the procedure for the determination of the proportions of living and dead cells in marine sediments coupled with a staining/destaining procedure, which allow a visible nucleoid region (the nucleoid containing cells, NuCC), which is utilized as an indication of presence of activity. The cells without a visible nucleoid region can be defined as inactive, dormant or dead (Gasol et al. 1995). An alternative approach for the quantification of the live/dead fraction of benthic prokaryotes is the use of FISH (Fluorescence *In Situ* Hybridization), which also described in this book and is based on the use of fluorescent primers. Ligating the RNA provides an indication of cells displaying metabolic activity.

34.2 FIELD PROCEDURES

Sediment samples are generally collected using multiple corers, which allow recovering undisturbed, intact sediment cores. Immediately after retrieval, sediment subsamples are gently removed from the corer inserting a cut-off 10 mL plastic syringe along the axis of the core. The syringe plunger is held fixed at the sediment surface while the barrel is pushed into the sediment for 1 cm, in an overall procedure that is analogous to piston coring. The syringes are then removed from the sediment and the sediment transferred into a sterile tube and resuspended in sterile water. Only fresh sediment samples can be processed [NOTE 1].

34.3 LABORATORY PROCEDURE

34.3.1 Equipment and Instruments

Sterile 15 mL tubes
Sterile 5 mL syringes
Sterile seawater [NOTE 2]
Tetrasodium pyrophosphate solution (SIGMA)
Anodisc filters, 0.2 μm pore size, 25 mm ø (Whatman)
Syringe filters, 0.2 μm pore size (Whatman Puradisc AS 25 mm ø)
Sterile Petri dishes
2–20 μL, 100–1000 μL, 1–5 mL pipettes and sterile tips
Sterilized forceps
Glass slides and coverslips
SYBR Green I stain (Invitrogen)
Propidium iodide [NOTE 3]
2-Propanol
Antifade solution

34.3.2 Solutions

Tetrasodium pyrophosphate solution: Prepare a stock solution of 500 mM tetrasodium pyrophosphate (SIGMA) in reagent-grade water prefiltered onto 0.2 μm filters (Whatman Puradisc).

SYBR Green I working solution: Original SYBR Green I stock (Invitrogen) is in concentrated form (10,000×), dissolved in DMS, and should be stored in the dark at –20°C. Before dilution, the stock solution must be de-frozen. Since the fluorochrome and the DMS have different cryoscopic points, be sure that both are completely melted. An ideal working solution of SYBR Green I has to be diluted 20 folds, in order to obtain 500× final concentration.

The stock solution (10,000×) must be diluted by sterile reagent-grade water in order to avoid contaminations. The working solution (500×) can be stored in the dark at 4°C, and is stable for a few weeks.

Propidium Iodide solution (PI) [NOTE 4]: Prepare a stock solution of 1 mg mL^{-1} (molecular probes) in reagent-grade water prefiltered onto 0.2 μm filters (Whatman Puradisc). Protect from light.

Antifade solution: Dissolve 50% glycerol, 50% phosphate buffered saline (FLUKA, pH 7.4–7.6), and 0.5% ascorbic acid (weight/volume) [NOTE 5].

Immersion oil for microscopy.

34.3.3 Procedure

1. Collect ca. 0.5–1 mL of sediment in a sterile tube.
2. Add sterile water for a final volume of 5 mL.
3. Incubate fresh subsamples (n = 3) with tetrasodium pyrophosphate at a final concentration of 5 mM (i.e., 50 μL of the 500 mM solution).
4. Incubate the samples in the dark and at room temperature for 15 min.
5. Sonicate the samples for 3 min with 30 s intervals between each minute for detaching the cells from the sediment particle. During intervals samples are manually shaken [NOTE 6].
6. Wait until the suspended sediment has deposed at the bottom of the tube. At the end of this step, a clear supernatant should be recognized [NOTE 7].
7. Dilute the supernatant with sterile seawater firstly 10×. Then, depending on the expected abundance of prokaryotes, dilute the samples from 100× to 1000× [NOTE 8].
8. Using forceps, place 0.2 μm anodisc filters on the filter holders. Place the funnel on the filter holder taking care that filter holders and funnels are perfectly overlapped.
9. Transfer the supernatant into the funnel and filter with vacuum pressure no greater than 20 kPa (or 150 mmHg) to avoid cells damage.
10. Remove the filter and place it on a sterile Petri dish.

34.3.4 Estimating the Living/Dead Prokaryote

11. Stain with 20 μL of SYBR Green I working solution and 20 μL of propidium iodide working solution.
12. Incubate in the dark for 20 min.
13. Wash the filter 3 times with 3 mL of sterile seawater.
14. Place a drop (20 μL) of antifade solution onto a clean slide, and mount the filter on the drop. Spot 20 μL antifade solution onto a clean coverslip, flip it over to cover the sample filter. Try to avoid the formation of bubbles.
15. Add one drop of immersion oil to the top of the coverslip.
16. The stained filters were examined by epifluorescence microscopy (1000× magnification) using the BP 450–490 filter set for SYBR Green I (yellow-green florescence) for total counting and the BP 546/12 filter for PI (red fluorescence) Figure 34.1 for dead cells.
17. Centrifuge at 800 × g speed for 1 min and discard the supernatant.
18. Place the tubes in an oven at 60°C for 24–48 h, until constant weight.
19. Weigh the dry sediment using an analytical balance.

34.3.5 Estimating the Active/Dormant Prokaryote [NOTE 9]

20. After the living/dead counting (as described earlier), the sediment bearing filters are placed another time in the filtration set-up.

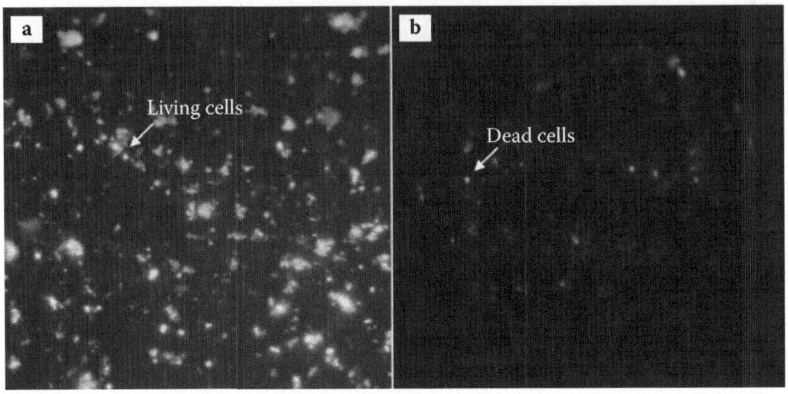

FIGURE 34.1 *(A color version of this figure follows page 234.)* Marine prokaryotes examined using epifluorescence microscopy (magnification 1000×; Zeiss Axioskop 2, BP 490 filter set for SYBR Green I for the yellow-green florescence and the BP 545 filter for propidium iodide). Figure illustrates prokaryotes stained with the SYBR Green I (total cells) (a) and prokaryotes stained with propidium iodide (dead cells) (b).

21. Add 10 mL of 2-propanol and incubate in the dark for 10 min (to remove the excess of stain). This step allows the estimation of the non-nucleoid-containing cells (non-NuCC) by washing the fraction of stain (SYBR Green I) not bound to the nucleoid.
22. Another time, place a drop (20 μL) of antifade solution onto a clean slide, and mount the filter on the drop. Spot 20 μL antifade solution onto a clean coverslip, flip it over to cover the sample filter. Try to avoid the formation of bubbles.
23. Add one drop of immersion oil to the top of the coverslip.
24. The stained filters are examined by epifluorescence microscopy, where the nucleoid (NuCC) of total cells appeared green due to the excitation of the SYBR Green I dye with which the cells have been stained.
25. After the counting of prokaryotes under blue excitation (for the NuCC counts), the sediment-bearing filters are immediately examined for the presence of red cells with a visible nucleoid region, and these nucleoid-containing cells are counted. This procedure is used for destaining the red fluorescent cells (i.e., those with compromised membrane integrity) on the filters stained using PI. This allowed a distinction to be made between actual dead cells (without a visible nucleoid region) and the red-stained cells with a still visible nucleoid region (Figure 34.1).

34.4 NOTES

1. Sample storage conditions have a critical effect on cell counts using PI staining. In fixed samples stained with PI, the number of living cells apparently decreased during storage, with a significant decrease already evident 2 h after fixation. After 40 days of preservation, the dead-cell percentage increased up to ca. 50% when compared with fresh cells (Manini and Danovaro 2006).
2. Sterile seawater (i.e., autoclaved and filtered through syringe filters 0.2 μm pore sizes) should be prepared before the beginning of the sample processing in order to avoid the blocking of important steps of the protocol.
3. PI is a double-charged phenanthridium derivative and is one of the most common stains for dead cells (Haugland 2002). PI has a molecular weight of 668.4 and is thus assumed to be unable to penetrate cell membranes. Evidence increasingly suggests that use of PI can lead to an overestimation of the size of the dead prokaryotic fraction, and can thus result in a bias between the living/dead prokaryotic counts (Williams et al. 1998; Barbesti et

al. 2000). The counts of the dead (red-stained) cells using PI increase progressively with increasing stain concentration (>20 μg mL^{-1}).

4. Do not use PI working concentrations >20 μg mL^{-1} as it is possible utilize another impermeant stains, the ethidium homodimer-2 (EthD-2, molecular probes). EthD-2 consists of two phenanthridinium fluorophores with a large spacer group, and it has four charges. Results of propidium iodide and ethidium homodimer-2 in the determination of the living/dead prokaryotic fractions in sediment samples do not display significant differences. However, at lower (e.g., 1.0–5.0 mg mL^{-1}) or higher (e.g., 30–40 mg mL^{-1}) concentrations, the living prokaryotic counts displayed significant differences between the two stains. Indeed, the counts of the dead (red-stained) cells using PI increase progressively with increasing stain concentration, whereas the counts using EthD-2 do not. Such differences are potentially related to the much higher molecular weight of EthD-2 (1292.71), which will prevent this compound from penetrating prokaryotic cells that do not have large pores or that have suffered membrane damage, which will hereafter be described as cells lacking in membrane integrity.
5. We recommend adding glycerol after PBS in order to avoid difficulties in the mixing of the reagents.
6. The treatment with ultrasounds does not alter the results obtained for the estimate of the living cells within 3 min. However, larger treatment could significantly alter the sample, determining the bursting of prokaryotic cells.
7. The deposition of the sediment typically requires from 30 s up to 15 min, depending on the grain size.
8. Generally, the sediment dilution ranges from 500× to 1000× for mud sediments of coastal and nutrient-rich samples, whereas it runs from 100× to 250× for deep-sea samples and sand sediments in coastal samples.
9. The living/dead staining protocol is coupled to the staining/destaining procedure, making it possible to gather additional synoptic information on the physiological state of benthic prokaryotes. The fraction of active prokaryotes can then be estimated as the number of prokaryotes displaying a visible nucleoid region (the nucleoid-containing cells, NuCC), whereas the cells without a visible nucleoid region can be defined as inactive/dormant or dead (Gasol et al. 1995).

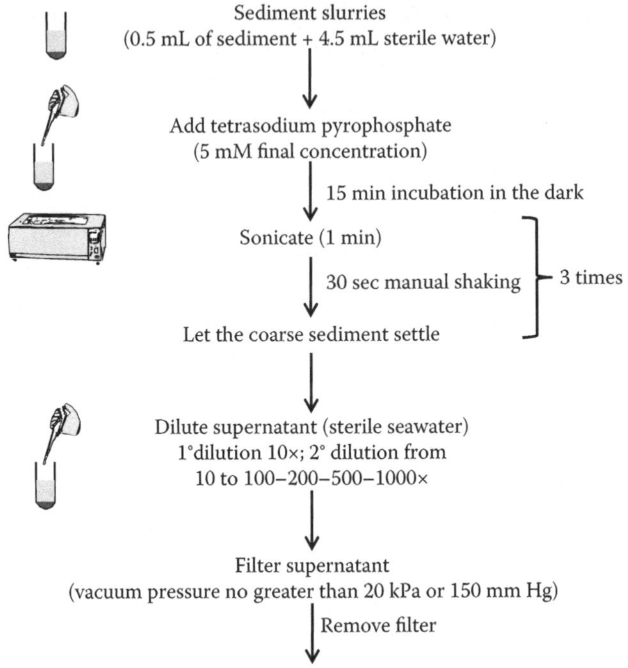

PROTOCOL 34.1 Protocol for extraction of prokaryotic cells from marine sediments.

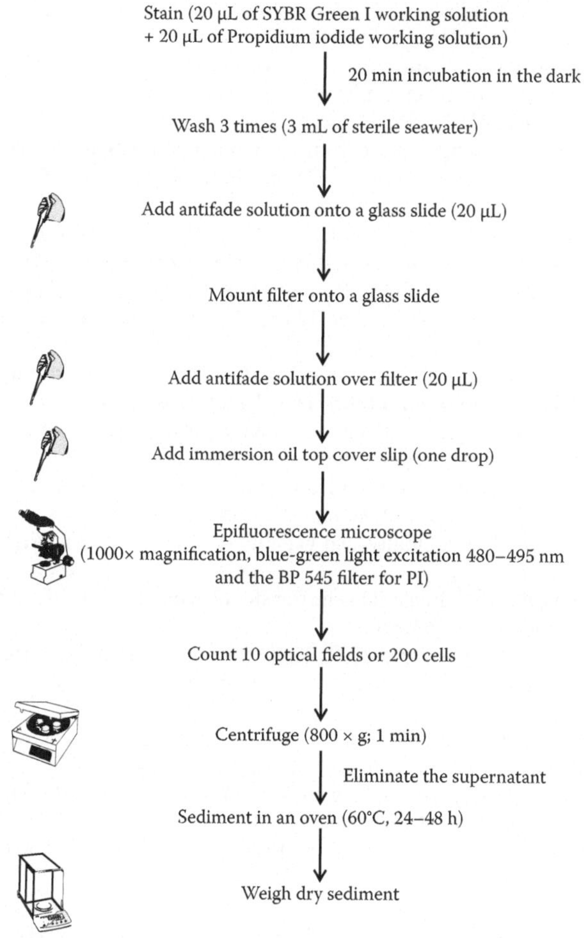

PROTOCOL 34.2 Protocol for estimating the living/dead prokaryotic fraction in marine sediments.

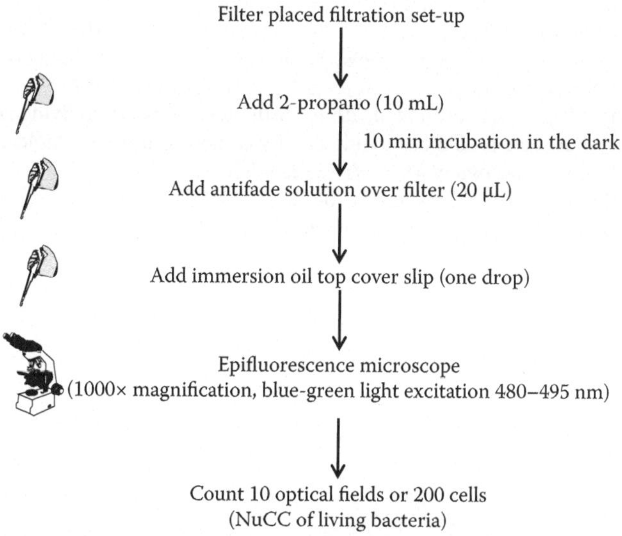

PROTOCOL 34.3 Protocol for quantifying the active versus dormant prokaryotic fraction in marine sediments.

REFERENCES

Boulos, L., M. Prevost, B. Barbeau, J. Coallier, and R. Desjardins. 1999. LIVE/DEAD Baclight: Application of a new rapid staining method for direct enumeration of viable and total bacteria in drinking water. *Journal of Microbiological Methods* 37: 77–86.

Carman, K. R. 1993. Microautoradiographic detection of microbial activity. In *Handbook of methods in aquatic microbial ecology*, ed. P. F. Kemp, B. F. Sherr, E. B. Sherr, and J. J. Cole, 397–404. Boca Raton: Lewis Publishers.

Gasol, J. M., P. A. Del Giorgio, R. Massana, and C. M. Duarte. 1995. Active versus inactive bacteria: Size-dependence in a coastal marine plankton community. *Marine Ecology Progress Series* 128: 91–97.

Gregori, G., S. Citterio, A. Ghiani et al. 2001. Resolution of viable and membrane compromised bacteria in freshwater and marine waters based on analytical flow cytometry and nucleic acid double staining. *Applied and Environmental Microbiology* 4662–4670.

Haugland, R. P. 2002. *Handbook of fluorescent probes and research chemicals*. http://www.invitrogen.com/site/us/en/home/References/Molecular-Probes-The-Handbook.html

Joux, F. and P. Lebaron. 1997. Ecological implications of an improved direct viable count method for aquatic bacteria. *Applied and Environmental Microbiology* 63: 3643–3647.

Lebaron, P., P. Catala, and N. Parthuisot. 1998. Effectiveness of SYTOX Green stain for bacterial viability assessment. *Applied and Environmental Microbiology* 64: 2697–2700.

Lloyd, D. and A. J. Hayes. 1995. Vigor, vitality, and viability of microorganisms. *FEMS Microbiology Letters* 133: 1–7.

Lorenz, M. and W. Wackernagel. 1994. Bacterial gene transfer by natural genetic transformation in the environment. *Microbiology Reviews* 58: 563–602.

Luna, G. M., E. Manini, and R. Danovaro. 2002. High fraction of dead and inactive bacteria in coastal marine sediments: Comparison of protocols and ecological significance. *Applied and Environmental Microbiology* 68: 3509–3513.

Manini, E. and R. Danovaro. 2006. Synoptic determination of living/dead and active/dormant bacterial fractions in marine sediments. *FEMS Microbiology Ecology* 55: 416–423.

McFeters, G. A., F. P. Yu, B. H. Pyle, and P. S. Stewart. 1995. Physiological assessment of bacteria using fluorochromes. *Journal of Microbiological Methods* 21: 1–13.

Porter, J., D. Deere, R. Pickup, and C. Edwards. 1996. Fluorescent probes and flow cytometry: New insights into environmental bacteriology. *Cytometry* 23: 91–96.

Proctor, L. M. and A. C. Souza. 2001. Method for enumeration of 5-cyano-2,3-dytolyl tetrazolium chloride (CTC)-active cells and cell-specific CTC activity of benthic bacteria in riverine, estuarine, and coastal sediments. *Journal of Microbiological Methods* 43: 213–222.

Quéric, N. V., T. Soltwedel, and W. E. Arntz. 2004. Application of a rapid direct viable count method to deep-sea sediment bacteria. *Journal of Microbiological Methods* 57: 351–367.

Rodriguez, G. G., D. Phipps, K. Ishiguro, and H. F. Ridgway. 1992. Use of a fluorescent redox probe for direct visualisation of actively respiring bacteria. *Applied and Environmental Microbiology* 58: 1801–1808.

Sherr, B. F., P. Del Giorgio, and E. B. Sherr. 1999. Estimating abundance and single-cell characteristics of respiring bacteria via the redox dye CTC. *Aquatic Microbial Ecology* 18: 117–131.

Williams, S. C., Y. Hong, D. C. A. Danavall et al. 1998. Distinguishing between living and nonliving bacteria: Evaluation of the vital stain propidium iodide and its combined use with molecular probes in aquatic samples. *Journal of Microbiological Methods* 32: 225–236.

Zweifel, U. L. and A. Hagström. 1995. Total counts of marine bacteria includes a large fraction of non-nucleoid-containing bacteria (ghost). *Applied and Environmental Microbiology* 61: 2180–2185.

35 Prokaryotic Biomass in Marine Sediments

CONTENTS

35.1 INTRODUCTION

A crucial step in the comprehension of the ecosystem functioning is the estimate of the magnitude of prokaryotic biomass as a particulate carbon and nitrogen pool relative to other major particulate pools of these elements, which participate in nutrient cycling and sinking flux. Essential elements for these estimates are the estimate of prokaryotic cell size and the use of appropriate conversion factors for estimating prokaryotic biomass from cell volume (biovolume).

At present, the most reliable approach for the estimate of prokaryotic cell size (and then the biovolume) is epifluorescence microscopy (EFM). Compared to other approaches (flow cytometry and transmission electron microscopy), EFM allow for the analysis of a larger number of samples with lower costs and greater dependability. The EFM drawback, namely, "fuzzy" edges of cell images (that hinder the measurement of the cell itself) is counterbalanced by the sensitive drastic shrinkage of cells in electron microscopy.

There are no generally accepted factors for converting biovolume into biomass. The first biovolume-to-biomass conversion factor reported in the literature (i.e., 121 fg of C μm^{-3}) was based on measurements of cultured *Escherichia coli* (Watson et al. 1977), which are 100-fold larger than native bacterioplankton. This value varies up to fivefold in the subsequent literature (Fuhrman and Azam 1980; Bjørnsen 1986; Bratbak 1985; Bratbak and Dundas 1984; Nagata 1986).

Bratbak and Dundas (1984) and Bratbak (1985) triggered a small revolution in bacterial appreciation with new estimates ranging from 160 and 930 fg C μm^{-3}. These estimates were based on pure cultures of marine bacteria grown on lab media and natural samples grown in enriched seawater. The first estimates of carbon per cell for native bacterioplankton grown on naturally occurring substrates were given by Lee and Fuhrman (1987), who grew natural assemblages from small (<0.8 µm) filtrates in particle-free seawater, and related microscopic volume estimates to C and N masses measured with a CHN analyzer. Interestingly, they determined that C per unit volume was itself inversely proportional to cell volume, such that C per cell was relatively constant at 20 fg over the observed size range of 0.036–0.077 μm^3. The value of 20 fg of C cell^{-1} has been assumed as a canonical status in marine microbiology (Cho and Azam 1990; Ducklow and Carlson 1992).

To date, the most reliable conversion factor used for biomass estimates is 310 fg C μm^{-3}, calculated by Fry (1990).

35.2 FIELD PROCEDURES

35.2.1 SAMPLING

Surface sediment samples may be collected by multiple corers or box corers, depending on the sampling depth.

Since vertical distribution of prokaryotes in sediments could be strongly heterogeneous (Mei and Danovaro 2004), in order to provide more realistic estimates it is strongly recommended that the sediment layers are not mixed during sampling and recovery of samples. This implies that sampling must be performed limiting sediment perturbation and maintaining an effective isolation of sediment samples from the surrounding waters. Slicing of sediment horizons must be performed using sterile utensils (i.e., tips, spatulas). Finally, all material and solutions used for the subsequent processing of sediment samples must be sterilized by autoclave in order to avoid any contamination.

35.2.2 SAMPLE STORAGE

The standard method to preserve samples for prokaryotic counting is the addition of sterile formaldehyde (2% final concentration; Luna et al. 2002) or freezing samples.

35.3 LABORATORY PROCEDURE

35.3.1 EQUIPMENT AND INSTRUMENTS

Autoclave
Laminar air flow
Filtration apparatus: Sterile filter holder for 25 mm diameter filters with 15 mL funnel (Millipore), vacuum pump, vacuum flask (for discharge)
Sonicator
Centrifuge
Epifluorescence microscopy

35.3.2 MATERIALS

Sterile 15 mL tubes
Sterile 5 mL syringes
Sterile seawater [NOTE 1]
Tetrasodium pyrophosphate solution (SIGMA)
Anodisc filters, 0.2 μm pore size, 25 mm ø (Whatman) if samples are stained with SYBR Green I
Black polycarbonate membrane filters (diameter 25 mm, pore size 0.2 μm; (Whatman) if samples are obtained with Acridine Orange
Syringe filters, 0.2 μm pore size
Sterile Petri dishes
2–20_ μL, 100–1000 μL, 1–5 mL pipettes and sterile tips
Sterilized forceps
Glass slides and coverslips

Antifade solution
Immersion oil for epifluorescence microscopy

35.3.3 Solutions

Tetrasodium pyrophosphate solution: Prepare 500 mM tetrasodium pyrophosphate (SIGMA) in reagent-grade water prefiltered onto 0.2 μm filters (Anotop, Whatman).

SYBR Green I working solution or Acridine Orange working solution. If samples are stained with SYBR Green I, use an appropriate antifade solution (see Chapter 12, "Prokaryotic Abundance" for details).

35.3.4 Procedure

1. Collect ca. 0.5–1g of sediment in a sterile tube.
2. Add sterile formaldehyde (2% final concentration) for a final volume of 5 mL [NOTE 2].
3. Add tetrasodium pyrophosphate at a final concentration of 5 mM (i.e., 20 μL of the 500 mM solution).
4. Incubate the samples in the dark and at room temperature for 15 min.
5. Sonicate the samples for 3 min with 30 s intervals between each minute. During intervals samples are manually shaken [NOTE 3].
6. Wait until the suspended sediment has deposed at the bottom of the tube. At the end of this step, a clear supernatant should be recognized [NOTE 4].
7. First dilute the supernatant with sterile seawater 10×. Then, depending on the expected abundance of prokaryotes, dilute the samples from 100× to 1000× [NOTE 5].
8. Using forceps, place 0.2 μm filters on the filter holders. Place the 15 mL funnel on the filter holder taking care that filter holders and funnels are perfectly overlapped.

35.3.4.1 Procedure with SYBR Green I

9. Transfer the supernatant into the funnel and filter with vacuum pressure no greater than 20 kPa (or 150 mmHg) to avoid destroying of cells.
10. When the sample has passed through, remove the filter and place it on a sterile Petri dish.
11. Stain with SYBR Green I (see Chapter 12, "Prokaryotic Abundance" for details on staining steps).
12. Incubate in the dark for 20 min at room temperature.

or

35.3.4.2 Procedure with Acridine Orange

9. Stain the sample diluted with 0.5 mL of Acridine Orange working solution.
10. Samples are manually shaken for 30 s.
11. Incubate in the dark at room temperature for a total of 5 min.
12. Transfer the stained sample into the funnel and filter, with vacuum pressure no greater than 20 kPa (or 150 mmHg) to avoid destroying prokaryotic particles.
13. Wash the filter 3 times with 3 mL of sterile seawater.
14. If samples are stained with SYBR Green I, use the appropriate antifade solution (see Chapter 12, "Prokaryotic Abundance" for details), whereas if the samples are stained with Acridine Orange, first add a drop of reagent-grade water under the coverslip before examining the slide (see Chapter 12 Prokaryotic Abundance for details).
15. Add one drop of immersion oil to the top of the cover slip, and examine the slide under blue-green light excitation (Fitch filter, excitation at 450–490 nm) using an epifluorescence

microscope (1000× magnification). Slides can be preserved up to 3 months at −20°C. After use, filter holders and funnels should be washed with 1N HCl.

16. Centrifuge at 800 × g speed for 1 min the original tube with sediment sample and eliminate the supernatant.
17. Place the tubes in an oven at 60°C for 24–48 h, until the sediment is dry.
18. Weigh the dry sediment using an analytical balance.

35.3.5 Estimate of Prokaryotic Biovolume and Biomass

1. By mean of an epifluorescence microscope equipped with a micrometric scale, examine the cellular dimensions in each replicate. Evaluate the diameter of cocci-like cells and the main dimensions of bacilli-like cells. We used three different sizes: small (<0.5 μm), medium (0.5< x <1.0 μm), and large (≥1 μm).
2. Examine all the cells of the slide for each replicate, dividing the cells in three groups (small, medium, and large).
3. After examinations, calculate the biovolume of each group of cell sizes (expressed as μm^3 $cell^{-1}$), considering each cell like a sphere $[(4/3)r^3 \times \pi]$.
4. Multiply the biovolume of each group of cells (small, medium, and large) for the conversion factor (310 fg C μm^{-3}) to obtain the biomass for each group of cells (expressed as fg C $cell^{-1}$).
5. Sum the biomass of the small, medium, and large cells (fg C $cell^{-1}$) and divide for sediment dry weight. Results are expressed as ng C g^{-1}.

35.4 NOTES

1. Sterile seawater (i.e., autoclaved and filtered through syringe filters (0.2 μm pore sizes) should be prepared before the beginning of the sample processing. Do not use reagent-grade water instead of sterile seawater: This could cause an osmotic shock and cell burst, leading to a considerable underestimation of prokaryotic abundance.
2. The formaldehyde solution should be made with sterile seawater in order to avoid the osmotic cell burst.
3. Do not increase the sonication time, since the consequent increase of temperature could alter the sample, determining the bursting of prokaryotic cells.
4. The deposition of the sediment typically requires from 30 s up to a few minutes, depending on the grain size. The deposition of small-size sediments could requires more time (maximum 15 min) in order to avoid prokaryote losses.
5. Generally, the sediment dilution ranges from 500× to 1000× for coastal and nutrient-rich samples, and from 100× to 250× for deep-sea samples.

Epifluorescence microscope (1000× magnification, blue-green light excitation 480–495 nm, micrometric scale)

↓

Evaluate diameter of cocci and bacilli-like cells (count 10 optical fields or 200 cells)

↓

Divide cells in size groups (small, medium, and large)

↓

Calculate biovolume (μm^3 $cell^{-1}$)

↓

Multiply biovolume cells (small, medium and large) for 310 fg C μm^{-3}

↓

Sum the biomass small, medium and large cells and divide sediment dry weight (ng C g^{-1})

PROTOCOL 35.1 Determination of the prokaryotic biomass in marine sediments.

REFERENCES

Bjørnsen, P. K. 1986. Automatic determination of bacterioplankton biomass by image analysis. *Applied and Environmental Microbiology* 51: 1199–1204.

Bratbak, G. and I. Dundas. 1984. Bacterial dry matter content and biomass estimations. *Applied and Environmental Microbiology* 48: 755–757.

Bratbak, G. 1985. Bacterial biovolume and biomass estimation. *Applied and Environmental Microbiology* 49: 1488–1493.

Cho, B. C. and F. Azam. 1990. Biogeochemical significance of bacterial biomass in the ocean's euphotic zone. *Marine Ecology Progress Series* 63: 253–259.

Ducklow, H. W. and C. A. Carlson. 1992. Oceanic bacterial production. In *Advances in microbial ecology*, ed. K. C. Marshall, 113–181. New York: Plenum Press.

Fry, J. C. 1990. Direct methods and biomass estimation. In *Methods in Microbiology*, ed. R. Grigorova and J. R. Norris, 22: 41–85. London: Academy Press Limited.

Fuhrman, J. A. and F. Azam. 1980. Bacterioplankton in the coastal waters of British Columbia, Antarctica and California. *Applied and Environmental Microbiology* 39: 1085–1095.

Lee, S. and J. A. Fuhrman. 1987. Reltionship between biovolume and biomass of naturally derived marine bacteriopolankton. *Applied and Environmental Microbiology* 53: 1298–1303.

Luna, G. M., E. Manini, and R. Danovaro. 2002. High fraction of dead and inactive bacteria in coastal marine sediments: Comparison of protocols and ecological significance. *Applied and Environmental Microbiology* 68: 3509–3513.

Mei, M. L. and R. Danovaro. 2004. Virus production and life strategies in aquatic sediments. *Limnology and Oceanography* 49: 459–470.

Nagata, T. 1986. Carbon and nitrogen content of natural planktonic bacteria. *Applied and Environmental Microbiology* 52: 28–32.

Watson, S. W., T. J. Novitsky, H. L. Quinby, and F. W. Valois. 1977. Determination of bacterial number and biomass in the marine environment. *Applied and Environmental Microbiology* 33: 940–946.

36 Benthic Prokaryotic Heterotrophic Production Using the Leucine Incorporation Method

CONTENTS

36.1 INTRODUCTION

Prokaryotic heterotrophic production (PHP) in marine sediments is a crucial variable for quantifying the contribution of prokaryotes to the carbon cycling in marine ecosystems. Prokaryotic heterotrophic production has been successfully measured in sediments with radioactive tracers since 1981 (Moriarty and Pollard 1981; Kirchman et al. 1986) and is still utilized (Manini et al. 2004; Danovaro et al. 2008, 2009). Biomass production in aquatic systems is now commonly estimated from the rate of incorporation of radioactively labeled leucine (i.e., ^{3}H-Leu) into prokaryotic proteins during short-term incubations (Kemp et al. 1993). Biomass production is calculated from the rate of protein synthesis because proteins comprise a large, fairly constant fraction of prokaryotic biomass. This procedure is used to estimate a conversion factor (cells or g C per mol of leucine incorporation) that converts leucine incorporation into production of biomass.

In this chapter, we describe the current method for measuring prokaryotic carbon production in marine sediments using the leucine incorporation method.

36.2 FIELD PROCEDURE

Sediment samples are preferentially collected using multiple corers, which allow recovering undisturbed, intact sediment cores. Immediately after retrieval, sediment subsamples are gently removed from the corer, inserting a cut-off 10 mL plastic syringe along the axis of the core. The syringe plunger is held fixed at the sediment surface while the barrel is pushed into the sediment for 1 cm, in an overall procedure that is analogous to piston coring. The syringes are then removed from the sediment and the sediment transferred into a sterile tube and resuspended in sterile seawater (collected from the same site) to form a homogeneous slurry (dilution 1:1).

36.3 LABORATORY PROCEDURE

36.3.1 Equipment and Instruments

Scintillation counter
Vacuum pump (Millipore or similar)
Heating block
Fridge
Vortex
Filter tower for 25 mm membrane filters
Bottle to collect radioactive fluid
Centrifuge for Eppendorf tubes (14–16,000 rpm)
Centrifuge for 15 mL tube (2,500 rpm)
Oven (60°C)
Analytical balance

36.3.2 Materials

Scintillation vials (20 mL)
Sterile pyrex tubes (15 mL)
Sterile plastic tubes (15 mL)
Sterile Eppendorf tubes
White polycarbonate membrane filters (diameter 25 mm, pore size 0.2 µm)
Plastic gloves

36.3.3 Solutions and Reagents

[4,5-^{3}H]-leucine solution (used at 0.5 µM final concentration)
Nonradioactive leucine Sigma; optional
Trichloroacetic acid (TCA) 5% (w/v)
Ethanol 80% (v/v)
Scintillation liquid (HIONIC FLUOR Packard or similar)
NaOH (2 N)

36.3.4 Procedure

1. Sediment subsamples (0.2 mL) are transferred to 2 mL sterile Eppendorf tubes. For each sediment sample, a total of three replicates and two blanks are analyzed. Sediment blanks are run as described for sediment samples but adding 1.7 mL of 80% ethanol immediately before 3H-Leucine addition.
2. Add 30 µL of an aqueous solution of L-[4,5-^{3}H] leucine solution. Leucine is generally used to reach at 0.5–1 µM final concentration.
3. Incubate in the dark for 1 h at *in situ* temperature.
4. After incubation, stop prokaryotic incorporation by adding 1.7 mL of cold 80% ethanol.
5. Vortex.
6. Centrifuge at 14,000–16,000 rpm for 5 min.
7. Remove gently supernatant.
8. Resuspend the sediment in 1.7 mL of cold 80% ethanol.
9. Vortex.
10. This centrifugation–resuspension procedure is carried out twice.
11. After the second procedure, the sediment–ethanol slurry is passed through one polycarbonate filter.

12. Filters are rinsed four times with 2 mL of cold 5% TCA, then transferred to sterile Pyrex® tubes.
13. Add 2 mL of 2 N NaOH.
14. Vortex.
15. Heat 2 h in the heating block at 100°C. At the end of this procedure, the polycarbonate filter must be dissolved.
16. Centrifuge the tubes at 2500 rpm for 5 min.
17. After centrifugation, transfer 1 mL of supernatant to scintillation vials containing 10 mL of scintillation liquid.
18. Place the vials in the liquid scintillation counter.
19. Record the DPM (disintegrations per minute).
20. Place the tubes in an oven at 60°C for 24–48 h, until the constant weight.
21. Weigh the dry sediment using an analytical balance.

36.3.5 Calculation of Rates of Biomass Production

^{3}H-leucine incorporation are converted into prokaryotic carbon production by using the following formula:

$$PHP\ (\mu g\ g^{-1}) = \left[nmol\ of\ leucine\ incorporated \cdot (100/7.3) \cdot M \cdot 0.86 \cdot R \cdot 2\right] \cdot g^{-1} \quad (36.1)$$

where

the nmol of leucine incorporated is calculated as follows:

$$DPM_{inc} / \left(S.A. \cdot 2.22 \cdot 10^6\right) \quad (36.2)$$

where

DPM_{inc} is disintegration per minute of the incorporated ^{3}H leucine and is calculated as "DPM sample–DPM blank."

S.A. is the specific activity (in Ci $mmol^{-1}$) of the ^{3}H Leucine.

The ratio 100/7.3 is the percentage of leucine into the total prokaryotic amino acid pool.

M is the molar weight of leucine.

0.86 is the conversion factor of bacterial protein production to prokaryotic C production.

R is the correction for radioactive decay of the ^{3}H-leucine.

2 is the intracellular isotope dilution.

g is sediment dry weight (grams).

36.4 NOTES AND PROBLEMS

There are several assumptions of the basis of this approach. The first is the incorporation is carried out at concentrations that maximize incorporation rates (substrate saturation). Substrate saturation can be assessed by performing incubations with increasing concentrations of tritiated leucine. Substrate saturation can widely vary among different areas but it is typically 50 μM.

The second is the intracellular isotope dilution as the presence of naturally-occurring nonlabeled leucine in sediments can dilute the added radioactive isotope. In the formula used here, the dilution of the isotope is assumed to be 2 (Simon and Azam 1989). Some authors suggest that, to correct for such isotope dilution, a kinetic approach can be utilized. This approach is based on measurements of protein synthesis in subsamples to which a number of subsaturated concentrations of radioactively labeled leucine had been added (Riemann and Azam 1992).

To do this some authors (e.g., van Duyl and Kop 1994) have utilized a mixture of radioactive ("hot") and nonradioactive ("cold") leucine. In this case, it is assumed that both labeled and unlabeled leucine are equally incorporated by prokaryotes.

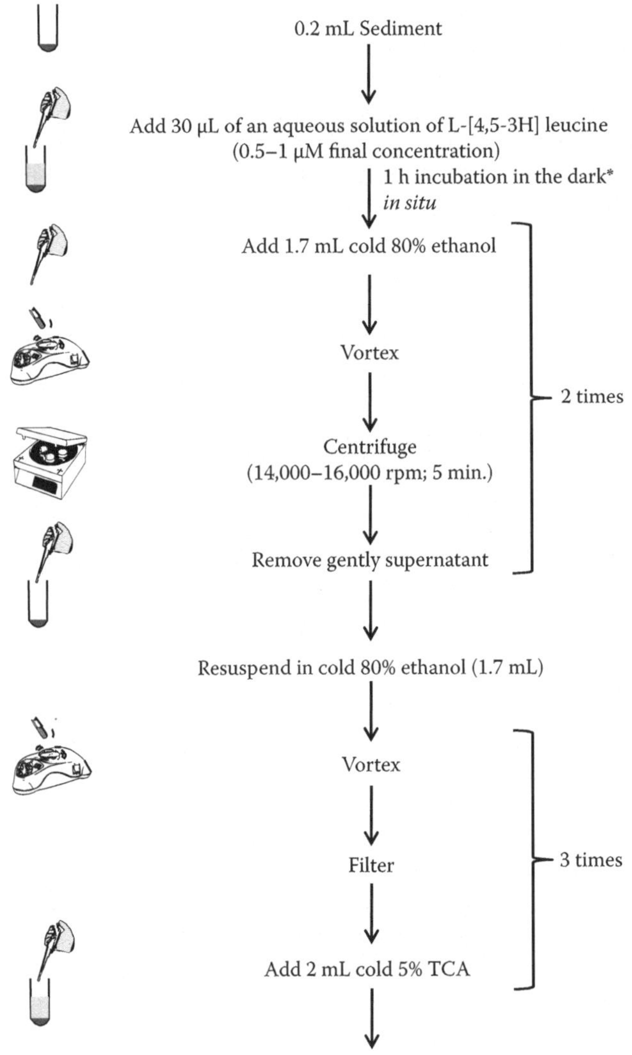

*Blank sample: to skip the step

(a)

PROTOCOL 36.1 Determination of benthic prokaryotic heterotrophic production. *Continued*

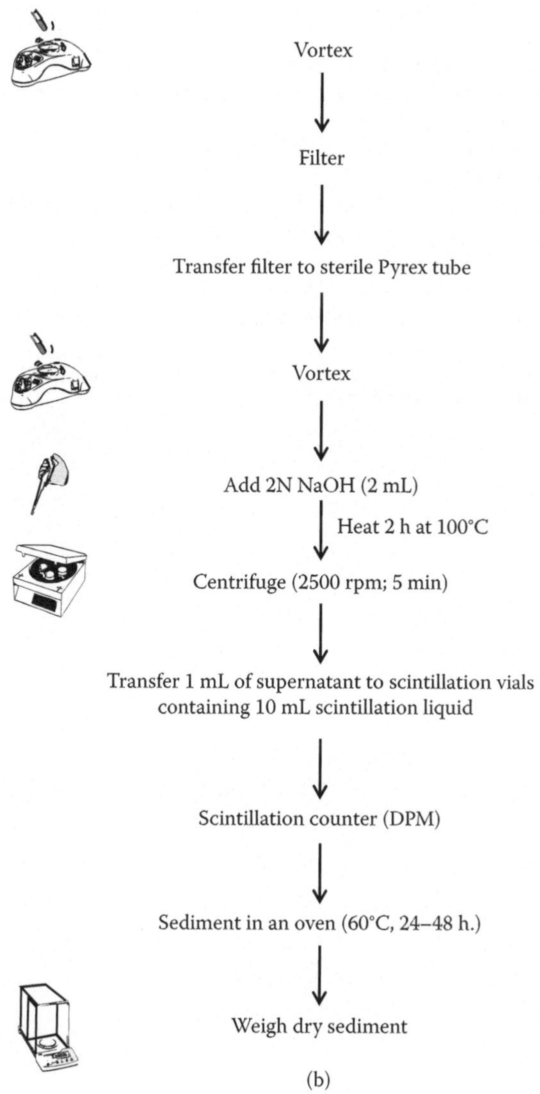

(b)

PROTOCOL 36.1 *Continued.*

REFERENCES

Danovaro, R., A. Dell'Anno, C. Corinaldesi et al. 2008. Major viral impact on the functioning of benthic deep-sea ecosystems. *Nature* 454: 1084–1087.

Danovaro, R., C. Corinaldesi, G. M. Luna, M. Magagnini, E. Manini, and A. Pusceddu. 2009. Prokaryote diversity and viral production in deep-sea sediments and seamounts. *Deep-Sea Research Part II* 56: 738–747.

Kemp, P. F., B. F. Sherr, E. B. Sherr, and J. J. Cole. 1993. *Handbook of methods in aquatic microbial ecology.* Boca Raton: Lewis Publishers.

Kirchman, D. L., S. Y. Newell, and R. E. Hodson. 1986. Incorporation versus biosynthesis of leucine: Implications for measuring rates of protein synthesis and biomass production by bacteria in marine systems. *Marine Ecology Progress Series* 32: 47–59.

Manini, E., G. M. Luna, and R. Danovaro. 2004. Benthic bacterial response to variable estuarine inputs. *FEMS Microbiology Ecology* 50: 185–194.

Moriarty, D. J. W. and P. C. Pollard. 1981. DNA synthesis as a measure of bacterial productivity in seagrass sediments. *Marine Ecology Progress Series* 5: 151–161.

Riemann, B. and F. Azam. 1992. Measurements of bacterial protein synthesis in eutrophic aquatic environments by means of leucine incorporation. *Marine Microbial Food Webs* 6: 91–105.

Simon, M. and F. Azam. 1989. Protein content and protein synthesis rates of planktonic marine bacteria. *Marine Ecology Progress Series* 51: 201–213.

van Duyl, F. C. and A. J. Kop. 1994. Bacterial production in North Sea sediments: Clues to seasonal and spatial variations. *Marine Biology* 120: 323–337.

37 Prokaryotic Chemoautotrophic Production in Marine Sediments

CONTENTS

37.1 INTRODUCTION

Recent investigations highlighted that a significant fraction of prokaryotic C production in deep waters may be derived from the uptake and assimilation of inorganic carbon (Pearson et al. 2001; Wuchter et al. 2003; Herndl et al. 2005). These findings open new questions on the potential quantitative relevance of prokaryotic chemoautotrophic production to the total C produced by prokaryotes also in deep-sea sediments.

The methodology utilized for determining prokaryotic chemoautotrophic production in deep waters has been adapted from the common ^{14}C incorporation technique developed in the early 1950s for photoautotrophic production (Steemann-Nielsen 1952). However, to date no standardized protocols still exist for the determination of prokaryotic chemoautotrophic production in marine sediments.

Grøntved (1960) was the first who utilized ^{14}C addition to estimate phototrophic primary production in shallow sediments. Subsequently, this procedure has been improved and applied for determining microphytobenthos production in coastal marine sediments worldwide (MacIntyre and Cullen 1995; Cibic et al. 2007). One of the problems in accurately estimating primary production based on ^{14}C incorporation is the dilution effect induced by the presence of ambient inorganic ^{12}C. Inorganic ^{12}C concentrations are, indeed, high in marine environments (both in the water column and in the sediments) and variable among systems, so that their concentrations must be determined.

Here we present a protocol for the determination of prokaryotic chemoautotrophic production in marine sediments lacking of photoautotrophic production (i.e., below photic zone and/or in subsurface sediment layers).

37.2 FIELD PROCEDURE

37.2.1 Equipment and Instruments

Eppendorf tubes

37.2.2 Solutions

[^{14}C] Bicarbonate solution
Buffered formalin (with sodium tetraborate, 20 g L^{-1})

37.2.3 Procedure

Sediment samples collected by a range of specific devices (e.g., multicorer/box corer for collecting undisturbed deep-sea sediments) must be immediately processed and analyzed at *in situ* temperature. Sediment subsamples are collected in cut-off plastic syringes and subsequently transferred into sterile tubes. Sediment samples are then diluted 1:1 (W/V) with 0.02 μm filtered seawater collected at the sediment–water interface. Sediment slurries [NOTE 1] are then processed as follows:

1. Prepare one set of sterile Eppendorf tubes for at least two blanks (Blk1, Blk2) and three replicate samples (R1, R2, R3).
2. Sediment subsamples (0.2 mL) are transferred to the sterile Eppendorf tubes.
3. Add 0.5–10 μCi of [^{14}C] bicarbonate [NOTE 2].
4. Stop immediately Blk1 and Blk2 by the addition of buffered formalin (2% final concentration).
5. Incubate sediment samples R1, R2, and R3 in the dark at *in situ* temperature from 4 up to 72 h [NOTE 3].
6. Stop the incubation by the addition of buffered formalin (2% final concentration).
7. Shake gently.
8. Store all samples in the dark at 4°C until further processing.

37.3 LABORATORY PROCEDURE

37.3.1 Equipment and Instruments

Autoclave
Scintillation counter
Fridge
Vortex
Centrifuge for Eppendorf tubes and for 15 mL tube
Oven
Analytical balance (±1 mg)
Scintillation vials
Glass tubes with tips
Sterile plastic tubes
Eppendorf tubes
Plastic gloves

37.3.2 Solutions and Reagents

[^{14}C] Bicarbonate solution
6 N HCl
Buffered formalin (with sodium tetraborate, 20 g L^{-1})
Scintillation liquid (Hionic Fluor Packard or similar)

37.3.3 Sample Processing after Incubation

1. Centrifuge the sample at 16,000 × g for 10 min.

2. Remove the supernatant.
3. Add 1.7 mL of phosphate saline buffer (PBS) and resuspend the sample.
4. Centrifuge at 16,000 × g for 10 min.
5. Remove the supernatant, and resuspend the sample again with 0.7 mL of PBS.
6. Transfer the sediment slurry into Pyrex glass tubes.
7. Add 0.25 mL of 6 N HCl and shaking for 2 h (or overnight) [NOTE 4].
8. Add 10 mL of scintillation liquid.
9. Vortex for 30 s.
10. Centrifuge at 3500 × g for 10 min.
11. Transfer the supernatant into a scintillation vial.
12. Read radioactivity with scintillation counter.
13. Record the DPM (disintegrations per minute).
14. Repeat again steps from 8 to 13 for two additional times [NOTE 5].

The liquid contained in the glass tube must be carefully removed and discharged in dedicated radioactive boxes. The sediment samples contained in the glass tubes must be dried for 24–48 h (60°C). Sediments are then weighed using an analytical balance.

37.3.4 Conversion into Prokaryotic Carbon Production

Data of [^{14}C] DPM are converted into g of carbon incorporated per hour per gram of dry sediment by using the following formula [NOTE 6]:

$$\frac{\left[DPM_{inc}\cdot 12\cdot 1.05\right]\cdot CO_2}{\left(S.A.\cdot 2.22\cdot 10^{12}\right)\cdot Y\cdot g\cdot H} \tag{37.1}$$

where

DPM_{inc} is the difference between the average DPM of the three sediment samples and the DPM of the blanks.
12 is the atomic weight of C.
1.05 is the isotope coefficient, since the uptake of ^{14}C is 5% lower than the uptake of ^{12}C.
CO_2 is the alkalinity of seawater used to resuspend sediment samples, expressed as molarity.
S.A. is the specific activity of [^{14}C]bicarbonate in Ci mol^{-1}.
2.22×10^{12} are the DPM emitted by 1 Ci.
Y is the concentration of added [^{14}C] bicarbonate, expressed as molarity.
g is the sediment dry weight expressed in grams.
H is the incubation time expressed in hours.

37.4 NOTES

1. The slurry procedure is generally used for the determination of heterotrophic prokaryotic C production in deep-sea sediments (Dixon and Turley, 2001; Danovaro et al., 2002), and it has been also used for determining microphytobenthos production in different shallow marine ecosystems (Jönsson, 1991; MacIntyre and Cullen, 1995; Barranguet et al., 1998; Cibic et al., 2007). This procedure allows the processing of many samples, typically collected during ecological studies. However, the preparation of a sediment slurry may potentially affect the actual estimates of both phototrophic and chemoautotrophic C production (e.g., by modifying the chemical micro-gradients within the samples). In this regards, some authors have observed a potential overestimation of primary productivity of

microphytobenthos which, however, has been attributed more to the light exposition during slurry preparation than to the effect of the slurry itself (Jönsson 1991; Barranguet et al. 1998).

2. The amount of spiked µCi in the sediment sample depends on the biomass and activity of the chemoautotrophic prokaryotes. However, since it is impossible either to distinguish the biomass of chemoautotrophic and heterotrophic prokaryotes or to estimate their activities a priori, it is necessary to perform site-specific experiments with increasing concentrations of [^{14}C] bicarbonate (i.e., increasing amounts of µCi). The use of increasing concentrations of radioactive can allow assessing of the threshold levels over which no further [^{14}C] bicarbonate uptake would be observed.
3. The appropriate incubation time for the estimate of prokaryotic chemoautotrophic production should be determined by time-course incubation experiments. These experiments define the time interval in which ^{14}C uptake is linear.
4. The step with 6 N HCl must be carried out with a fume cupboard.
5. The DPM obtained by the three countings is used for the calculation of the prokaryotic C production.
6. We have adapted the formula used for the estimation of phototrophic primary production to convert DPM incorporated into organic C produced by prokaryotes through chemosynthetic processes. Therefore, we have assumed that inorganic C has been used to produce prokaryotic biomass. CO_2 fixation by chemoautotrophic aerobic prokaryotes using ammonium or other reduced compounds as electron donors and energy sources (Francis et al. 2005; Könneke et al. 2005) has been documented in different marine ecosystems including the deep sea (Wuchter et al. 2003; Herndl et al. 2005; Ingalls et al. 2006). However, since the metabolic pathways for converting inorganic C into biomass by benthic deep-sea chemoautotrophic prokaryotes are still largely unknown, the formula proposed here for estimating chemoautotrophic prokaryotic production could be modified and/or implemented in the future. We suggest combining the present procedure with independent analyses (e.g., micro autoradiography coupled with fluorescent *in situ* hybridization; Lee et al. 1999) to provide direct indication of inorganic C fixation into the prokaryotic biomass.

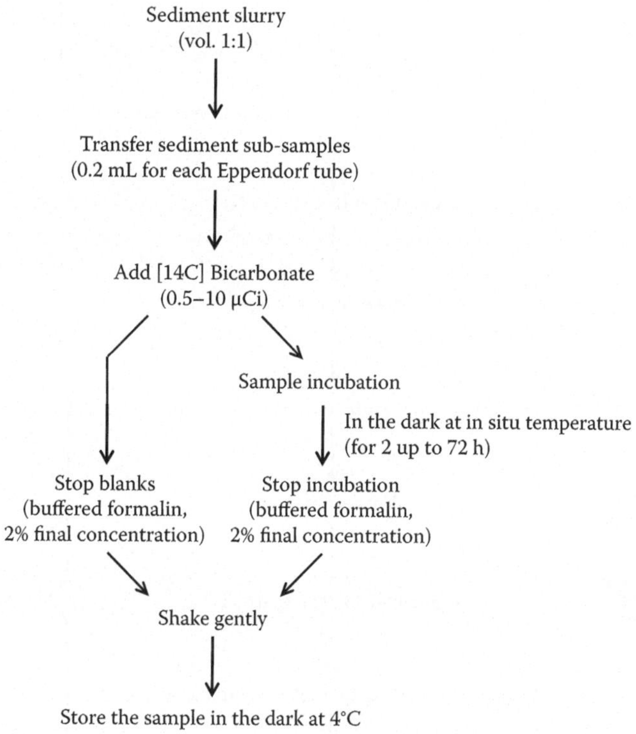

PROTOCOL 37.1 Determination of prokaryotic chemoautotrophic production in marine sediments on board.

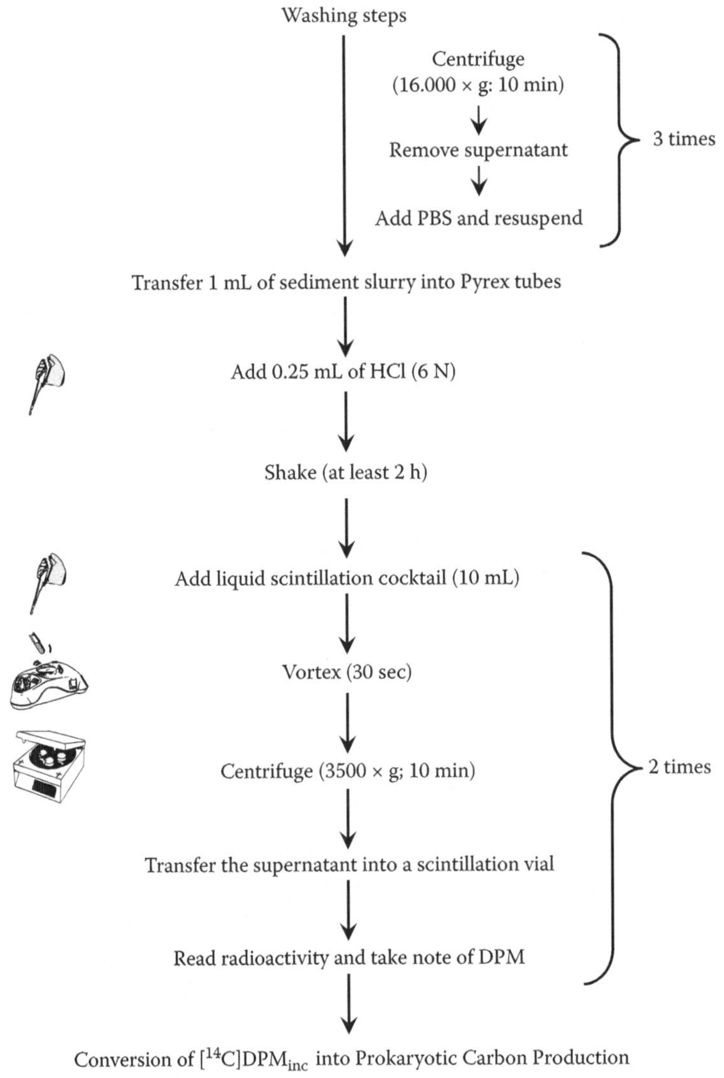

PROTOCOL 37.2 Determination of prokaryotic chemoautotrophic production in marine sediments in laboratory.

REFERENCES

Barranguet, C., J. Kromkamp, and J. Peene. 1998. Factors controlling primary production and photosynthetic characteristics of intertidal microphytobenthos. *Marine Ecology Progress Series* 173: 117–126.

Cibic, T., O. Blasutto, N. Burba, and S. Fonda Umani. 2007. Microphytobenthic primary production as ^{14}C uptake in sublittoral sediments of the Gulf of Trieste (northern Adriatic Sea): Methodological aspects and data analyses. *Estuarine, Coastal, and Shelf Science*75: 50–62.

Danovaro, R., M. Manini, and A. Dell'Anno. 2002. Higher abundance of bacteria than of viruses in deep Mediterranean sediments. *Applied and Environmental Microbiology* 68: 1468–1472.

Dixon, J. L. and C. M. Turley. 2001. Measuring bacterial production in deep-sea sediments using 3H-thymidine incorporation: Ecological significance. *Microbial Ecology* 42: 549–561.

Francis, C. A., K. J. Roberts, J. M. Berman, A. E. Santoro, and B. B. Oakley. 2005. Ubiquity and diversity of ammonia-oxidizing archaea in water columns and sediments of the ocean. *Proceedings of the National Academy of Sciences*14683–14688.

Grøntved, J. 1960. On the productivity of microphytobenthos and phytoplankton in some Danish fjords. *Meddelelser fra Danmarks Fiskeri-Og Havundersogelser* 3: 1–17.

Herndl, G. J., T. Reinthaler, E. Teira et al. 2005. Contribution of archaea to total prokaryotic production in the deep Atlantic Ocean. *Applied and Environmental Microbiology* 71: 2303–2309.

Ingalls, A. E., S. R. Shah, R. L. Hansman et al. 2006. Quantifying archaeal community auotrophy in the mesopelagic ocean using natural radiocarbon. *Proceedings of the National Academy of Sciences* 103: 6442–6447.

Jönsson, B. 1991. A ^{14}C-incubation technique for measuring microphytobenthic primary productivity in intact sediment cores. *Limnology and Oceanography* 36: 1485–1492.

Könneke, M., A. E. Bernhard, J. R. De la Torre, C. B. Walker, J. B. Waterbury, and D. A. Stahl. 2005. Isolation of an autotrophic ammonia-oxidizing marine aechaeon. *Nature* 437: 543–546.

Lee, N., P. H. Nielsen, K. A. Andreasen et al. 1999. Combination of fluorescent *in situ* hybridizaion and microautoradiography a new tool for structure-function analyses in microbial ecology. *Applied and Environmental Microbiology* 65: 1289–1297.

Maclntyre, H. L. and J. J. Cullen. 1995. Fine-scale vertical resolution of chlorophyll and photosynthetic parameters in shallow-water benthos. *Marine Ecology Progress Series*.122: 227–237.

Pearson, A., A. P. McNichol, B. C. Benitez-Nelson, J. M. Hayes, and T. I. Eglington. 2001. Origin of lipid biomarkers in Santa Monica Basin surface sediment: A case study using compound-specific Δ14C analysis. *Geochimica et Cosmochimica Acta* 65: 3123–3137.

Steemann-Nielsen, E. 1952. The use of radioactive carbon (^{14}C) for measuring organic production in the sea. *Journal du Conseil Permanent International d'Exploration de la Mer* 16: 117–40.

Wuchter, C., S. Schouten, H. T. S. Boschker, and J. S. S. Damsté. 2003. Bicarbonate uptake by marine Crenarchaeota. *FEMS Microbiology Letter* 219: 203–207.

38 Meiofaunal Biomass and Secondary Production

CONTENTS

38.1 INTRODUCTION

Meiofauna represent the dominant numerical and biomass component of deep-sea benthic metazoans (Vincx et al. 1994; Cook et al. 2000; Danovaro et al. 2002a), and play an important ecological role as the main pathway of energy transfer to higher trophic levels. High pressure, unpredictable arrival of energy inputs, long-term starvation, and the refractory composition of food sources have been suggested as responsible for the decrease of faunal abundance and biomass with increasing water depth (Vincx et al. 1994; Thurston et al. 1994; Soetaert et al. 1997; Stora et al. 1999; Gooday et al. 2001; Smith et al. 2002; Gutzmann et al. 2004; Vanhove et al. 2004). However, the recent discovery of hotspots of benthic biomass in the deep sea (Danovaro et al. 2002a; Valdovinos et al. 2003) is challenging the paradigm of decreasing abundance and biomass of benthic organisms with increasing depth. Determining the benthic biomass is rather important to obtain a good description of the status of deep-sea ecosystems since it can be considered an indicator of ecosystem functioning. Due to several difficulties (i.e., research vessel availability, high cost of equipment and ship time, loss of research data due to weather conditions) to perform *in situ* measurements of deep-sea ecosystem functioning (Snelgrove 1999), the estimation of meiofaunal biomass (relatively easy to perform in a laboratory) can be a useful descriptor of ecosystem functioning.

An estimate of the production of meiofaunal assemblages is highly suitable for assessing the trophic efficiency of marine ecosystems as (1) meiofauna are characterized by life cycles spent entirely in the sediment (i.e., lack of pelagic dispersion); (2) their production is not masked by allochthonous inputs; (3) they respond promptly to all different organic matter sources; and (4) they display high abundance and turnover rates (often higher than macrofaunal production), thus playing a key role in benthic energy flows (Heip et al. 1982; Feller and Warwick 1988; Higgins and Thiel 1988; Ceccherelli and Mistri 1991; Danovaro and Fabiano 1996; Albertelli et al. 1999). Meiofauna represent a direct link between producers and higher trophic levels because they are efficient in exploiting organic detritus, and at the same time are predated upon by both macrofauna and juvenile fish (Watzin 1983; De Morais and Bodiou 1984; Lindquist et al. 1994). An estimate of the meiofaunal secondary production can be considered a good descriptor of ecosystem efficiency and the capability of the ecosystem to transfer energy and material to higher trophic levels (Möller et al. 1985; Danovaro et al. 2002b). This is important with deep-sea sediments where meiofauna can represent the dominant component of the benthos (Gambi et al. 2003). Here, we report the different methods to measure the meiofaunal biomass and secondary production.

38.2 FIELD PROCEDURES

In deep-sea soft sediments a multicorer is used to collect simultaneously several meiofaunal samples (e.g., 4–12 corers; Elmgren and Radziejewska 1989). The instrument allows the collection of undisturbed sediment samples of a given surface area for analysis of the horizontal and vertical distribution of benthic organisms. Another common sampling instrument for soft sediments is the box corer (Fleeger et al. 1983). The weight of a box corer after sampling can range from 150 to >750 kg and the collection of surface sediment amounting from 0.02 to 1.0 m^2 down to a depth of 50 cm. From the box corer, meiofaunal samples can be collected using manual Plexiglas® corers (internal diameter 3.6 cm), one replicate from each different deployment. However, during sampling recovery most box corers allow a surface sediment mixing, which might alter the quality of the sample. Bett et al. (1994), Shirayama and Fukushima (1995), and Bett (2000), comparing sampling efficiency of multicorers and box corers in deep-sea environments, pointed out that meiofaunal abundance collected using box corers was significantly lower. However, some studies conducted in the Eastern Mediterranean demonstrated that the use of two different samplers (USNEL-type box corer and multicorers) provide identical results in terms of benthic parameters, suggesting an adequate sampling efficiency of both sampling devices (Danovaro et al. 1999).

38.3 MEIOFAUNAL BIOMASS

The procedures to collect, store, extract, and sort meiofaunal samples for the determination of biomass are the same reported in Chapter 17, Abundance of Metazoan Meiofauna.

38.3.1 Laboratory Procedures

38.3.1.1 Equipment and Instruments

Stereomicroscope equipped with a micrometer scale
Sieve 20 μm
Delfuss cuvette
Wash bottle
Handling needle

38.3.1.2 Solution

Fresh water
4% Buffered formalin
Rose Bengal

38.3.1.3 Sample Preparation for the Determination of Meiofaunal Taxa Biomass

Samples were treated using the same methods reported for the meiofaunal counting and sorting. After the extraction, the sample is transferred into the Delfuss cuvette, using the 20 μm sieve. The sample, washed with fresh water to eliminate the residuals of formalin, is transferred into the Delfuss cuvette. Before starting the sorting and determining biomass, wait a few minutes to allow the sedimentation of all organisms in the bottom of the cuvette. Start the analyses of the sample from the first cell on the left site of the first layer of the cuvette. At the end of the sorting and biomass determination, the sample is transferred again in the 4% buffered formalin and stained with Rose Bengal.

38.3.1.4 Procedure

1. Transfer the samples fixed with 4% buffered formalin into a sieve with a mesh net of 20 μm [NOTES 1–3].
2. Wash the sample to eliminate the 4% buffered formalin.
3. Transfer the sample in a Delfuss cuvette with water using a wash bottle.
4. Estimate the body length (L) and maximal body width (W) of each specimen using a micrometer-scale (magnification 25× and 40×, Figure 38.1) [NOTES 4–12].
5. At the end of the determination of the biomass, transfer the sample through a 20 μm mesh net [NOTE 13].
6. Transfer the samples into a 50 mL tube using 4% buffered formalin using a wash bottle.
7. Add a few drops of 0.5 g L^{-1} Rose Bengal solution with a Pasteur pipette.
8. Store the samples at room temperature.

38.3.2 Notes

1. It is recommended to sort the meiofaunal sample and to determine the biomass during the same analysis to avoid several transfers of the sample from the tube to the sieve and the cuvette. All these steps are rather critical and offer the opportunity to lose some specimens.
2. Caution: Use the formalin carefully, dressing in all the safety devices (laboratory coat, gloves, glass, respiratory protection), and handle under an extractor fan.
3. The solutions containing the formalin have to be collected and eliminated using the safety procedures.
4. Wait a few minutes before starting the sorting.
5. To avoid the mixing of the sample, do not knock the side of the cuvette. If this does happen, restart the sorting and the estimation of biomass from the beginning.
6. Measure the body length (L) and the maximal body width (W) of each specimen using the micrometer scale along the main axis of the organism, following a hypothetical layer

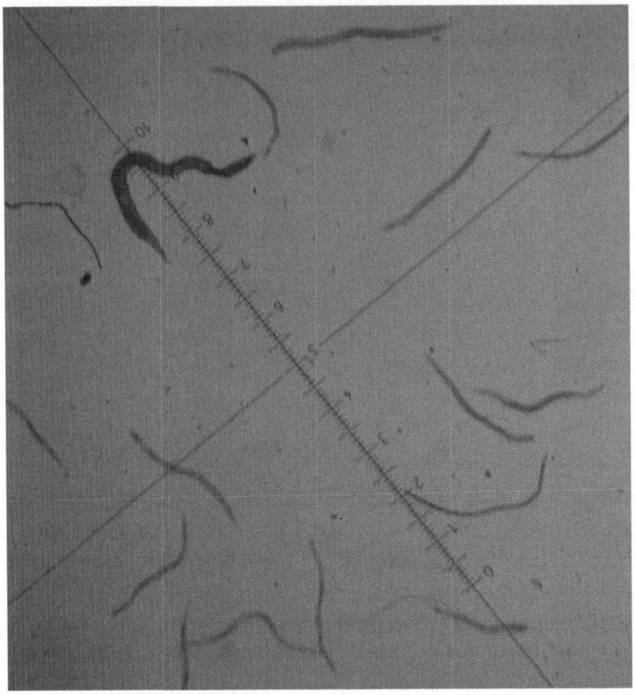

FIGURE 38.1 Prescreening for the subsequent measurement of body length (L) and body width (W) using a micrometer scale.

along the center of the body. For specimens characterized by long length, move the ocular equipped with the micrometer scale to maintain the position along the main axis of the body (Figures 38.2 and 38.3).

7. Use a handling needle to move the specimens in the optimal position to measure of the body length and width.
8. Use the magnification 25× or 40×. The magnification is chosen on the basis of the size of each specimen.
9. Annotate the magnification used to measure the body length and width to convert the measures (obtained using the micrometer scale) in μm or mm. The conversion factor is strictly related to the model of the stereomicroscope. Take care to estimate this conversion factor using a micrometer scale on a standard micrometer slide.
10. For the determination of the biomass, do not consider specimens that do not have an intact body.
11. The number of individuals used for the determination of the biomass is strictly related to the abundance of each taxon (i.e., 50 specimens for each taxon). In deep-sea samples, where abundance is rather low, all specimens should be measured.
12. At the end of the sorting and biomass determination, check the surface of the water in the cuvette to ensure specimens that are suspended are counted.
13. Collect all specimens from the tubes, from the sieve, and from the cuvette. Clean the sieve very well at each step during the preparation of the sample for sorting and after the end of the counting.

38.3.3 Calculations

The determination of meiofaunal biomass can be performed using the volumetric method that consists in indirect estimates of biomass extrapolating organism weight from a volume. This method is

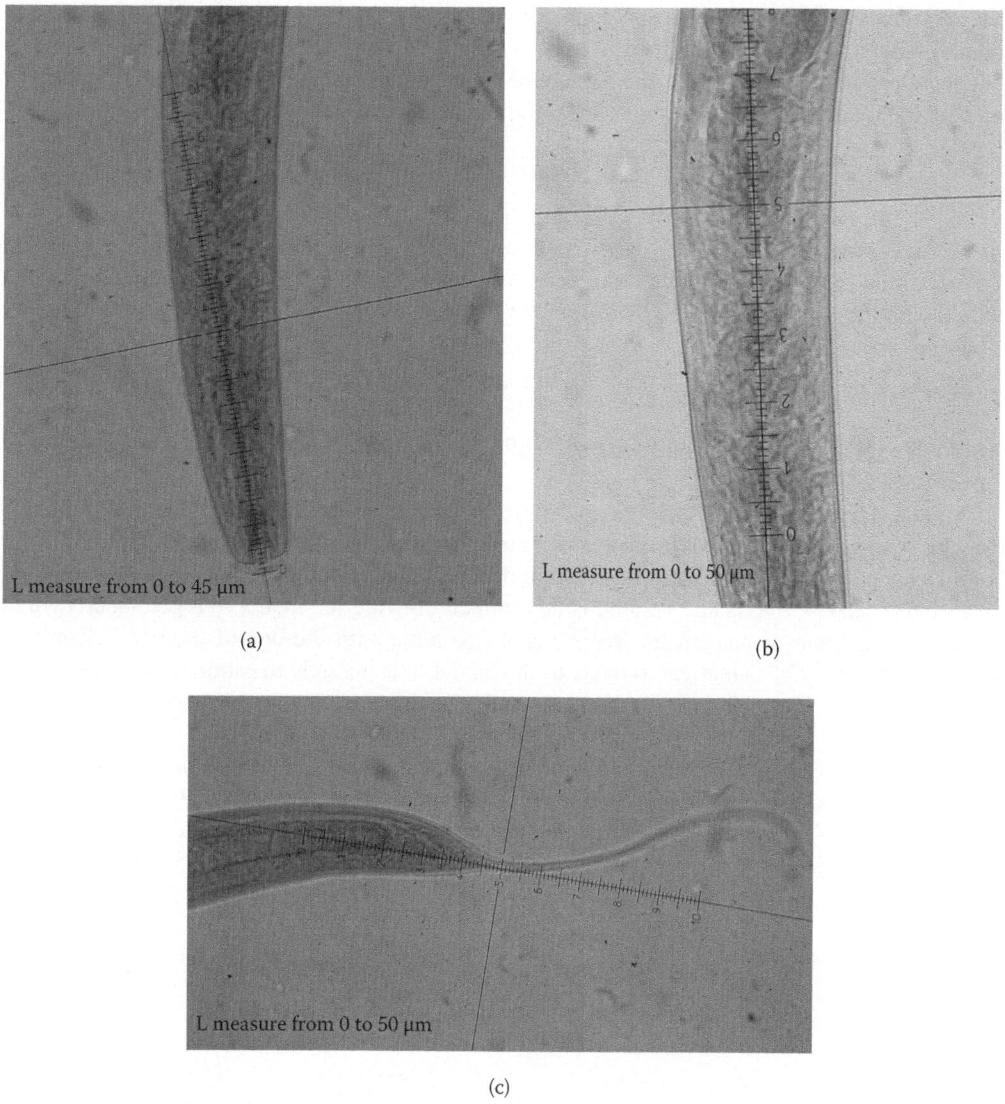

FIGURE 38.2 Some examples of measures of the body length (L) of nematodes performed using a micrometer scale along the main axis of the body.

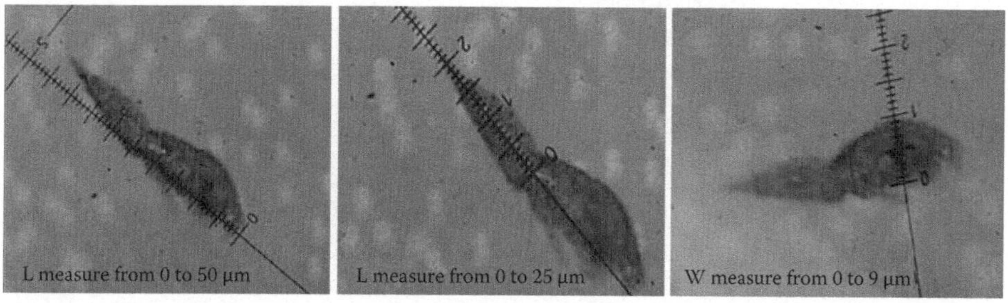

FIGURE 38.3 Some examples of measures of the body length (L) and maximal body width (W) of copepods using a micrometer scale.

based on the association of the morphology of meiofaunal organisms to geometric shapes, and the volume estimates (V) can be obtained from body length (L) and body width (W) at the microscope or stereomicroscope. For instance, the biovolume of nematodes can be approximated using a cylinder, with the following formula proposed by Andrassy (1956):

$$V = L \cdot W^2 \cdot 0.063 \cdot 10^5 \tag{38.1}$$

where V is expressed in nL (10^{-9} L) whereas W and L are expressed in μm. W and L are measured in a microscope equipped with a micrometer scale or with an image analyzer. Alternatively, it can be used with the formula proposed by Warwick and Price (1979):

$$V = C \cdot L \cdot W^2 \tag{38.2}$$

where V is expressed in nL, L and W in mm, while C is the conversion factor specific for each taxon (for nematodes C = 530).

This later formula is derived from a procedure applicable also to organisms characterized by an irregular shape (i.e., copepods). It involves constructing models in scale based on shapes drawn at the microscope or based on photos taken from different perspectives in order to respect the body dimension (L and W max). The volume is determined drawing the model in a graduated cylinder and measuring water displacement. Reporting this volume with the one of the parallelepiped of length (L) and base (W^2) identical to those of the model, it is possible to estimate the C conversion factor (Table 38.1; Warwick and Gee 1984). For all other taxa, except for nematodes, L and W can be measured using a stereomicroscope equipped with a micrometer scale.

In general, the volume can be approximately converted into wet weight assuming a specific gravity of 1.13 (Warwick and Gee 1984). For nematodes, dry weight is generally 20%–25% of the wet weight (Myers 1967; Wieser 1960). For further conversions it is useful to consider that C content of the biomass of meiobenthic organisms is ca. 40% of its dry weight, while organic N is ca. 10% of

TABLE 38.1
List of the C Conversion Factors

Taxon	C
Nematodes	530
Ostracods	450
Kinorhynchs	295
Turbellarians	550
Tardigrades	614
Idroids	385
Polychaetes	530
Oligochaetes	530
Tanaidaceans	400
Isopods	230
Copepods	[a]

[a] For copepods the factor C varies depending on the shape of the animal, as suggested by Warwick and Gee (1984), from a minimum of 230 for discoid forms (scutelliforms) to a maximum of 750 for cylindrical forms.

dry weight (Feller and Warwick 1988). Jensen (1984) proposed to convert nematode wet weight to organic carbon using the conversion factor 12.4%.

38.4 NEMATODE BIOMASS

38.4.1 Laboratory Procedures

38.4.1.1 Sample Preparation for the Determination of Nematode Biomass

For the determination of nematode biomass using a microscope, specimens have to be transferred on slides to measure body length and width. For the determination of nematode biomass the measurements cannot be obtained using the stereomicroscope. Nematodes should be prepared following a procedure to obtain a permanent preparation that can be also used for the nematode genera and species identification.

38.4.1.2 Slide Preparation

The following procedures allow the preparation of permanent slides that can be use for the determination of the nematode biomass and identification. The microscope slides should be prepared with a drop of glycerin and a ring of paraffin; 10–15 nematodes for each sample can be fixed on a slide. This method is based on the clearing and—when combined with alcohols—dehydrating properties of glycerine (Figure 38.4). Dehydration must be gradually reached, otherwise soft-bodied animals might be damaged and made irremediably useless for observation (Vincx 1996).

Nematodes should be transferred using a handling needle from the cuvette to a staining block containing a solution of formalin (4%) and glycerine (100 mL solution: 99 mL formalin 4% and 1 mL glycerin). After 12 h, nematodes are transferred in an oven at 36°C for a night. During the permanence in the oven, a few drops of a solution containing ethanol (95%) and glycerin are added in order to prevent the desiccation of the nematodes. After the treatment in the oven, nematodes can be transferred onto the slides. Glycerin makes the nematodes transparent. Water results in slow penetration of the glycerin into the nematode tissue whereas ethanol gives the best penetration. The ethanol (95%) and glycerine solutions prevent the desiccation of the nematodes.

38.4.1.3 Equipment and Instruments

Light microscope, equipped with a micrometer scale
Sieve with a mesh net of 20 μm
Delfuss cuvette
Wash bottle
Fresh water

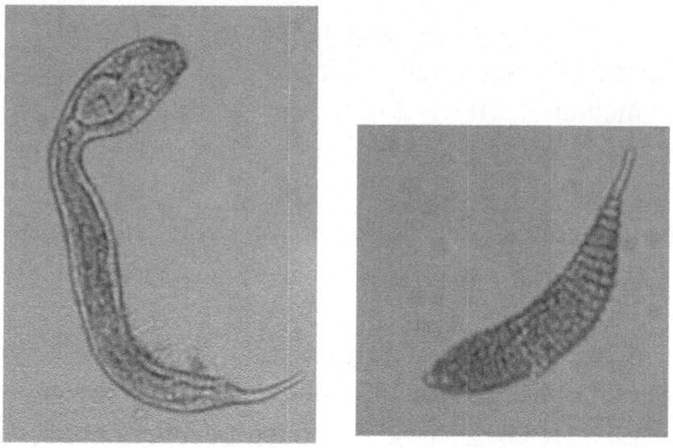

(a) (b)

FIGURE 38.4 Specimens belonging to the genus (a) *Draconema* and (b) *Richertsia*.

4% Buffered formalin
Rose Bengal
Staining block
Slides
Cover glass
Paraffin in Petri dish
Glycerin
Lighter or hotplate
Oven
Handling needle
Glass tube

38.4.1.4 Methods for Preparation of Permanent Slides

Transfer the samples into a cuvette using the procedures reported for meiofaunal counting and sorting [NOTES 1, 2].
Pick out the nematodes using a handling needle from the cuvette to a staining block containing a few drops of the solution containing formalin 4% and glycerin 1% [NOTE 3].
Add a few drops of Rose Bengal.
Place the staining block in a beaker containing a thick layer of ethanol (95%) overnight [NOTE 4].
The day after, place the staining block containing the nematodes (washed in the solution containing formalin 4% and glycerin 1%) in the oven at 36°C.
Add a few drops of the solution containing 95% ethanol and 5% glycerin every 2 h.
Leave the staining block in the oven overnight to let the ethanol evaporate until the nematodes are in pure glycerin.

At this step, the nematodes are ready to be transferred onto permanent slides.

38.4.1.5 Methods for Preparing the Slides

Prepare a few slides with a paraffin ring [NOTES 5–7].
Heat a glass Pirex tube (internal diameter 1.5–2 cm) and insert it in the Petri dish containing the paraffin.
Place the glass tube at the center of the slide to produce a smooth and thin paraffin ring.
Place a small drop of glycerin at the center of the paraffin ring.
Transfer a few nematodes (10–15 specimens) from the staining block to the glycerin drop on the slide using a stereomicroscope [NOTES 8, 9].
Arrange the nematodes, using a handling needle to avoid any overlap.
Gently place a cover glass on the paraffin ring and the glycerin drop.
Heat the slide on a hotplate or using a lighter for few seconds, keeping the slide horizontal.
Wait for the paraffin to melt and, after 1–2 min, seal the slide [NOTE 10].
Write down all informations relating to the sample (project, station, depth, replicate) with a permanent waterproof ink.
Place the slides in cardboard or plastic slide trays.

38.4.1.6 Notes

This procedure is a mix of two techniques suggested by Seinhorst (1959) and De Grisse (1969). The first procedure is based on 96% ethanol and glycerin. It is relatively rapid, labor saving (2–3 d), and gives very good results. De Grisse modified the procedure described by Seinhorst, combing the killing and fixing of fresh samples. This procedure is based on the use of three solutions: (1) 4% formalin and glycerin (99:1), (2) 96% ethanol and glycerin (95:5), and (3) 96% ethanol and glycerin (50:50). The first solution is made hot to fix the sample. This method requires 3 days.

1. The procedure described here is a combination of these techniques, but results are obtained relatively rapidly and the quality of the permanent slides is high.
2. It is recommended that nematodes be sorted and picked out at the same time as the counting and sorting of meiofauna to avoid several transfers of the sample from the tube to the sieve and cuvette. All these steps are rather critical in that there is the danger of losing some specimens.
3. Care should be taken in the transfer of nematodes using the handling needle; check to see if some nematodes remain on the needle. A quantity of 100–120 nematodes for each replicate can be transferred onto permanent slides. When abundance is rather low, try to pick out all the specimens present in the sample.
4. Close the beaker containing the thick layer of ethanol and the staining block using Parafilm to avoid the total evaporation of the ethanol. The presence of Parafilm allows the accumulation of the ethanol from the bottom of the beaker to the staining block to avoid desiccation of the nematodes.
5. Exercise care in producing the paraffin rings. Avoid large paraffin blocks that can wash out the drop of glycerin containing the nematodes during the heating.
6. Avoid large drops of glycerin that can wash out the melting paraffin and prevent the closing of the slide during the heating.
7. Other seals to close the slides are "Zut" cement; nail varnish; brown cement; silicone rubber; araldite resin, and a local putty. All sealing agents are affected by immersion oil. At the end of analysis, clean your slides after use by means of soft paper.
8. If, after the treatment of the staining block into the oven, the solution containing the nematodes is too dense, add a few drops of pure glycerin to facilitate the transfer of the nematodes onto the slides using the handling needle.
9. Alternative to the glass slides, the use of aluminum double-cover glass slides (Cobb 1917) is recommended, which offer the following advantages: (1) specimens can be viewed from either side, (2) the slides stack horizontally without one touching its neighbors, and (3) the slides can withstand being dropped on a hard floor.
10. A more rapid method for preparing microscopy slides implies the use of Hoyer's medium. First, wash the specimen with a drop of distilled water on a slide; then, add a drop of Hoyer's medium (30 g extremely fine gum arabic dissolved in 50 mL distilled water with 20 mL glycerin and 125 g chloral hydrate, filtered through cotton wool) that acts as a dehydrating, clearing, and sealing agent. The organism is transferred onto this drop and a cover glass is carefully lowered onto them to prevent the formation of air bubbles. This method, despite having the advantage of being rapid, might generate an excessive clearing of the specimen, making the observation of internal organs quite difficult (Danovaro et al. 2004).

38.4.2 Nematode Biomass Determination

When permanent slides containing nematodes are prepared, we can proceed to the determination of biomass. The procedures for the measurements of nematode length and width are similar to those reported for the main meiofaunal taxa.

38.4.2.1 Materials

Microscope equipped with a micrometer scale
Nematodes on permanent slides

38.4.2.2 Methods

Measure the body length (L) and maximal body width (W) of each specimen using a micrometer scale (minimum magnification 40×) [NOTES 1–6].

38.4.2.3 Notes

1. Using the micrometer scale, measure the body length (L) and the maximal body width (W) of each specimen along the main axis of the organism, following an hypothetical layer along the center of the body. For specimens characterized by long length, move the ocular equipped with the micrometerscale to maintain the position along the main axis of the body.
2. Do not include the filiform tails in the measurement of the body length.
3. Use the magnification 40× to obtain more precise measurements.
4. Annotate the magnification used to measure the body length and width to convert the measures (obtained using the micrometer scale) in micrometers. The conversion factor is strictly related to the model of the microscope. Take care to estimate this conversion factor using a micrometer scale on a standard micrometer slide.
5. For the determination of the biomass, do not consider specimens without an intact body.
6. For faster determinations of nematode biomass, it is possible to acquire the image using a software for image elaboration. The measurements can be performed using different models of image analyzer: Semiautomatic images system (analySIS®2.1; Garcia et al. 2007) and Quantimet 500+ (Flach et al. 2002).

38.5 MEIOFAUNAL BIOMASS

The difficulties of obtaining accurate biological estimates of fecundity, natural mortality, predation, and population dynamics of meiofauna generally make investigations into the energetic of meiofaunal assemblages difficult and time consuming. The problems of obtaining such information for meiofaunal assemblages increase due to sampling problems when we work in deep-sea ecosystems (*i.e.*, with expensive special equipment; Snelgrove 1999) and reduced individual body mass (Schewe and Soltwedel 1998). Since meiofauna are characterized by short generation time, selection of the temporal scale of observations is rather difficult. Also in the shallower depths, estimates of production are generally not accurate due to an inability to follow the development of a single cohort. This is very difficult for meiofaunal assemblages because of the reduced size and the overlapping size of individuals of different cohorts. The adoption of more sophisticated methods based on population dynamics models (predator–prey interaction and strong assumptions on the age structure of the population) would be required to estimate meiofaunal secondary production (Downing and Rigler 1984). Meiofaunal secondary production can be grossly estimated by the production to biomass ratio (P/B), which is a rather simple indirect method. P/B expresses the turnover of one species or of a group of organisms and indicates the rate at which an unit of biomass is renewed in one year. In practice, this indirect method requires only determination of the total meiofaunal biomass.

38.5.1 Direct Methods

Direct methods for estimating total meiofaunal production (i.e., species-specific growth increments and identifiable cohorts) are difficult to apply at the entire assemblage, because of the taxonomic difficulty in identifying all meiofaunal taxa to species level and their small size (Arntz 1971; Waters 1977). Moreover, methods based on *in situ* estimates of oxygen consumption (Shirayama 1992) are again difficult to apply because meiofaunal respiration cannot be efficiently distinguished from prokaryote oxygen consumption. For the limits of the direct methods, indirect approaches are the most applied and utilized methods to estimate meiofaunal secondary production (Danovaro et al. 2002b; Heip et al. 1983; Escaravage et al. 1989).

38.5.2 Indirect Methods

Indirect methods for the estimation of meiofaunal secondary production are based on the determination of meiofaunal biomass in terms of total assemblages and different taxa.

38.5.2.1 Determination of Meiofaunal Biomass

All procedures to determinate meiofaunal biomass (nematodes and higher taxa) are described in details in the section Meiofaunal Biomass and Production. When total meiofaunal biomass and the biomass of the main taxa have been measured, different indirect methods can be applied to measure the secondary production.

38.5.2.2 Minimal Production

This method for estimating secondary production based on summation of biomass increments of a population (Arntz 1991). Values of production based on the summation of biomass increments at each sampling period during a year's sampling interval can be calculated as follows

$$P_{tot} = \Delta B_{t_1-t_0} + \Delta B_{t_2-t_1} + \ldots \Delta B_{t_n-t_{n-1}} \tag{38.3}$$

where

P_{tot} = total annual meiofaunal production.
ΔB_{t1-t0} = the total biomass increment between two subsequent samplings.
$t_n - t_0$ = one year.

The primary limitation of this method is that it underestimates the production since the fraction of biomass consumed by predation is not considered. This approach is utilized in coastal ecosystems (Danovaro et al. 2002b) but results are rather difficult to apply in deep-sea ecosystems where intensive studies on the temporal variability of benthic assemblages are very difficult to carry out and are expensive.

38.5.2.3 Production to Biomass Ratio (P/B)

This method is useful for immediate estimations of potential secondary production by multiplying the biomass (estimated using the volumetric method or image analyzers) by the P/B value. The production to biomass ratio is based on the assumption that the P/B ratio for the higher taxa is known. The value of secondary production can be calculated as follows:

$$P_{tot} = P / B \cdot B_{annual} \tag{38.4}$$

where

P_{tot} = total meiofaunal production.
P/B = production to biomass ratio.
B_{annual} = mean annual meiofaunal biomass.

McIntyre (1964) and Gerlach (1971) suggested a mean P/B ratio of 9 and 10, respectively, for the entire meiofaunal assemblage, obtained by considering different turnover rates for different taxa. The P/B ratios allow an approximate estimation of meiofaunal secondary production because only within nematodes P/B ratio can vary from 4 to 63. However, further experiments are required to validate these P/B ratios.

38.5.2.4 Production to Biomass Specific Ratio (P/B for the Higher Meiofaunal Taxa)

P/B specific ratios for each taxon were proposed by Lassere et al. (1976) and Warwick and Price (1979). Warwick and Price (1979) suggested: P/B = 8.4 for nematodes, 18 for harpacticoids, and 5 for annelids and turbellarians, whereas Lassere et al. (1976) reported: P/B = 9.75 for nematodes, 21 for copepods, 11.3 for turbellarians, 9.7 for annelids, and 9.12 for other taxa. The application of the production to biomass specific ratio for each higher taxon allows us to estimate the meiofaunal secondary production as follows:

$$P_{tot} = P_{nem} + P_{cop} + P_{anel} + P_{turb} + P_{oth} \tag{38.5}$$

and

$$P_{taxon} = P/B_{taxon} \cdot B_{taxon} \tag{38.6}$$

where

P_{tot} = total meiofaunal production.
P_{nem} = total nematode production.
P_{cop} = total harpacticoid copepod production.
P_{anel} = total annelid production.
P_{turb} = total turbellarian production.
P_{oth} = total production of all others meiofaunal taxa.
P_{taxon} = annual production of each taxon.
P/B_{taxon} = annual biomass turnover of each taxon.
B_{taxon} = annual mean biomass of each taxon.

The limit of such a method is due to the fact that the P/B ratio of several taxa is so far unknown and is often *nonapplicable* to a pool of organisms belonging to different groups [NOTES 1 and 2].

38.5.3 Notes

All of the approaches reported here have obvious limitations.

1. Methods based on P/B ratios are subjected to problems of accuracy since it is not easy to establish a single P/B ratio for the entire assemblage (Lasserre et al. 1976; Warwick and Price 1979; Warwick 1980).
2. Although the P/B ratio methods need further validation, these still remain the best approach to estimate the total meiofaunal secondary production (Arntz 1971; Ankar and Elmgren 1976; Waters 1977; Warwick 1980; Heip et al. 1983; Vranken and Heip 1986; Escaravage et al. 1989; Chardy and Dauvin 1992; Chardy et al. 1993; Danovaro and Fabiano 1996).

38.5.4 Temporal Variability of Meiofaunal Secondary Production

The analysis of meiofaunal secondary production (MSP) can be subjected to temporal analysis for identifying which sampling intervals display the highest biomass increments and which factors are responsible for such changes (Danovaro et al. 2002b). In such a case the Warwick's fractionated P/B ratios can be applied to each sampling period, as follows:

$$MSP = \sum_{i=1}^{n}\left[MBM_t \cdot P/B_t\right]/T_1 + \sum_{i=1}^{n}\left[MBM_t \cdot P/B_t\right]/T_2 + \ldots \sum_{i=1}^{n}\left[MBM_t \cdot P/B_t\right]/T_n \tag{38.7}$$

and

$$MSP_{T_1} + MSP_{T_2} + \ldots + MSP_{T_n} = \text{one sampling year} \tag{38.8}$$

where

MSP = meiofaunal secondary production.

n = number of taxa.

MBM_t = the average meiofaunal biomass of each taxon at each sampling interval.

P/B_t = is the production to biomass ratio utilized for each taxon on a daily basis and calculated by dividing the annual P/B ratio by the number of days of each sampling period.

T = the number of days between two subsequent samplings.

T_1, T_2, T_n = is the sequential number of sampling intervals.

MSP_{T1} = is meiofaunal production determined for each sampling interval.

This method allows us to estimate the monthly production and to identify which taxon is responsible for such biomass variation. This approach can be utilized in coastal or transitional ecosystems (Danovaro et al., 2002b), but is rather difficult to apply in deep-sea ecosystems in which studies on the temporal (on monthly basis) variability of benthic assemblages are practically not existent [NOTES 1 and 2].

38.5.4.1 Notes

1. Alternative approaches to estimate the secondary production are based on allometric equations by which it is possible to estimate P/B ratio as a function of some biotic and abiotic parameters that influence the organisms' growth and production. Banse and Mosher (1980) proposed a simple model correlating the benthic organisms P/B ratio with the adult biomass. Schwinghamer et al. (1986) calculate the meiofaunal production as follows:

$$P/B = 0.073 \cdot M^{-0.337} \tag{38.9}$$

where

P = annual production (Kcal m^{-2} a^{-1}).

B = average annual biomass (Kcal m^{-2}).

M = mean individual body mass (Kcal).

Tumbiolo and Downing (1994) also included in the model abiotic parameters able to affect benthic faunal production in marine environments (i.e., temperature and depth). The proposed equation is:

$$\log P = 0.24 + 0.96 \cdot \log B - 0.21 \log Wm + 0.03T - 0.16(Z+1) \tag{38.10}$$

where

P = annual production (g dry weight m^{-2} a^{-1}).

B = mean annual biomass (g dry weight m^{-2}).

Wm = highest individual body weight (mg dry weight).

T = temperature (°C) possibly at the bottom.

Z = depth (m).

2. This model is highly unspecific (useful for comparisons at different latitudes and depths) and flexible (for instance, the variable (Z + 1) allows the application in intertidal environments where depth is assumed to be 0). The main limit of this equation is that it has been calculated analyzing only one meiobenthic group (i.e., copepods). Further investigations are needed to validate this method.

Transfer the residual sample into a Delfuss cuvette

↓

Estimate the body length (L) and maximal body width (W)
of each specimen using a micrometer- scale (magnification 25×, 40×)

↓

At the end of the sorting, transfer the sample on a 20 μm sieve

↓

Transfer the samples to a 50 mL tube using 4% buffered formalin

↓

At the end of the sorting, transfer the sample on a 20 μm sieve

↓

Transfer the samples to a 50 mL tube using 4% buffered formalin

↓

Add some drops of 0.5 g L^{-1} Rose Bengal
and store at room temperature

PROTOCOL 38.1 Determination of meiofaunal biomass.

REFERENCES

Albertelli, G., A. Covazzi-Harriague, R. Danovaro, M. Fabiano, S. Fraschetti, and A. Pusceddu. 1999. Differential responses of bacteria, meiofauna, and macrofauna in a shelf area (Ligurian Sea, NW Mediterranean): Role of food availability. *Journal of Sea Research* 42: 11–26.

Andrassy, I. 1956. Die rauminhalts-und gewichtsbestimmüng der fadenwurmer (Nematoden). *Acta Zoologica Academiae Scientarum Hungaricae* 2: 1–15.

Arntz, W. E. 1971. Biomasse und production des makrobenthos in den tierferen Teilen der Kieler Bucht im Jahr 1968. *Kiel Meersforsch* 27: 36–72.

Banse, K. and S. Mosher. 1980. Adult body mass and annual production/biomass relationships of field populations. *Ecological Monograph* 50: 355–379.

Bett, B., A. Vanreusel, M. Vincx et al. 1994. Sampler bias in the quantitative study of deep-sea meiobenthos. *Marine Ecology Progress Series* 104: 197–203.

Bett, B. J. 2000. Comparative benthic ecology of the Rockall Trough and Faeroe-Shetland Channel. In *Atlantic margin environmental surveys of the seafloor 1996 and 1998 (Atlantic Frontier Environmental Network)* [CDROM], ed. Geotek Ltd, Section 4.3.2. London, UKOOA.

Ceccherelli, V. U. and M. Mistri. 1991. Production of the meiobenthic harpacticoid copepod Canuella perplexa. *Marine Ecology Progress Series* 68: 225–234.

Chardy, P. and J. C. Dauvin. 1992. Carbon flows in a subtidal fine sand community from the western English Channel: A simulation analysis. *Marine Ecology Progress Series* 81: 147–161.

Chardy, P., P. Gros, H. Mercier, and Y. Monbet. 1993. Benthic carbon budgets for the Bay of Saint Brieuc (Western Channel). Application of inverse method. *Oceanologica Acta* 16: 687–694.

Cobb, N. A. 1917. Notes on nemas. *Contribution to a Science Nematology* 5: 117–128.

Cook, A. A., P. J. D. Lambshead, L. E. Hawkins, N. Mitchell, and L. A. Levin. 2000. Nematode abundance at the oxygen minimum zone in the Arabian Sea. *Deep-Sea Research Part I* 47: 75–85.

Danovaro, R. and M. Fabiano. 1996. Meiofaunal production in subtidal sediments of the Ligurian Sea: Potential contribution of bacteria and microphytobenthos to the food demand. In *Proceedings 11th Conference of the Italian Association of Oceanology and Limnology*, ed. G. Albertelli, A. De Maio, and M. Picazzo, 603–613. Genova: AIOL.

Danovaro, R., C. Gambi, and N. Della Croce. 2002a. Meiofauna hotspot in the Atacama Trench (Southern Pacific Ocean). *Deep-Sea Research Part I* 49: 843–857.

Danovaro, R., C. Gambi, and S. Mirto. 2002b. Meiofaunal production and energy transfer efficiency in a seagrass bed (*Posidonia oceanica*) of the Western Mediterranean. *Marine Ecology Progress Series* 234: 95–104.

Danovaro, R., D. Marrale, N. Della Croce, A. Dell'Anno, and M. Fabiano. 1999. Heterotrophic nanoflagellates, bacteria, and labile organic compounds in continental shelf and deep-sea sediments of the eastern Mediterranean. *Microbial Ecology* 35: 244–255.

De Grisse, A. T. 1969. Redescription ou modifications de quelques techniques utilisées dans l'étude des nématodes phytoparasitaires. *Mededelingen Rijks Faculteit Landbouww Et Gent* 34: 351–369.

De Morais, L. T. and J. Y. Bodiou. 1984. Predation on meiofauna by juvenile fish in a western Mediterranean flatfish nursery ground. *Marine Biology* 82: 209–215.

Downing, J. A. and F. H. Rigler. 1984. *A manual on methods for the assessment of secondary productivity in fresh waters.* IBP Handbook No. 17. Oxford: Blackwell Scientific Publications.

Elmgren, R. and T. Radziejewska. 1989. Recommendations for quantitative benthic meiofauna studies in the Baltic. *Baltic Marine Biologists Publication* 12: 1–23.

Escaravage, V., M. E. Garcia, and J. Castel. 1989. The distribution of meiofauna and its contribution to detritic pathway in tidal flats (Arcachon Bay, France). In *Topics in marine biology, Scientia Marina*, ed. J. D Ros, 53: 551–559.

Feller, R. J. and R. M. Warwick. 1988. Energetics. In *Introduction to the study of meiofauna*, ed. R. P. Higgins and H. Thiel, 181–196. London: Smithsonian Institution Press, London.

Flach, F., A. Muthumbi, and C. Heip. 2002. Meiofauna and macrofauna community structure in relation to sediment composition at the Iberian margin compared to the Goban Spur (NE Atlantic). *Progress in Oceanography* 52: 433–457.

Fleeger, J. W., W. B. Sikora, and J. P. Sikora. 1983. Spatial and long-term temporal variation of meiobenthic-hyperbenthic copepods in lake Pontchartrain, Lousiana. *Estuarine, Coastal, and Shelf Science* 16: 441–453.

Gambi, C., A. Vanreusel, and R. Danovaro. 2003. Biodiversity of nematode assemblages from deep-sea sediments of the Atacama Slope and Trench (Southern Pacific Ocean). *Deep-Sea Research Part I* 50: 103–117.

Garcia, R., K. A. Koho, H. C. De Stigter, E. Epping, E. Koning, and L. Thomsen. 2007. Distribution of meiobenthos in the Nazaré canyon and adjacent slope (western Iberian Margin) in relation to sedimentary composition. *Marine Ecology Progress Series* 340: 207–220.

Gerlach, S. A. 1971. On the importance of marine meiofauna for benthos communities. *Oecologia* 6: 176–190.

Gooday, A. J., J. A. Huges, and L. A. Levin. 2001. The foraminiferan macrofauna from three North Carolina (USA) slope sites with contrasting carbon flux: A comparison with the metazoan macrofauna. *Deep-Sea Research Part I* 48: 1709–1739.

Gutzmann, E., P. Martìnez Arbizu, A. Rose, and G. Veit-Köhler. 2004. Meiofauna communities along an abyssal depth gradient in the Drake Passage. *Deep-Sea Research Part I* 51: 1617–1628.

Heip, C., P. M. Herman, and M. Vincx. 1983. Variability and productivity of meiobenthos in the Southern Bight of the North Sea. *Rapport et Procès-Verbaux des Réunions du Conseil Permanent International Pour l'Exploration de la Mer* 183: 51–56.

Heip, C., P. M. J. Herman, and A. Coomans. 1982. The productivity of marine meiobenthos. *Academiae Analecta (Klasse der Wetenschappen)* 44: 1–20.

Higgins, R. P. and H. Thiel. 1988. *Introduction to the study of meiofauna*. Washington, DC: Smithsonian Institution Press.

Jensen, P. 1984. Measuring carbon content in nematodes. *Helgoländer Meeresuntersuchungen* 38: 83–86.

Lasserre, P., J. Renaud-Mornand, and J. Castel. 1976. Metabolic activities of meiofaunal communities in a semi-enclosed lagoon. Possibilities of trophic competition between meiofauna and mugil fish. In *Proceeding 10th European Marine Biology Symposium*, ed. G. Persoone and E. Jasper, 393–414. Wettener: University Press.

Lindquist, D. G., L. B. Cahoon, I. E. Clavijo et al. 1994. Reef fish stomach contents and prey abundance on reef and substrata associated with adjacent artificial and natural reefs in Onslow Bay, North Carolina. *Bulletin of Marine Science* 55: 308–318.

McIntyre, A. D. 1964. Ecology of marine meiobenthos. *Biological Reviews* 44: 245–290.

Möller, P., L. Pihl, and R. Rosenberg. 1985. Benthic faunal energy flow and biological interaction in some shallow marine soft bottom habitats. *Marine Ecology Progress Series* 27: 109–121.

Myers, R. F. 1967. Osmoregulation in Pangrellus redivivus and Aphelenchus avanae. *Nematologica* 12: 579–586.

Schewe, I. and T. Soltwedel. 1998. Deep-sea meiobenthos of the central Arctic ocean: Distribution patterns and size-structure under extreme oligotrophic conditions. *Vie et Milieu* 49: 79–92.

Schwinghamer, P., B. Hargrave, D. Peer, and C. M. Hawkins. 1986. Partitioning of production and respiration among size groups of organisms an intertidal benthic community. *Marine Ecology Progress Series* 31: 131–142.

Seinhorst, J. W. 1959. A rapid method for the transfer of nematodes from fixative to unhydrous glycerine. *Nematologica* 4: 67–69.

Shirayama, Y. and T. Fukushima. 1995. Comparison of deep-sea sediments and overlying water collected using multiple corer and box corer. *Journal of Oceanography* 51: 75–82.

Shirayama, Y. 1992. Respiration rates of bathyal meiobenthos collected using a deep-sea submersible SHINKAI 2000. *Deep-Sea Research Part A. Oceanographic Research Papers* 39 (5): 781–788.

Smith, K. L. Jr., R. J. Baldwin, D. M. Karl, and A. Boetius. 2002. Benthic community responses to pulses in pelagic food supply: North Pacific Subtropical Gyre. *Deep-Sea Research Part I* 49: 971–990.

Snelgrove, P. V. R. 1999. Getting to the bottom of marine biodiversity: Sedimentary habitats. *BioScience* 49: 129–138.

Soetaert, K., J. Vanaverbeke, C. Heip et al. 1997. Nematode distribution in ocean margin sediments of the Goban Spur (northeast Atlantic) in relation to sediment geochemistry. *Deep-Sea Research Part I* 44: 1671–1683.

Stora, G., M. Bourcier, A. Arnoux et al. 1999. The deep-sea macrobenthos on the continental slope of the northwestern Mediterranean Sea: A quantitative approach. *Deep-Sea Research Part I* 46: 1339–1368.

Thurston, M. H., B. J. Bett, A. L. Rice, and P. A. B. Jackson. 1994. Variations in the invertebrate abyssal megafauna in the North Atlantic Ocean. *Deep-Sea Research Part I* 41: 1321–1348.

Tumbiolo, M. L and J. A. Downing. 1994. An empirical model for the prediction of secondary production in marine benthic invertebrate population. *Marine Ecology Progress Series* 114: 165–174.

Valdovinos, C., S. A. Navarrete, and P. A. Marquet. 2003. Mollusk species diversity in the southern eastern Pacific: Why are there more species towards the pole? *Ecography* 26: 139–144.

Vanhove, S., H. Vermeeren, and A. Vanreusel. 2004. Meiofauna towards the South Sandwich Trench (750–6300m), focus on nematodes. *Deep-Sea Research Part II* 51: 1665–1687.

Vincx, M. 1996. Meiofauna in marine and freshwater sediments. In *Methods for the examination of organismal diversity in soils and sediments*, ed. G. S. Hall, 187–195. Cab International.

Vincx, M., B. J. Bett, A. Dinet et al. 1994. Meiobenthos of the deep Northeast Atlantic. *Advance in Marine Biology* 30: 2–88.

Vranken, G. and C. Heip. 1986. The productivity of marine nematodes. *Ophelia* 26: 429–442.

Warwick, R. M. 1980. Population dynamics and secondary production of benthos. In *Marine benthic dynamics*, ed. K. R. Tenore and C. B. Coull, 1–24. Columbia: Univ. South Carolina Press.

Warwick, R. M. and J. M. Gee. 1984. Community structure of estuarine meiobenthos. *Marine Ecology Progress Series* 18: 97–111.

Warwick, R. M. and R. Price. 1979. Ecological and metabolic studies on free-living nematodes from an estuarine mud-flat. *Estuarine, Coastal, and Shelf Science* 9: 257–271.

Waters, T. F. 1977 Secondary production in inland waters. *Advances in Ecological Research* 10: 91–164.

Watzin, M. C. 1983. The effects of meiofauna on settling macrofauna meiofauna may structure macrofaunal communities. *Oecologia* 59: 163–166.

Wieser, W. 1960. Benthic studies in Buzzards Bay. II. The meiofauna. *Limnology and Oceanography* 5: 121–137.

39 Macrofaunal and Megafaunal Biomass and Biochemical Composition

CONTENTS

39.1 INTRODUCTION

Benthic standing stock represents a time-averaged consequence of energy input (Smith et al. 1997; Herman et al. 2001) that is largely dependent on the input of particulate organic carbon for source systems (Rowe et al. 1991; Cosson et al. 1997; Danovaro et al. 1999; Smith and Demopoulos 2003). Variations in benthic standing stock can provide important indications of ecological changes in the deep-sea ecosystem. Since Rowe's (1983) classic review of bathymetric biomass trends in the macrofauna, much literature has appeared documenting standing stock on local and regional scales (Figure 39.1). However, there is still little information concerning the biomass, trophic strategies, and biochemical composition of macrofauna and megafauna inhabiting the deep-sea benthos (Barnes et al. 1976; Haedrich and Rowe 1977).

While the average size of organisms decreases with depth, large megafaunal species are well represented at abyssal depths and have an important ecological role (Billett et al. 2001; Ruhl and Smith 2004). Scavenging amphipods living at abyssal depths actually exhibit gigantism (Hessler et al. 1972); they are much larger than typical peracarid crustaceans (Perrone et al. 2002). These organisms are highly mobile, and are adapted to quickly locate and consume large food falls. Their unusual size may be a response to a different set of selective pressures than those affecting smaller sedentary members of the community.

The dichotomy between gigantism in a small number of species and size reduction in most deep-sea benthic species represents an interesting example of the multiple ability of deep-sea species to adapt to extreme conditions.

Holothuroids are one of the dominant invertebrate megafauna in the deep sea characterized by soft bottoms (Zenkevitch and Birstein 1956; Laubier and Sibuet 1979). However, deep-sea hard bottoms contain several species, with apparently limited geographic distribution. The deep-sea canyons or the seamounts for instance host very diverse assemblages, characterized by the presence of corals, sponges, crinoids, that dominate the biomass and create new habitats that host several invertebrate species (Pitcher et al. 2007).

The community standing stock of macrofauna and megafauna decrease with depth; this is a universal phenomenon that involves a complex transition in the relative importance of the different size groups. Bacterial abundance and biomass show no decline with depth. All of the animal-size groups experience significant exponential decreases in both abundance and biomass. The abundance of larger animals is significantly lower and decreases more rapidly than for smaller groups. The upper continental slope is dominated by macrofaunal biomass, and the abyss by bacterial and meiofaunal biomass. The steeper regressions for both abundance and biomass of the macrofauna

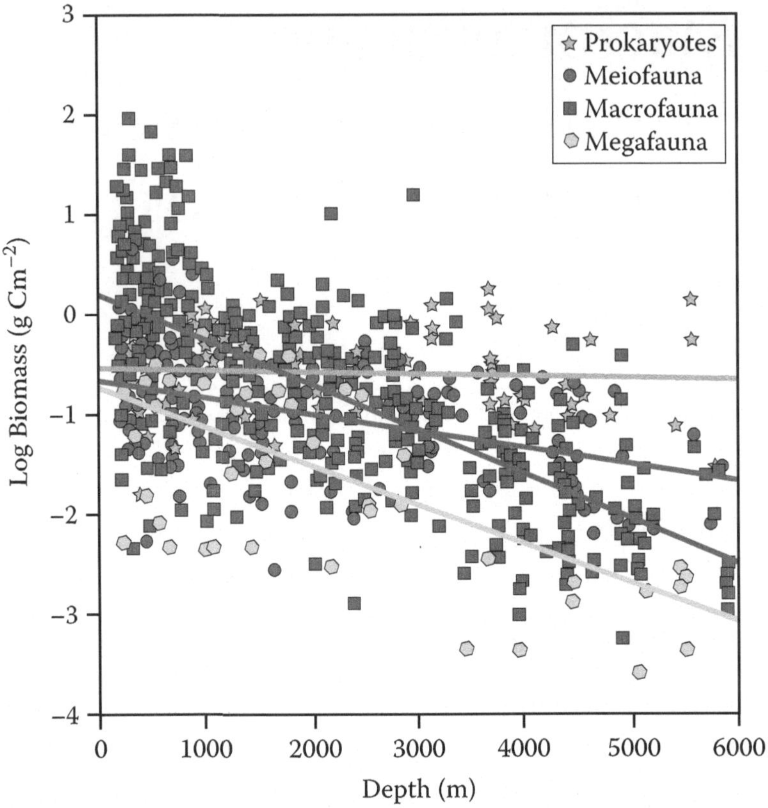

FIGURE 39.1 Relationships of biomass to depth for bacteria, metazoan meiofauna, macrofauna, and megafauna. Data for the four different classes are provided in Rex et al. 2006 (Appendices 2 and 3 available at www.int-res.com/articles/suppl/m317p001_app.pdf).

and megafauna, compared to smaller size groups, suggest that large organisms are more affected by the reduction in available energy with depth. Larger organisms have more efficient metabolism per unit weight, but still require more energy than small organisms. At great depths, the population densities of larger organisms may become too low, due to energy restrictions, to remain reproductively viable (Rex 1973; Thiel 1975; Rex et al. 2005).

Most of deep-sea benthic organisms feed upon detrital organic matter present in the sediment and phytodetritus that reaches the bottom via sedimentation or lateral advection. The bioenergetic strategy of macro- and megafauna is likely to be strongly governed by environmental conditions, and the biochemical composition of these organisms is likely to reflect changes in food availability. Lipid accumulation is the most widespread long-term energy storage strategy in aquatic crustaceans and their reproductive potential is largely dictated by lipid content (e.g., review by Sargent and Henderson 1986; also Gatten et al. 1980). Environmental nutritional conditions directly regulate the breeding success of individuals and, subsequently, population dynamics. The seasonal cycle in the composition of the benthic, deposit-feeding amphipod is largely determined by the brief period of spring phytoplankton bloom sedimentation and the long-lasting deficiency of good-quality nutrition during late autumn–early spring. Carbon and nitrogen levels followed closely the patterns of lipid and protein, respectively. Biochemical changes (proximate chemical composition, total lipids, glycogen, and cholesterol contents) during the reproductive cycle can be utilized for investigating benthic life strategies of *Aristeus antennatus* (Crustacea: Penaeidea), *Parapenaeus longirostris* (Crustacea: Penaeidea), and *Nephrops norvegicus* (Crustacea: Astacidea) and their response to biological and environmental factors (sex, maturation, reproduction, food availability, and depth).

Nektobenthic species (*A. antennatus* and *P. longirostris*) typically show higher protein, lipid, cholesterol, and glycogen contents, and lower moisture content in the muscle than the benthic–endobenthic species (*N. norvegicus*). The energy content of the nektobenthic species is typically higher than that of endobenthic species. Depth can have a key role in the observed changes in the biochemical composition, as depth can have a significant interspecific effect on food intake. It was also evident that reproductive cycle has profound effects upon the biochemistry of the three species. Gonadal maturation has large associated energy costs and the biochemical composition can be influenced by or synchronized with seasonal feeding activity or food availability

39.2 PROCEDURE

A number of basic parameters that measure functional aspects, or can represent surrogates of functional variables, are typically measured in all collections of deep-sea macrofauna and megafauna. Usually, immediately after the macrofaunal sample is retrieved on board, this is sieved and subsequently sorted. In the case of megafauna, this is just sorted once on board. The functional variables measured for macrofauna include, after the taxonomic identification to species level, the determination of the biomass by weighting the specimens individually after desiccation. The most practical method for obtaining an estimate of the metabolically active biomass is the calculation of the Ash-Free-Dry-Weight (AFDW) which is the value of the dry weight from which the ashes are subtracted. To determine the biomass, organisms are weighed after drying at 60°C–80°C for 24–48 h when reaching the constant weight, higher temperatures can lead to a significant loss of the most volatile compounds of the organic matter. Organisms are then ignited in a muffle furnace at 450°C for 4–5 h to estimate ash content. Higher temperature can lead to a significant carbonate loss. AFDW is calculated including shells. Measurements should be performed using an analytical balance with an accuracy of 0.1 mg. Macrofaunal taxa too numerous or too small to weigh individually can be weighed collectively and a mean individual wet weight calculated.

The functional variables measured for megafauna include:

1. Length (in cm)
2. Width of each specimen (cm)
3. Individual wet weight (g)

After all these measurements have been made, the data of abundance (number of individuals) and biomass (total weight of each species or taxon) are combined to calculate total macrofaunal or megafaunal biomass.

Biomass has to be standardized by area sampled in order to allow for comparisons between samples, areas, and studies. Typically macrofaunal biomass is standardized to gC m^{-2}. In order to do so, the wet weight is measured on freshly collected (wet) material in which the excess of water is removed from the organisms prior to weighing. After the wet weight is determined, there will be an accuracy of ± 0.0001 g for macrofauna and ± 0.01 g for megafauna.

To standardize by area, the distance fished on the seafloor is calculated from the navigation data on board (latitude and longitude) and the touch-down and lift-up times of the net. This distance is transformed into area fished by taking into account the trawl mouth opening. The biomass data are divided by the area fished (km^2), providing the area corrected data in kg.km^2 (Company et al. 2004; Ramírez-Llodra et al. 2008).

39.3 DETERMINATION OF THE BIOCHEMICAL COMPOSITION

For the analysis of the biochemical composition of deep-sea organisms, the samples are treated as follows: For each biochemical analysis (lipids, proteins, and carbohydrates) and for each sampling occasion, at least 10 mg of tissue (or the entire organism, if smaller) in 3–5 replicates is placed on a Whatman GF/F fiberglass filters, previously treated in a muffle furnace (450°C for 2 h), washed with prefiltered seawater, and stored at −20°C. Alternatively, the organisms are directly placed in a test tube and stored at −20°C until analysis.

It is possible also to determine the biochemical composition of organisms fixed in prefiltered seawater containing buffered formalin (4% final concentration) (Dell'Anno et al. 1999).

For the analyses of the protein, carbohydrate, and lipid content of deep-sea organisms, all glassware used for the biochemical composition of deep-sea animals is soaked in 1 M NaOH and subsequently rinsed with MilliQ water to avoid any organic contamination. In order to increase extraction efficiency, prior to all analyses the tissue is homogenized in 1 mL of MilliQ water and then sonicated for 2 h at 4°C, frozen twice for 30 min (−20°C), and sonicated again for 10 min at 4°C. All crude material (i.e., the entire homogenate) is used separately for each biochemical analysis.

All biochemical analyses are carried out with 1 mL of homogenate, after appropriate dilution. To determine the detection limits of the analyses, internal standards can be added to replicate samples. The accuracy of the methods was tested against prepared standards obtained from dilution of 100 μg mL^{-1} stock solution for proteins and carbohydrates, and of 40 μg mL^{-1} stock solution for lipids; final concentrations were 2.5, 5, 10, and 20 μg mL^{-1} for all standards.

Lipid extraction is carried out according to Bligh and Dyer (1959) by direct elution with chloroform and methanol (1:2 v:v). The extracted lipids were then dried at 80°C (20 min) and determined spectrophotometrically after carbonization at 180°C in concentrated sulfuric acid, according to Marsh and Weinstein (1966); tripalmitine solutions are used as standard. Various techniques are applicable for lipid analysis (Barnes and Blackstock 1973; Folch et al. 1957) but the method of Marsh and Weinstein (1966) was selected because of its capability to detect a large range of lipid species, and because of its adequate sensitivity and accuracy (± 1.5 μg). Protein concentration is determined according to Hartree (1972) using bovine albumin serum solution as standard. The determination of proteins based on Folin-reacting peptides is carried out spectrophotometrically. Carbohydrates are analyzed according to Gerchakov and Hatcher (1972), and glucose solution was used as standard. The method is based on the same principles as the common phenol-sulfuric protocol of Dubois et al. (1956). In order to avoid the interference of formalin in carbohydrate analysis, filters were accurately rinsed with 10 mL artificial seawater.

For each biochemical analysis, blanks are made using the same filters previously treated in muffle furnace (450°C, 2 h). To evaluate the possible interference of the preservative and stain on the biochemical analyses, additional blanks, on three replicates, are analyzed after treatment with formalin solution (4%), Rose Bengal (final concentration 0.5 mg mL^{-1}), and a combination of both. The blanks are then rinsed with prefiltered artificial seawater and processed as previously described for the other samples.

The individual concentration of lipids, proteins, and carbohydrates is obtained by dividing the total concentration of each biochemical compound by the dry weight of the organism (or part there of) analyzed.

Organic matter content is then referred to as the sum of lipids, proteins, and carbohydrates. Biochemical parameters were converted to energy equivalents as follows: lipids = 39.95 J mg^{-1} DW, proteins = 23.63 J mg^{-1} DW, and carbohydrates = 17.18 J mg^{-1} DW.

The calorific value can be calculated using the Winberg (1971) equation as

$$(cal\ mg^{-1})\text{cal mg}^{-1} = 0.041 \cdot \%carbohydrates + 0.055 \cdot \%proteins + 0.095 \cdot \%lipids \qquad (39.1)$$

The protocols for the analysis of the biochemical composition of the organisms, after the pretreatment described here, are identical to those reported for the analysis of sediment organic matter.

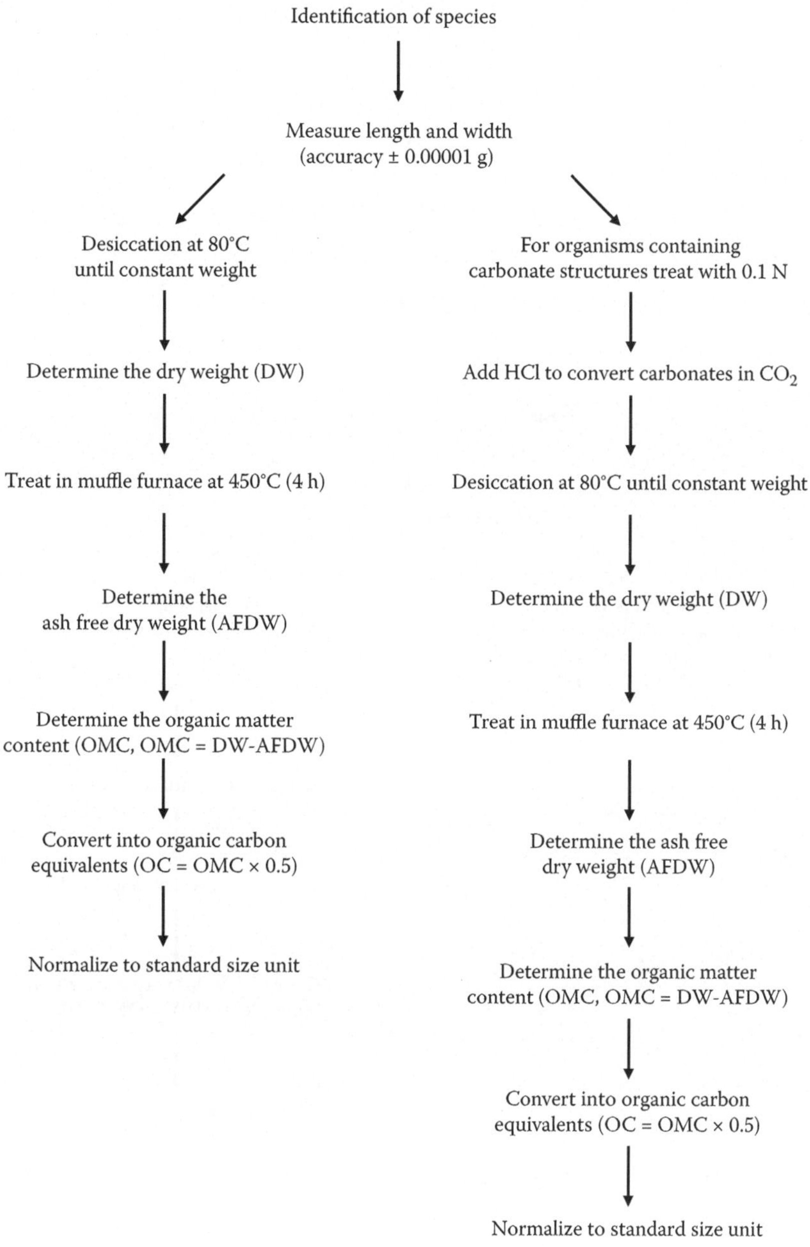

PROTOCOL 39.1 Determination of macrofaunal biomass.

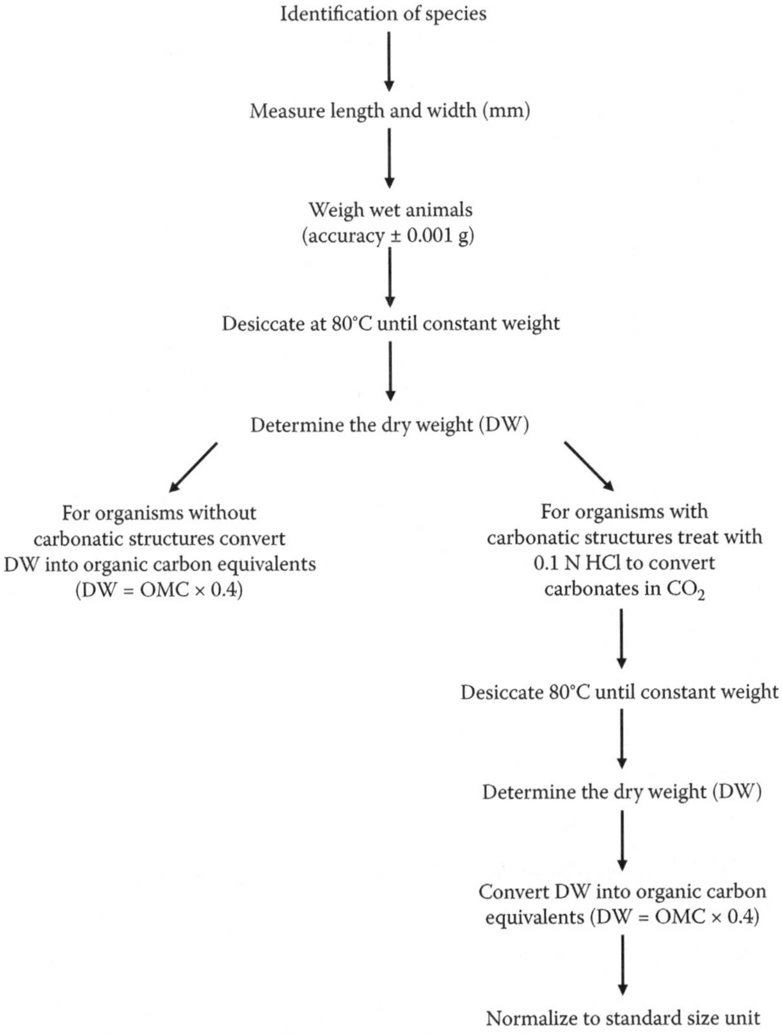

PROTOCOL 39.2 Determination of megafaunal biomass.

REFERENCES

Barnes, A. T., L. B. Quetin, J. J. Childress, and D. L. Pawson. 1976. Deep-sea macroplanktonic sea cucumbers: Suspended sediment feeders captured from deep submerged vehicle. *Science* 194: 1083–1085.

Barnes, H. and J. Blackstock. 1973. Estimation of lipids in marine animals and tissues: Detailed investigation of the sulphophosphovanillin method for "total" lipids. *Journal of Experimental Marine Biology and Ecology* 12: 103–118.

Bligh, E. G. and W. Dyer. 1959. A rapid method for total lipid extraction and purification. *Canadian Journal of Biochemistry and Physiology* 37: 911–917.

Company, J. B., P. Maiorano, A. Tselepides et al. 2004. Deep-sea decapod crustaceans in the western and central Mediterranean Sea: Preliminary aspects of species distribution, biomass, and population structure. *Scientia Marina* 68 (Suppl. 3): 73–86.

Cosson, N., M. Sibuet, and J. Galeron. 1997. Community structure and spatial heterogeneity of the deep-sea macrofauna at three contrasting stations in the tropical northeast Atlantic. *Deep Sea Research Part I* 44: 247–269.

Danovaro, R., A. Dell'Anno, D. Martorano, P. Parodi, D. Marrale, and M. Fabiano. 1999. Seasonal variation in the biochemical composition of deep-sea nematodes: Bioenergetic and methodological considerations. *Marine Ecology Progress Series* 179: 273–283.

Dubois, M., K. A. Gilles, J. K. Hamilton, P. A. Rebers, and F. Smith. 1956. Colorimetric method for determination of sugars and related substances. *Analytical Chemistry* 28: 350–356

Folch, J., M. Lees, and G. H. S. Stanley. 1957. A simple method for the isolation and purification of total lipides from animal tissues. *Journal of Biological Chemistry* 226: 497–509.

Gatten, R. R., J. R. Sargent, T. E. V. Forsberg, S. C. M. O'Hara, and E. D. S. Corner. 1980. On the nutrition and metabolism of zooplankton. XIV. Utilization of lipid by *Calanus helgolandicus* during maturation and reproduction. *Journal of Marine Biological Association of United Kingdom* 60: 391–399.

Gerchakov, S. M. and P. G. Hatcher. 1972. Improved technique for analysis of carbohydrates in sediments. *Limnology and Oceanography* 17: 938–943.

Haedrich, R. L. and G. T. Rowe. 1977. Megafaunal biomass in the deep sea. *Nature* 269: 141–142.

Hartree, E. F. 1972. Determination of proteins: A modification of the Lowry method that gives linear photometric response. *Analytical Biochemistry* 48: 422–427.

Herman, P. M. J., K. Soetaert, J. J. Middelburg et al. 2001. The seafloor as the ultimate sediment trap using sediment properties to constrain benthic-pelagic exchange processes at the Goban spur. *Deep-Sea Research Part II* 48: 3245–3264.

Hessler, R. R., J. D. Isaacs, and E. L. Mills. 1972. Giant amphipod from the abyssal Pacific Ocean. *Science* 175: 636–637.

Laubier, L. and M. Sibuet. 1979. Ecology of the benthic communities of the deep northeast Atlantic. *Ambio Special Report* 6: 37–42.

Marsh, J. B. and W. J. Weinstein. 1966. A simple charring method for determination of lipids. *Journal of Lipid Research* 7: 574–576.

Perrone, F. M., A. Dell'Anno, R. Danovaro, N. Della Croce, and M. H. Thurston. 2002. Population biology of Hirondellea sp. nov. (Amphipoda: Gammaridea: Lysianassoidea) from the Atacama Trench (south-east Pacific Ocean). *Journal of the Marine Biological Association of the UK* 82(3): 419–425.

Pitcher, T. J., T. Morato, P. J. B. Hart, M. R. Clark, N. Haggan, and R. S. Santos. 2007. *Seamounts: Ecology, fisheries, and conservation.* Fish and Aquatic Resources Series, 12. Oxford: Blackwell Publishing.

Ramírez-Llodra, E., M. Ballesteros, J. B. Company, L. Dantart, and F. Sardà. 2008. Spatio-temporal variations of biomass and abundance in bathyal non-crustacean megafauna in the Catalan Sea (North-western Mediterranean). *Marine Biology* 153: 297–309.

Ruhl, H. A. and K. L. Smith. 2004. Shifts in deep-sea community structure linked to climate and food supply. *Science* 305: 513–515.

Sargent, J. R. and R. J. Henderson. 1986. Lipids. In *The biological chemistry of marine copepods,* Ed. E. D. S. Corner and S. C. M. O'Hara, 59–108, Oxford: Oxford University Press.

Smith, C. R. and A. W. J. Demopoulos. 2003. Ecology of the deep ocean floor. In *Ecosystems of the world, Volume 28: Ecosystems of the deep ocean,* Ed. P. A. Tyler, 179–218. Amsterdam: Elsevier.

Winberg, G. G. 1971. *Methods for the estimation of production of aquatic animals*. London and New York: Academic Press.

Zenkevitch, L. A. and J. A. Birstein. 1956. Studies of the deep water fauna and related problems. *Deep Sea Research* 4: 54–64.

40 Protistan Grazing on Benthic Prokaryotes and Viruses

CONTENTS

40.1 INTRODUCTION

Protozoans are important consumers of prokaryotes, and are able to facilitate remineralization of detritus, thus playing a pivotal role in nutrient recycling (Sherr et al. 1982; Sherr and Sherr 1993; Patterson et al. 1989; Hondeveld et al. 1992). In addition, protozoans are consumed by meio- and macrofauna, providing a trophic link between prokaryotes and the upper trophic levels of the food webs (Sherr et al. 1986a). Direct estimates of bacteria losses by grazing can provide an estimate of the carbon being transferred to higher trophic levels (Carrick et al. 1991; Dolan and Gallegos 1991). Also, virus–protozoa interactions may be important in the material flow in aquatic systems. Phagotrophic nanoflagellates consume and digest, either directly or indirectly (via ingestion of virus-infected bacteria), a variety of marine viruses, thereby deriving nutritional benefit and serving as a natural sink for marine viral particles (González and Suttle 1993; Maranger et al. 2002). Information on viruses–prokaryotes–protozoans interactions in benthic environments and the biomass flow in the sediment are almost completely lacking (Fischer et al. 2003; Wieltschnig et al. 2003). The reasons for the general lack of benthic studies are the specific characteristics of the sediment, presenting several problems including the difficulty of quantitative extraction of the organisms from sediment.

Estimate of the impact of protistan grazing has been studied using various approaches such as a dilution technique (Landry et al. 1984; Weisse 1991), specific eukaryotic inhibitors (Sherr et al. 1986; Caron et al. 1991; Hondeveld et al. 1999), grazers exclusion by prefiltration (Bongiorni et al. 2005; Jacquet et al. 2005), uptake of radiolabeled bacteria (Jürgens and Güde 1991; Simon et al. 1998), and uptake of fluorescently labeled and immuno-labeled particles (González 1999; Kemp 1988; Sherr et al. 1987; Gonzàlez and Suttle 1993; Christofferson et al. 1997). Recently, fluorescent *in situ* hybridization (FISH) has been applied to determine ingested natural bacteria within ciliates. This direct a posteriori technique has the advantage of specifically staining phylogenetic lineages (e.g., eubacteria) after ingestion and thereby a false determination of the number of ingested bacteria is avoided (for details see Diederichs et al. 2003; Cleven and Königs 2007).

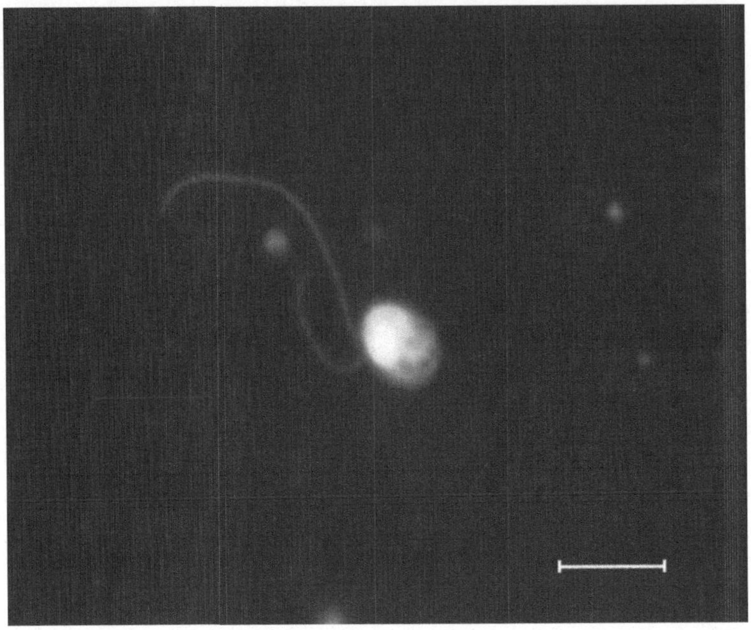

FIGURE 40.1 *(A color version of this figure follows page 234.)* Nanoflagellate and prokaryote cells stained with DAPI. Scale bar is 5 μm.

Most grazing studies have used fluorescently labeled tracer bacteria (FLB; Carlough and Meyer 1990; Epstein 1997; González 1999; Kemp 1988; Sherr et al. 1987; Hondeveld et al.1992) or direct staining of sediments (Starink et al. 1994), fluorescently labeled viruses (FLV, Gonzàlez and Suttle 1993; Noble and Fuhrman 2000), or inert plastic fluorescently labeled microspheres (FLM, Wilks and Sleigh 1998; Carrias et al. 1996). These techniques have diverse drawbacks, which include the selectivity of grazers for or against the tracer particles, the differences between fluorescently labeled particles and natural prey, change in the natural abundance of preys by addition of FLP, and the amount of time and labor required (Starink et al. 1994; Vaqué et al. 1994). However, the essential advantages of the use of fluorescently labeled preys/particles is that prey and predators do not have to be separated (which is not achievable for pico- and nano-sized organisms in sediments); direct identification of most active consumers in the protistan assemblage; the minimum manipulation of natural samples; and the determination of grazing rates via short time experiments (Hondeveld et al. 1992; Gonzàlez and Suttle 1993; Sherr and Sherr 1993; Starink et al. 1994; Wieltschnig et al. 2003; Fischer et al. 2006; Figure 40.1).

Here, we report protocol for the estimation of protist grazing with FLB and FLV (Hondeveld et al. 1992; Sherr and Sherr 1993; Gonzàlez and Suttle, 1993).

40.2 FIELD PROCEDURES

Sediment samples are collected with box corers (subsampled with Plexiglas® core, syringes) or multicorers and immediately used for grazing experiments, as explained.

40.3 LABORATORY PROCEDURE

40.3.1 Equipment and Instruments

Autoclave
Centrifuge

Sonicator
Water bath
Laminar air flow
Filtration apparatus: Sterile filter holder for 25 mm diameter filters with 15 mL funnel Millipore, vacuum pump, vacuum flask (for discharge)
Epifluorescence microscope

40.3.2 Materials

Sterile 50 mL falcon tubes
Sterile 10 mL test tubes
Sterile 2 mL Eppendorf tubes
Nucleopore polycarbonate filters, 0.2, 0.8, or 2.0 μm pore size, 25 mm Ø (Whatman)
2–20 μL, 100–1000 μL, 1–5 mL pipettes, and sterile tips.
Glass slides and coverslips
Disodium hydrogen phosphate (Na_2HPO_4)
Na_4 pyrophosphate (PPi)
Sodium chloride (NaCl)
Phosphate-buffered saline solution
DTAF stain (5-([4,6-dichlorotriazin-2-yl]amino)fluorescein hydrochloride; Sigma)

40.3.3 Solutions

DTAF solution is prepared by dissolving DTAF powder in 0.05 M of Na_2HPO_4 (final concentration of 4 mg mL^{-1}); filter through a 0.02 μm pore-size polycarbonate filter before use.
PPi buffer (0.02 M tetrasodium pyrophosphate (PPi)–0.85% NaCl solution) is prepared by dissolving 0.89 g of tetrasodium pyrophosphate (PPi) in 100 mL of deionized water and adding NaCl (final concentration 0.85%).
Phosphate-buffered saline solution (pH 9).

40.3.4 Procedure

40.3.4.1 Suspension of Fluorescently Labeled Viruses (FLVs)

1. Collect ca. 20 L of seawater (with 5 L Niskin bottles).
2. Filter sample through a 0.2 μm (pore-size) filter to remove bacteria and protists.
3. Concentrate this virus-sized fraction using a spiral cartridge concentration system (Amicon, Inc.) to a final volume of ca. 150 mL.
4. Further concentrate using Centriprep-30 centrifugal concentration units (Amicon, Inc.), to a final volume of ca. 5 mL.
5. Add 0.5 μL of the DTAF-Na_2HPO_4 solution to 1 mL of viral suspension (ca. 10^{10} viruses).
6. Mix gently and incubate overnight at 4°C in the dark.
7. Sonicate (1 min) subsamples of the DTAF stained virus suspension [NOTE 1].
8. Filter an aliquot through a 0.2 μm polycarbonate filter.
9. Count under epifluorescence microscope (as explained in Chapter 15).
10. Process immediately for grazing experiment or store at −20°C until analysis.

40.3.4.2 Suspension of Fluorescently Labeled Bacteria (FLB)

1. Harvest a bacterial culture (in its log phase) (3000 g for 20 min) [NOTE 2].
2. Rinse resuspended the pellet with 10 mL cold phosphate-buffered saline solution (PBS, pH 9).
3. Add 2 mg of DTAF [NOTE 3].

4. Incubate in a water bath at 60°C for 2 h.
5. Centrifuge the cells and washed ca.3 times with PBS to remove excess DTAF.
6. Resuspend the cells in PPi buffer in order to reach the cells density requested and sonicate two times (several seconds).
7. Check bacterial density by DAPI stain and epifluorescence microscopy.
8. Store the bacterial suspension frozen at −20°C.

40.3.4.3 Set up of Grazing Experiments

9. Thown and sonicate (1 min) subsamples of the DTAF-stained bacteria or virus suspension [NOTE 4].
10. Place ca. 10 g of the sediment slurry in a whirlpack bag or beaker immersed in a water bath at *in situ* temperature.
11. Incubate sample for 1 h in order to allow the microbial assemblage to recover from handling stress.
12. Gently mixed with a pipette, sediment with a FLB or FLV suspension giving a final concentration of 5%–30% of natural bacterial abundance and 40%–50% natural viral abundance [NOTE 5].
13. Collect subsamples of ca.1–2 mL sediment at 5 or 15 min intervals (e.g., 5, 10, 15, 20, 30, and 60 min).
14. Fix sediment with ice-cold glutaraldehyde (1.5%–2% final concentration) and store at 4°C [NOTE 6].
15. Extract protists from sediments as explained in Chapter 15.
16. Filter extracts onto 0.8 or 2.0 μm Nuclepore pore-size polycarbonate filters.
17. Stain with DAPI as previously described for protists enumerations (Chapter 15).
18. Count cells by epifluorescence microscopy using UV light to spot protists, and shifting to blue light excitation to check and count ingested bacteria/viruses inside. In order to determine the average number of fluorescent particles per cell, a minimum of 30–50 phagotrophic protists should be inspected for each time point and a minimum of 100 consumed preys should be enumerated per filter.

40.3.4.4 Estimate of Ingested Rates

Ingestion rates of FLP by the protistan taxon examined is given by the linear regression of the linear portion of the uptake curve.

As an example, the calculation of grazing rates for bacterivorous nanoflagellates is:

Per cell-clearance rate of FLBs = cell specific uptake rate of FLB/ FLB g^{-1}
Where cell specific uptake of FLBs = FLBs nanoflagellate^{-1} h^{-1}
Ingestion rate (bacteria cell^{-1} h^{-1}) = FLBs nanoflagellate^{-1} h^{-1} × ratio of total bacteria to FLB
Community grazing rate (bacteria g^{-1} h^{-1}) = ingestion rate × nanoflagellate
Concentration

40.3.5 Notes

1. Sonication reduced clumping of the viruses; however, longer sonication times do not improve the results.
2. Bacterial clumping often increase after the stationary phase; alternatively, a bacteria-culture concentrate of a natural bacterial population can be obtained by serial filtrations through 100, 10 μm-mesh net and 1.0 μm-polycarbonate filters (Nuclepore).
3. This solution is 0.2 μm prefiltered and stored frozen (Sherr et al. 1987).

4. If the culture or suspension is too dense, it is possible to sonicate the suspensions or filter it to remove clumps.
5. Tracer concentration higher than the observed natural prokaryotic assemblage result in a prey-saturated environment (Fenchel 1982). In such an environment, rates of prey uptake become independent from prey density. FLVs are added to the samples at a final concentration of about 10^7 mL^{-1}, whereas the FLB are added at about 10^6 mL^{-1}.
6. For minimizing the egestion of particles (Sanders et al. 1989).

Place 10 g sediment in a water bath (T room)

↓ Incubate 1 h

Add DTAF stained bacteria (FLB)
or stained virus (FLV)

↓

Sonicate (1 min)

↓

Add FLB or FLV to sediment
and gently mix

↓

Collect 1–2 mL sediment
at 5 min intervals to 60 min

↓

Fix subsamples with ice cold
glutaraldehyde (store at 4°C)

↓

Extract protists as explained

↓

Filter on polycarbonate filter

↓

Add 2 mL DAPI working solution

↓

Wash filter 3 times with 5 mL PFSW

↓ Vacuum filter, <35 mmHg

Mount the filter on microscope slide
with a drop of immersion oil

↓

Count cells under
epifluorescence microscope (100× magnifications)

PROTOCOL 40.1 Protocol for estimation of benthic protistan grazing.

REFERENCES

Bongiorni, L., M. Magagnini, M. Armeni, R. Noble, and R. Danovaro. 2005. Viral production, decay rates, and life strategies along a trophic gradient in the north Adriatic Sea. *Applied and Environmental Microbiology* 71: 6644–6650.

Carlough, L. A. and J. L. Meyer. 1990. Rates of protozoan bacterivory in three habitats of a southeastern blackwater river. *Journal of the North American Benthological Society* 9: 45–53.

Caron, D. A., E. L. Lim, G. Miceli, J. B. Waterbury, and F. W. Valois. 1991. Grazing and utilization of chroococcoid cyanobacteria and heterotrophic bacteria by protozoa in laboratory cultures and a coastal plankton community. *Marine Ecology Progress Series* 76: 205–217.

Carrias, J. F., C. Amblard, and G. Bourdier. 1996. Protistan bacterivory in an oligomesotrophic lake: Importance of attached ciliates and flagellates. *Microbial Ecology* 31: 249–268.

Carrick, H. J., G. L. Fahnenstiel, E. F. Stroemer, and R. G. Wetzel. 1991. The importance of zooplankton-protozoanntrophic couplings in Lake Michigan. *Limnology and Oceanography* 36: 1335–1345.

Christofferson, K., O. Nybroe, K. Jürgens, and M. Hansen. 1997. Measurement of bacterivory by heterotrophic nanoflagellates using immunofluorescence labeling of ingested cells. *Aquatic Microbial Ecology* 13: 127–134.

Cleven, E. J. and S. Königs. 2007. Growth of interstitial ciliates in association with ciliate bacterivory in a sandy hyporheic zone. *Aquatic Microbial Ecology* 47: 177–189.

Diederichs, S., C. Beardsley, and E. J. Cleven. 2003. Detection of ingested bacteria in benthic ciliates using fluorescence *in situ* hybridization. *Systematic and Applied Microbiology* 26: 624–630.

Dolan, J. R. and C. L. Gallegos. 1991. Trophic coupling of rotifers, microflagellates, and bacteria during fall months in the Rhode River Estuary. *Marine Ecology Progress Series* 77: 147–156.

Epstein, S. S. 1997. Microbial food webs in marine sediments. I. Trophic interactions and grazing rates in two tidal flat communities. *Microbial Ecology* 34: 188–198.

Fenchel, T. 1982. Ecology of heterotrophic microflagellates: IV. Quantitative occurrence and importance as bacterial consumers. *Marine Ecology Progress Series* 9: 35–42.

Fischer, U. R., C. Wieltschnig, A. K. T. Kirschner, and B. Velimirov. 2003. Does virus-induced lysis contribute significantly to bacterial mortality in the oxygenated sediment layer of shallow oxbow lakes? *Marine Ecology Progress Series* 69: 5821–5289.

Fischer, U. R., C. Wieltschnig, A. K. T. Kirschner, and B. Velimirov. 2006. Contribution of virus-induced lysis and protozoan grazing to benthic bacterial mortality estimated simultaneously in microcosms. *Environmental Microbiology* 8: 1394–1407.

González, J. M. 1999. Bacterivory rate estimates and fraction of active bacterivores in natural protist assemblages from aquatic systems. *Applied and Environmental Microbiology* 65: 1463–1469.

González, J. M. and C. A. Suttle. 1993. Grazing by marine nanoflagellates on viruses and virus-sized particles: Ingestion and digestion. *Marine Ecology Progress Series* 94: 1–10.

Hondeveld, B. J. M., R. P. M. Bak, and F. C. Van Duyl. 1992. Bacterivory by heterotrophic nanoflagellates in marine sediments measured by the uptake of fluorescently labeled bacteria. *Marine Ecology Progress Series* 89:63–71.

Hondeveld, B. J. M., R. P. M. Bak, W. Van Raaphorst, and F. C. Van Duyl. 1999. Impact of grazing by benthic eukaryotic organisms on the nitrogen sediment-water exchange in the North Sea. *Journal of Sea Research* 41: 255–268.

Jürgens, K. and H. Güde. 1991. Seasonal changes in the grazing impact of phagotrophic flagellates on bacteria in lake Constance. *Marine Microbial Food Webs* 5: 27–37.

Jacquet, S., I. Domaizon, S. Personnic et al. 2005. Estimates of protozoan- and viral-mediated mortality of bacterioplankton in Lake Bourget (France). *Freshwater Biology* 50: 627–645.

Kemp, P. F. 1988. Bacterivory by benthic ciliates: Significance as a carbon source and impact on sediment bacteria. *Marine Ecology Progress Series* 49: 163–169.

Landry, M. R., L. W. Haas, and V. L. Fagerness. 1984. Dynamics of microbial plankton communities: Experiments in Kaneohe Bay, Hawaii. *Marine Ecology Progress Series* 16: 127–133.

Maranger, R., P. A. Del Giorgio, and D. F. Bird. 2002. Accumulation of damaged bacteria and viruses in lake water exposed to solar radiation. *Aquatic Microbial Ecology* 28: 13–227.

Noble, R. T. and J. A. Fuhrman. 2000. Rapid virus production and removal as measured with fluorescently labeled viruses as tracers. *Applied and Environmental Microbiology* 66: 3790–3797.

Patterson, D. J., J. Larsen, and J. O. Corliss. 1989. The ecology of heterotrophic flagellates and ciliates living in marine sediments. *Progress in Protistology* 3: 185–277.

Sanders, R. W., K. G. Porter, S. J. Bennett, and A. E. DeBiase. 1989. Seasonal patterns of bacterivory by flagellates, cilliates, rotifers, and cladocerans in a freshwater plankton community. *Limnology and Oceanography* 34: 673–687.

Sherr, B. F., E. B. Sherr, and T. Berman. 1982. Decomposition of organic detritus: A selective role for Microflagellate protozoa. *Limnology and Oceanography* 27: 765–769.

Sherr, E. B., B. F. Sherr, and G. A. Paffenhofer. 1986a. Phagotrophic protozoa as food for metazoans: A "missing" troph~clink in marine pelagic food webs? *Marine Microbial Food Webs* 1: 61–80.

Sherr B. F., E. B. Sherr, T. L. Andrew, R. D. Fallon, and S. Y. Newell. 1986. Trophic interactions between heterotrophic protozoa and bacterioplankton in estuarine water analyzed with selective metabolic inhibitors. *Marine Ecology Progress Series* 32: 169–179.

Sherr B. F., E. B. Sherr, and R. D. Fallon. 1987. Use of monodispersed, fluorescently labeled bacteria to estimate *in situ* protozoan bacterivory. *Applied and Environmental Microbiology* 53: 958–965.

Sherr, E. B. and B. F. Sherr. 1993. Protistan grazing rates via uptake of fluorescently labeled prey. In: *Handbook of methods in aquatic microbial ecology*, ed. P. F. Kemp, B. F. Sherr, E. B. Sherr, and J. J. Cole, 695–702. Boca Raton: Lewis Publishers.

Simon, M., C. Bunte, M. Schulz, M. Weiss, and C. Wünsch. 1998. Bacterioplankton dynamics in Lake Constance (Bodensee): Substrate utilization, growth control, and long-term trends. *Archiv für Hydrobiologie Special Issues Advances in Limnology* 53: 195–221.

Starink, M., M.-J. Bär-Gilissen, R. P. M. Bak, and T. E. Cappenberg. 1994. Quantitative centrifugation to extract benthic protozoa from freshwater sediments. *Applied and Environmental Microbiology* 60: 167–173.

Vaqué, D., J. M. Gasol, and C. Marrassé. 1994. Grazing rates on bacteria: The significance of methodology and ecological factors. *Marine Ecology Progress Series* 109: 263–274.

Weisse, T. 1991. The annual cycle of heterotrophic freshwater nanoflagellates: Role of bottom-up and top-down control. *Journal of Plankton Research* 13: 167–185.

Wieltschnig, C., U. R. Fischer, A. K. T. Kirschner, and B. Velimirov. 2003. Benthic bacterial production and protozoan predation in a silty freshwater environment. *Microbial Ecology* 46: 62–72.

Wilks, S. A. and M. A. Sleigh. 1998. Grazing rates in Euplotes mutabilis: Relationships between particle size and concentration. *Microbial Ecology* 36: 165–174.

41 Meiobenthic Predation on Prokaryotes

CONTENTS

41.1 INTRODUCTION

Meiofauna have the ability to absorb dissolved organic matter, eat particulate organic matter, and predate other temporary and permanent meiofauna (Giere 1993), but the behavior of most metazoan meiofauna is adapted to pick specific microbial food items (e.g., microalgae, prokaryotes, and protozoans; Montagna et al. 1983). Such predation on benthic microbes has been estimated in terms of microbial biomass consumed by meiofauna per unit of time. Benthic prokaryote abundance and distribution might, in turn, affect meiofauna predation rates (Montagna and Yoon 1991). Despite the potential relevance of meiofaunal predation rates for the functioning and energy transfer within the benthic food web, little information is available on the impact of meiofaunal predation on natural microbial assemblages. This is largely due to the difficulty of applying *in situ* techniques to measure predation rate. The labeling technique is actually the best tool for such kinds of studies since radioactive tracer is introduced to measure synoptically microbial uptake of the tracer and meiofaunal predation (Montagna 1993). In addition, the use of dual labels allow measurement of the predation on both autotrophic and heterotrophic prokaryotes.

41.2 FIELD PROCEDURE

Surface sediment samples may be collected both manually by means of Plexiglas® cores or through equipments such as multicorers, depending on the sampling depth. Immediately after retrieval, sediment subsamples should be gently removed from the corer. Only the top 1 cm, where more than

50%–70% of total meiofauna abundance is present, is removed and transferred into a sterile falcon tube (50 mL).

41.3 LABORATORY PROCEDURE

41.3.1 Equipment and Instruments

Autoclave
Laminar air flow
Filtration apparatus: Sterile filter holder for 25 mm diameter filters with 15 mL funnel (Millipore), vacuum pump, vacuum flask (for discharge)
Sonicator
Centrifuge for falcon
Epifluorescence microscopy
Scintillation counter
Vacuum pump (Millipore or similar)
Bottle to collect radioactive fluid
Analytical balance
Room temperature
Stereomicroscope for meiofaunal sorting

41.3.2 Materials

Sterile 50 mL Falcon tube
Sterile 15 mL tubes
Sterile 10 mL syringes
Sterile Eppendorf tubes
Sieves for meiofaunal extraction
Wash bottle
Petri dish
Sterile seawater
Tetrasodium pyrophosphate solution (SIGMA)
Syringe filters, 0.2 μm pore size (Whatman Puradisc AS 25 mm ø)
Sterilized forceps
3,6-bis[dimethylamino]acridinium chloride (acridine orange [AO])
Glass slides and coverslips
Formalin
Nalidix acid
Leucine
$MgCl_2$
HCl
Polycarbonate filters black (Whatman) 0.2 μm pore size, 25 mm ø
2–20 μL, 100–1000 μL, 1–5 mL pipettes and sterile tips
Scintillation vials (20 mL)
White polycarbonate membrane filters (diameter 25 mm, pore size 0.2 μm)
LUDOX

41.3.3 Solutions and Reagents

[4,5-^{3}H]-leucine solution (used at 0.5 μM final concentration)
Nonradioactive leucine (Sigma—optional)

4 μCi of a solution of ^{14}C bicarbonate
HCl 1%
TCA 5% (w/v)
Scintillation liquid (HIONIC FLUOR Packard or similar)
Formalin 2% (v/v) final concentration in artificial seawater, sterilized and prefiltered onto 0.2 μm filters (Whatman Puradisc) and add Borax (sodium tetraborate, SIGMA) to formalin pure (20 g l^{-1}) so that the solution is at pH 8.8 to 9.0.
Tetrasodium pyrophosphate solution (work solution 5 mM): Prepare a stock solution of 500 mM tetrasodium pyrophosphate (SIGMA) in reagent-grade water prefiltered onto 0.2 μm filters (Whatman Puradisc)
3,6-bis[dimethylamino]acridinium chloride (acridine orange [AO]) working solution 0.025% (w/v) in reagent-grade water. Prepare 50 mL solution and dark preserved at 4°C for 3 weeks.

41.3.4 Predation Experiments (See Scheme)

For each sediment sample collected there should be at least nine Plexiglas cores: Three replicate sediment cores for experiment meiofaunal predation, three replicate sediment cores for inhibitor control sediments, and three replicate sediment cores for adsorption effect or killed control [NOTE 1].

41.3.4.1 Experiment Meiofaunal Predation

Sediment subsamples (three replicates) are gently removed from the cores (Ø 3.6 cm) and only the top 1 cm collected, then transferred into a sterile falcon tube (50 mL).
Resuspend in 12 mL of sterile seawater [NOTE 2] (collected from the same site) to form an homogeneous slurry (dilution 1:1).
Samples are manually shaken.
Add two radiolabeled substrates: 6 μCi of an aqueous solution of L-[4,5-3H] leucine (for prokaryotic incorporation), which is used to reach 0.5–1 μM final concentration, and 4 μCi of a solution of ^{14}C bicarbonate (for microphytobenthos incorporation).
Samples are manually shaken.
Incubate for 1 h at *in situ* temperature and ambient light conditions.
After incubation, to add 4 g of $MgCl_2$ for 5 min [NOTE 3].
Stop meiofaunal predation on prokaryotes and microalgae incorporation by adding 11 mL of seawater and 1 mL of formalin (to reach at 4% final concentration).
Samples are manually shaken.
Take an aliquot of 1 mL for heterotrophic prokaryotic C production analysis (incorporation of ^{3}H-leucine) and an aliquot of 1 mL for photosynthetic fixation (incorporation of ^{14}C).
Sediment subsamples (1 mL) are transferred to 2 mL sterile Eppendorf tubes and preserved at 4°C until laboratory procedure. For each sediment sample a blank is analyzed.
Falcon sediment is preserved at 4°C and used for meiofaunal sorting.

41.3.4.2 Inhibitor Control Sediments

Sediment subsamples (three replicates) are gently removed from the cores (Ø 3.6 cm) and only the top 1 cm collected, then transferred into a sterile falcon tube (50 mL).
Resuspend in 12 mL of sterile seawater (collected from the same site) to form an homogeneous slurry (dilution 1:1).
Samples are manually gently mixed.
Add two radiolabeled substrates (6 μCi of an aqueous solution of L-[4,5-^{3}H] leucine and 4 μCi of a solution of ^{14}C bicarbonate).

For measure the passive uptake by meiofauna, saturate with a solution of nalidix acid plus leucine to inhibit prokaryote incorporation of ^{3}H-leucine.
Samples are manually gently mixed.
New procedure: Incubate for 1 h at an *in situ* temperature in the dark to inhibit photosynthetic fixation of $^{14}CO_2$.
After incubation, add 4 g of $MgCl_2$ for 5 min.
Samples are gently manually mixed.
Stop the incorporation reaction by adding 11 mL of seawater and 1 mL of formalin (to reach at 4% final concentration).
Samples are gently manually mixed.
Take an aliquot of 1 mL for heterotrophic prokaryotic C production analysis (incorporation of ^{3}H-leucine) and an aliquot of 1 mL for primary production (incorporation of ^{14}C).
Sediment subsamples (1 mL) are transferred to 2 mL sterile Eppendorf tubes and preserved at 4°C until laboratory procedure. For each sediment sample a blank is analyzed.
Falcon sediment is preserved at 4°C and used for meiofauna sorting.

41.3.4.3 Adsorption Effect or Killed Control

Sediment subsamples (three replicates) are gently removed from the cores (Ø 3.6 cm) and only the top 1 cm is collected, then transferred into a sterile falcon tube (50 mL).
Resuspend in 24 mL of sterile seawater (collected from the same site) plus 1 mL of pure formalin, and then add the two radiolabeled substrates (6 μCi of an aqueous solution of L-[4,5-3H] leucine and 4 μCi of a solution of ^{14}C bicarbonate).
Samples are gently mixed manually.
Take an aliquot of 1 mL for heterotrophic prokaryotic C production analysis (incorporation of ^{3}H-leucine) and an aliquot of 1 mL for primary production analysis (incorporation of ^{14}C).
Sediment subsamples (1 mL) are transferred to 2 mL sterile Eppendorf tubes and preserved at 4°C until laboratory procedure. For each sediment sample a blank is analyzed.
Falcon sediment is preserved at 4°C and used for meiofaunal sorting.

41.3.4.4 Prokaryotic Production

Heterotrophic prokaryotic production and chemoautotrophic primary production is undertaken as follows:

1. After the field procedure, 1 mL of slurry sediment is passed through a polycarbonate filter both for heterotrophic prokaryotic production and for microphytobenthos primary production.
2. Filters for heterotrophic prokaryotic production are rinsed four times with 2 mL of cold 5% TCA, whereas filters for microphytobenthos primary production are rinsed four times with 2 mL of HCl 1%. [NOTE 4].
3. Transfer the filters to scintillation vials containing 5 mL reagent-grade water and 15 mL of scintillation liquid.
4. Place the vials in the liquid scintillation counter.
5. Take note of DPM (disintegrations per minute) for ^{3}H and ^{14}C.

41.3.4.5 Meiofaunal Sorting [NOTE 5]

1. Before starting with the standard protocol for meiofaunal extraction, eliminate the water above the sediment as much as possible, paying attention to avoid the resuspension of the sediment. Filter through the 20 μm mesh net. If necessary the sample can be centrifuged for 2 min at 3000 rpm to remove the water.

2. After this preliminary step, check and follow the standard procedure reported in Chapter 4, Section 4.5 Meiofaunal Diversity.
3. For meiofaunal sorting, check the procedures for samples preparation reported in the previous chapter.
4. Pick up all nematodes, copepods, and polychaetes observed in the samples.
5. Transfer each specimen in different vials (one for each group) containing 1 mL of reagent-grade water.
6. Place the vials into the oven at 60°C for 24–48 h to allow the evaporation of all the water.
7. Add 1 mL of Packard Soluene 350 to destroy the tissues, and keep again the vials in the oven at 60°C for 24 h.
8. After 24 h in the oven, add 10 mL of Packard Hionic-flour; samples can be analyzed with the scintillation counter.

41.3.4.6 Measurements of Prokaryotic Biomass

Prokaryotic biomass is measured on subsamples collected from the same cores used for predation experiments: Three replicate subsamples (about 0.5 g) and preserve with 4.5 mL of buffered formalin solution (2% final concentration). For direct cell count see Chapter 12, "Prokaryotic Abundance" and Chapter 35, "Prokaryotic Biomass in Marine Sediments."

41.3.5 Calculations

41.3.5.1 Rate of Biomass Production [NOTE 6]

Data of ^{3}H-leucine incorporation are converted into heterotrophic prokaryotic carbon production by using the formula reported in this book in Chapter 36, "Benthic Prokaryotic Heterotrophic Production Using the Leucine Incorporation Method."

41.3.5.2 Rate of Primary Production [NOTE 6]

The procedure to measure the rates of prokaryotes autotrophic production is reported in this book in a previous chapter.

Meiofaunal predation rates (G) [NOTE 7] are estimated as follows:

$$G = 2F / t \tag{41.1}$$

$$F = M / B \tag{41.2}$$

where

F is the fraction of label uptake in Meiofauna (M) relative to each microbial component (B) in a time t = 1 h. Both M and B are corrected using controls.

Meiofauna counts were reported to number of individuals per sample.

41.4 NOTES

1. The killed control is used to distinguish between label uptake due to absorption and adsorption by meiofauna in the control treatment. The killed control is subtracted to the inhibitor control to quantify meiofaunal absorption of radiolabeled substrates.
2. Sterile seawater (i.e., autoclaved and filtered through syringe filters 0.2 μm pore sizes) should be prepared before the beginning of the sample processing in order to avoid the interruption important steps of the protocol.
3. $MgCl_2$ for meiofauna anesthetic.

4. Filters for prokaryotic secondary production are rinsed with cold 5% TCA, while filters for microphytobenthos primary production are rinsed with HCl 1% to remove excess of leucine and sodium bicarbonate radioactive, respectively.
5. Meiofaunal sorting is laborious and time consuming.
6. Primary and secondary production can be expressed as DPM sample^{-1} h^{-1}.
7. Predation rates can be underestimated (Moens et al. 1999).

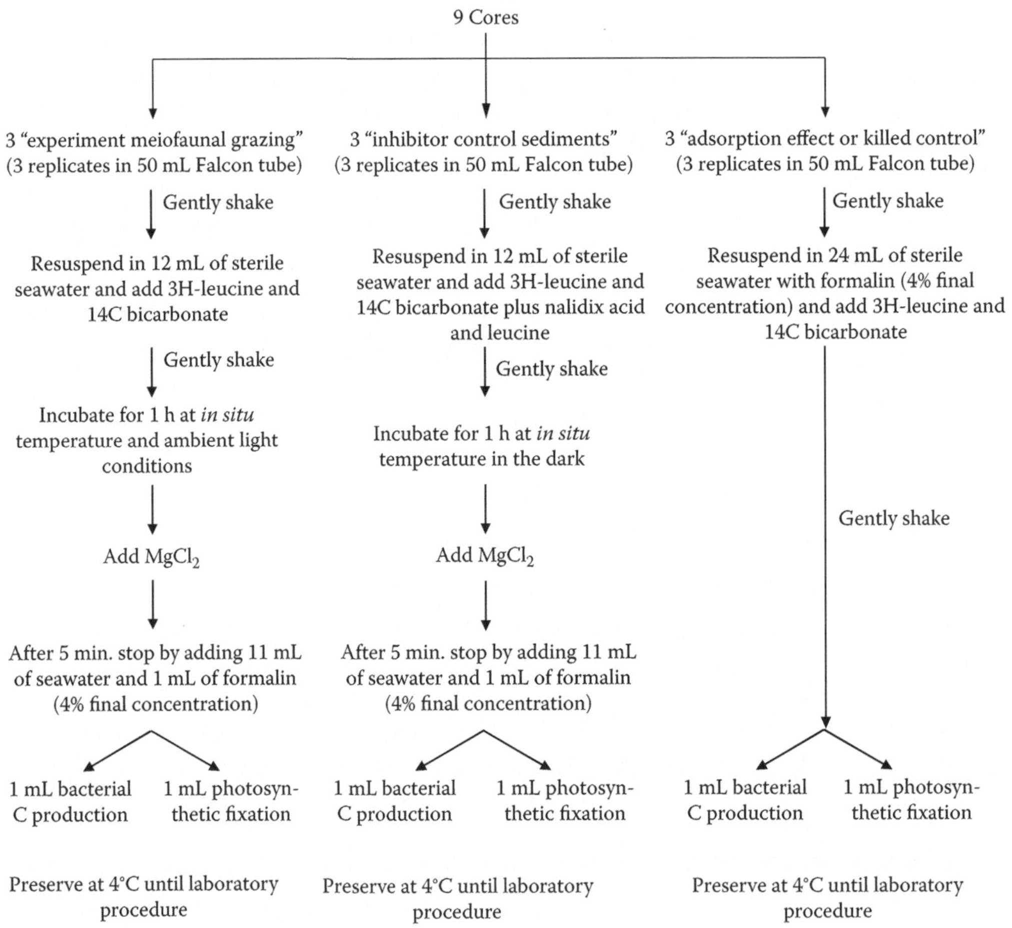

PROTOCOL 41.1 Determination of meiofaunal predation rates.

REFERENCES

Giere, O. 1993. *Meiobenthology. The microscopic fauna in aquatic sediments.* Berlin Heidelberg: Springer-Verlag.

Moens, T., L. Verbeeck, and M. Vincx. 1999. Preservation and incubation time-induces bias in tracer aided grazing studies on meiofauna. *Marine Biology* 133: 69–77.

Montagna, P. A. 1993. Radioisotope technique to quantity *in situ* microbivory by meiofauna in sediments. In *Handbook of methods in aquatic microbial ecology*, ed. P. F. Kemp, B. F. Sherr, E. B. Sherr, J. J. Cole, 745–753. Boca Raton, FL: Lewis Publishers.

Montagna, P. A., B. C. Coull, T. L. Herring, and B. W. Dudley. 1983. The relationship between abundances of meiofauna and their suspected microbial food (diatoms and bacteria). *Estuarine, Coastal, and Shelf Science* 17: 381–394.

Montagna, P. A. and W. B. Yoon. 1991. The effect of freshwater inflow on meiofaunal consumption of sediment bacteria and microphytobenthos in San Antonio Bay, Texas, U.S.A. *Estuarine, Coastal, and Shelf Science* 33: 529–547.

42 Analysis of Benthic Food Webs and Benthic Trophodynamics

CONTENTS

42.1 INTRODUCTION

The diverse origin of organic matter makes it difficult to discriminate what the most important food sources are in the animal diet and to quantitatively estimate their proportions. The different food sources contribute to an important detrital organic matter pool (Pocklington and Tan 1987; Mann 1988) and may be used differently by different animals due to their digestibility or because of hydrology that influences their production and availability to consumers (Schwinghamer et al. 1983; Monbet 1992). Studies of the feeding relationships of deep-sea organisms are very limited and have concentrated primarily on gut-content data that display substantial limitations (Polunin et al. 2001). For example, gut contents reflect diets at particular points in time and space, and severely neglect certain types of dietary materials such as gelatinous plankton and detritus that may nevertheless be very important in the sustenance of marine food webs (Polunin et al. 2001). Other limitations include problems associated with the voiding of gut contents upon capture and the very sporadic feeding of many carnivorous species. Furthermore, many animals crush or grind their food, such that the identification of prey becomes very difficult, and it is not often clear which components of their diets are actually assimilated, particularly when refractory or amorphous foods are ingested (Pinnegar and Polunin 2000). The use of stable nitrogen and carbon isotopes for the study of trophic interactions is now common in aquatic ecosystems (Pinnegar and Polunin 2000) also in the deep sea (Schubert and Calvert 2001; Bosley et al. 2004; Mohtadi et al. 2005; Levesque et al. 2006; Usui et al. 2006).

42.2 USE OF STABLE ISOTOPES (∂ ^{13}C AND ∂ ^{15}N) FOR THE DETERMINATION OF FOOD PREFERENCES

Stable isotopes can provide an indication of the prey assimilated over a considerable period of time (Hesslein et al. 1993) and elucidate aspects of both the origin and transformation of organic matter ingested (Hyndes and Lavery 2005; Usui et al. 2006). Stable nitrogen isotopes have been used to help define the trophic level of organisms (Vander Zanden et al. 1997). Carbon isotopes may be useful indicators of sources of production (Harrigan et al. 1989). Given the constrains and paucity of gut-content data, stable isotopes can clearly help to elucidate some aspects of the structure of deep-sea food webs (Polunin et al. 2001). Food-web analysis is made by measuring stable isotope ratios (∂ ^{13}C and ∂ ^{15}N) based on a stepwise enrichment of the heavier stable isotope between the prey and consumers tissue during assimilation processes (Rau 1982; Fry 1988; Griffin and Valiela 2001). The enrichment of ∂ ^{13}C and ∂ ^{15}N at each trophic level is reported to be about 1‰ and 3.3‰, respectively (Hobson and Welch 1992), although these values may differ due to the enzymatic assimilation activity of the consumer. By measuring isotopic ratios of as many as possible components of the food web, it is possible to establish relative trophic positions of species within the community (Polunin et al. 2001). Here, we present the methods for the preparation of samples for stable isotope analyses and for stomach content analysis.

42.3 FIELD PROCEDURES

42.3.1 Sampling

In deep-sea soft sediments a multicorer is used to collect simultaneously several meiofaunal samples (e.g., 4–12 corers; Elmgren and Radziejewska 1989). The instrument allows for the collecting of undisturbed sediment samples of a given surface area to analyze the horizontal and vertical distribution of benthic organisms. Another common sampling instrument for soft sediments is the box corer (Fleeger et al. 1983; Shirayama and Fukushima 1995). The weight of a box corer after sampling can range from 150 to >750 kg, and it collects a surface sediment of 0.02–1.0 m^2 down to a depth of 50 cm. From the box corer, meiofaunal samples could be collected using manual Plexiglas® corers (internal diameter 3.6 cm), one replicate from each different deployment. The use of transparent corers is recommended to analyze the redox potential discontinuity depth on the basis of changes in sediment color. The corer is inserted into the sediment to a certain depth and subsequently gently extracted from the sediment to keep it undisturbed, avoiding the mixing and redistribution of organisms along the vertical profile. Closing the corer *in situ* allows the transport in laboratory in undisturbed conditions.

Typically in these experiments, fishes, brachyuran crabs, shrimps, octopus, etc., are captured in different zones using a suction device attached to a rotating multichamber collector, mesh nets, or benthic cages. Also the classical Agassiz trawls and other tools can be used to collect megafaunal specimens (Bett 2000).

42.3.2 Material and Methods

The methods used for preserving samples for stable isotopes analysis should be carefully chosen. In general, samples (sediments and biota) for stable isotopes analyses should be immediately prepared for the mass spectrometer after their collection. When this is not possible, *that is*, during an oceanographic cruise, samples should be frozen at −20°C until the preparations for analysis in the laboratory.

Previous studies (Kaehler and Pakhomov 2001) on three unrelated marine species (fish, octopus, and kelp) revealed that the ∂ ^{15}N signatures tend to increase slightly but not significantly during both formalin and ethanol preservation. Data from previous studies suggest that preserved samples should not be used for ∂ ^{13}C analysis as the effects of preservatives on carbon isotope signatures vary greatly between species and studies (Kaehler and Pakhomov 2001). In contrast, the effect of preservatives on $\partial\ ^{15}N$ signatures is comparatively small, which may facilitate the use of preserved samples in studies of stable nitrogen isotopes. Samples intended for isotope analysis should be frozen, freeze-dried, or oven-dried; preservatives may be used only for $\partial\ ^{15}N$ analysis and organisms should be sacrificed immediately after collection (Fleeger et al. 1983; Kaehler and Pakhomov 2001).

Bosley and Wainright (1999) and Kaehler and Pakhomov (2001) conclude that freezing is the only preservation method that does not significantly influence stable isotopes. After collection, organisms should be immediately frozen until further preparation becomes possible.

42.3.3 Storage

Samples for the isotopes stables should be keep frozen (−20/−72°C) until ready for examination by the mass spectrometer in a laboratory. Specimens of macrofauna, megafauna, and fish should be kept frozen in Eppendorf vials or sterile plastic bags. Sediment organic matter samples should be kept frozen in a Petri dish. Meiofaunal samples for meiofauna should be kept frozen in a Plexiglas corer.

42.3.3.1 Equipment and Instruments

Plexiglas corer internal diameter 3.6 or 4.5 cm
Stopper for the corer
Parafilm
Adhesive tape

42.3.3.2 Procedure

1. Keep the corer vertical [NOTE 1].
2. Remove stopper from the lower opening [NOTE 2].
3. Close the lower opening with Parafilm [NOTE 3].
4. Fix the Parafilm using the adhesive tape [NOTE 3].
5. Remove stopper from the upper opening.
6. Close the upper opening with the Parafilm.
7. Fix the Parafilm using the adhesive tape.
8. Write all information relative to the samples (project, station, depth, replicate) using a permanent marker.
9. Cover the notes with a transparent tape.

42.3.4 Notes

1. It is important to keep the corer vertical to avoid the mix of the sediment layers and the alteration of the vertical distribution of meiofaunal organisms.
2. Take care to remove the stopper from the lower opening; be sure that the upper opening is well closed. The upper stopper plays a key role in keeping sediments from falling down.
3. Close the corer with Parafilm and adhesive tape within a few minutes after removal from the collection to avoid the alteration of the vertical distribution of meiofaunal organisms.

42.4 LABORATORY PROCEDURES

All procedures are based on the idea of converting the sample to a pure gas. This is then introduced into the isotope ratio mass spectrometer for analysis (Michener and Lajtha 2007). Some aspects should be kept in mind for the stable isotopes analysis; according to Michener and Lajtha (2007), the minimum sample sizes are normally 1 μmol C, 2 μmol N, and 20 μmol S (12 μg C, 28 μg N, and 640 μg S), respectively. For smaller samples size, the stable isotope analysis becomes rather difficult, and it is important to contact the laboratory in which the analyses will be performed to ask for the detection limit of the mass spectrometer.

42.4.1 Sediment Samples

In the laboratory, sediment is freeze-dried and ground to powder using a pestle and mortar [NOTE 1]. Freeze-dried sediment samples are analyzed for organic carbon and nitrogen stable isotopes after acidification, according to the procedure of Nieuwenhuize et al. (1994) [NOTE 2].

42.4.1.1 Notes

1. Collect 5–15 g sample sieve to remove large rocks, shells. Dry at 60°C and grind to a fine powder in a mortar and pestle. Store the fine powder in tin capsules for solid samples.
2. Give an approximate C and N content of samples and pass on this information to the laboratory where the analysis at mass spectrometer will be performed.

42.4.2 Meiofaunal Samples

Live nematodes should be extracted from the sediment following the procedures described in the previous chapters, but in this case fresh material is immediately extracted from sediment without using any fixative. After the extraction, meiofauna is immediately sorted and picked up. If this is not possible, meiofaunal samples can be stored in grade reagent at 4°C until the sorting, which should be performed within 12 h. Also, from frozen samples, meiofauna should be immediately sorted after extraction, and specimens picked up. If this is not possible, meiofaunal samples can be stored in reagent-grade water at 4°C until sorting, which should be performed within 12 h. The sorting and pick up of meiofaunal organisms should be performed using the procedures reported in previous chapters. Live nematodes should be starved for 12 h in 0.7 μm filtered seawater to facilitate cleaning the gut contents, and then rinsed cautiously prior to further removal of any remaining contaminants, which is done manually using a dissecting microscope (Chouch 1989). About 80–100 nematodes are finally collected on precombusted Whatman GF/F filters (450°C for 4 h) [NOTES 1–3]. Filter should be transferred in an oven (60°C) for one night and then kept frozen until the analysis with the mass spectrometer (Danovaro et al. 1999).

42.4.2.1 Notes

1. 80 to 100 individuals are pooled together in the same filter to obtain a sufficient biomass for the isotopic analysis. The stable isotopes ratio of a single individual could be not detected by the mass spectrometer.
2. Place the filter in the oven on aluminum foil. Transfer each filter in an Eppendorf when the filters are at room temperature. This step is important because the relative high temperature can melt plastic containers and alter $\partial\ ^{13}C$ measurements.
3. Place the filter in the Eppendorf and bend the filter keeping the nematodes in the internal part of the filter, using tweezers.

42.4.3 Macrofaunal Samples

All individuals are cleaned of epibionts and incubated overnight in filtered seawater from the sampling site to allow evacuation of gut contents [NOTE 1]. The specimens are then killed by freezing. After dissection from the shell (mollusks) or from the entire cuticle (crabs), the tissue is acidified (10% HCl) to remove all carbonates debris and rinsed with distilled water. The material is subsequently freeze-dried and ground to powder using a mortar and pestle (Riera et al. 1999). Samples are kept frozen until analysis with the mass spectrometer. One whole organism is typically sufficient for the analysis [NOTES 2 and 3].

42.4.4 Megafaunal Samples and Fishes

Megafauna and fishes are held in tanks of filtered sea water for 36 to 48 h to prevent unassimilated food from affecting the measurement of stable isotope (Griffin and Valiela 2001) [NOTE 1]. White muscle tissue was sampled from each fish, freeze-dried and ground to a fine homogeneous powder; most invertebrate samples consisted of whole organisms [NOTE 2]. The crustacean samples have to be decalcified to remove the inorganic carbon from the exoskeleton since inorganic carbonates tend to be less negative in $\partial\ ^{13}C$ than other body fractions (DeNiro and Epstein 1978). A small quantity of 1 M hydrochloric acid is added to the samples and left for 3 h. The samples are then rinsed with distilled water, dried at 60°C for 12 h, and reground. Since this process results in enrichment of $\partial\ ^{15}N$ (Pinnegar and Polunin 1999), crustacean samples required analysis before (for $\partial\ ^{15}N$) and after (for $\partial\ ^{13}C$) the decalcification (Polunin et al. 2001) [NOTE 3].

42.4.5 Notes

1. Collect at least 5 mg of biomass. For small animals (less than 1 mg), use the whole animal, but remove gut contents by dissection or by holding the animals alive until gut contents are cleared.
2. You may have to pool many individual small animals to meet the 5 mg minimum (i.e., nematodes). For smaller samples, the validity of the stable isotopes analyses could be compromised (Michener and Lajtha 2007). For larger animals use a scrap of muscle tissue from a carcass or liver tissue. Dry at 60°C and grind to a fine powder with mortar and pestle.
3. For the analyses of the stable isotopes, samples should be transferred in glass or plastic vials and aluminum foil (small animals). It is important to give an approximate C and N content of samples and to pass this information to the laboratory where these analyses will be performed. Any carbonates (shell, $CaCO_3$ precipitates) should be removed before $\partial\ ^{13}C$ analysis of organic fractions.

42.5 DATA TREATMENT

$\partial\ ^{15}N$ and $\partial\ ^{13}C$ are determined using a continuous flow isotope ratio mass spectrometry. Isotope ratios are expressed as parts of thousand (‰) differences from a standard reference material:

$$\partial X = \left[\left(R_{sample}/R_{standard}\right) - 1\right] \cdot 10^3 \qquad (42.1)$$

where

X is ^{15}N or ^{13}C, R is the ratio $^{15}N/^{14}N$, or $^{13}C/^{12}C$ for N or C isotopes in the sample.

The standard reference materials are Vienna Pee Dee Belemnite (vPDB) for carbon and atmospheric N_2 for nitrogen (Pinnegar and Polunin 1999). Experimental precision based on standard deviation of replicates of the internal standard is 0.12 ‰ for $\partial\ ^{15}N$ and 0.09 ‰ for $\partial\ ^{13}C$.

The values of ∂ X obtained from the analyses should be negative or positive. Values are positive when the value of the $R_{standard}$ is higher than the R_{sample}, suggesting that the standard contains more heavy isotopes than the sample. Vice versa is true of the positive values. In general, the values of $\partial\ ^{15}N$ and $\partial\ ^{13}C$ increase, moving from the potential food sources to the higher trophic levels of the food web. This is due to the enrichment of the heavier isotope content (^{13}C and ^{15}N) during the transfer from one trophic level to the next level. This enrichment happens in the metabolic process and respiration that favor the accumulation of heavier isotopes in the organisms.

In one example of the stable isotope analyses performed in estuarine ecosystems, the different values of $\partial\ ^{13}C$ were obtained from different potential food sources and consumers. The values of $\partial\ ^{13}C$ allows discrimination of the preferential diet for different taxa (Figure 42.1). It is recommended that to perform the stable isotope analyses, two or more stable isotopes should be combined. The use of $\partial\ ^{15}N$ can complement other isotopic data (i.e., $\partial\ ^{13}C$) for the determination of food sources, but it can also contribute to understanding the mechanisms involved in trophic transfer of the organic matter and the trophic structure of the food web (Figure 42.2).

The trophic mediation between a food source and a consumer cannot be clearly detected on the basis of $\partial\ ^{13}C$ analyses because $\partial\ ^{13}C$ only shows an average enrichment of 1‰ per trophic level (Rau et al. 1983; Fry and Sherr 1984). The enrichment in ^{13}C allows a direct correspondence between a food source and its consumer to be established (Fry and Sherr 1984), but the enrichment is not large enough for following the trophic mediation between primary production at the base of the food chain and the organisms. However, $\partial^{15}N$ has been successful in investigating the trophic structure within aquatic ecosystems since an increase in $\partial\ ^{15}N$ by about 3.51‰ per trophic level occurs as nitrogen is transferred (Minagawa and Wada 1984; Wada et al. 1987). The values of $\partial\ ^{15}N$ can be interpreted as a function of both food sources and trophic level.

42.6 STOMACH CONTENT ANALYSES

To help infer trophic relations among deep-sea species, stomach contents of the most abundant consumers in the study area can be collected and analyzed. Animals are dissected as soon as they were brought to the surface, and the contents of their stomachs and intestines are examined under a binocular-dissecting microscope. Recognizable food items are identified to species level wherever is possible.

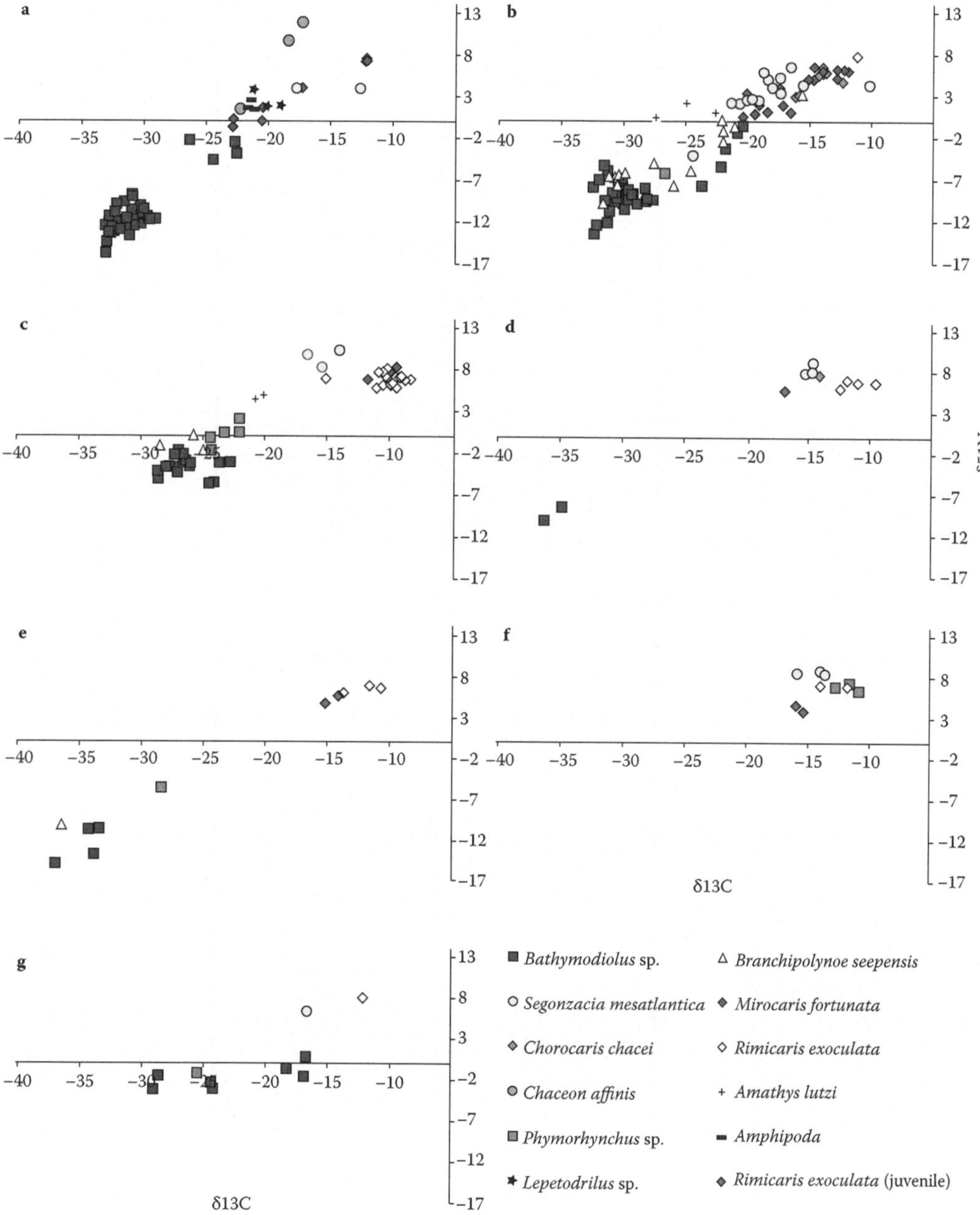

FIGURE 42.1 ***(A color version of this figure follows page 234.)*** ∂ ^{13}C and ∂ ^{15}N values identifying the different main groups of organisms sampled at Mid-Atlantic Ridge deep-sea hydrothermal vent fields: (a) Menez Gwen; (b) LuckyStrike; (c) Rainbow; (d) Broken Spur; (e) Snake Pit; (f) TAG; (g) Logatchev. (Data from Colaço, A., F. Dehairs, and D. Desbruyeres. 2002. *Deep-Sea Research Part I* 49: 395–412.)

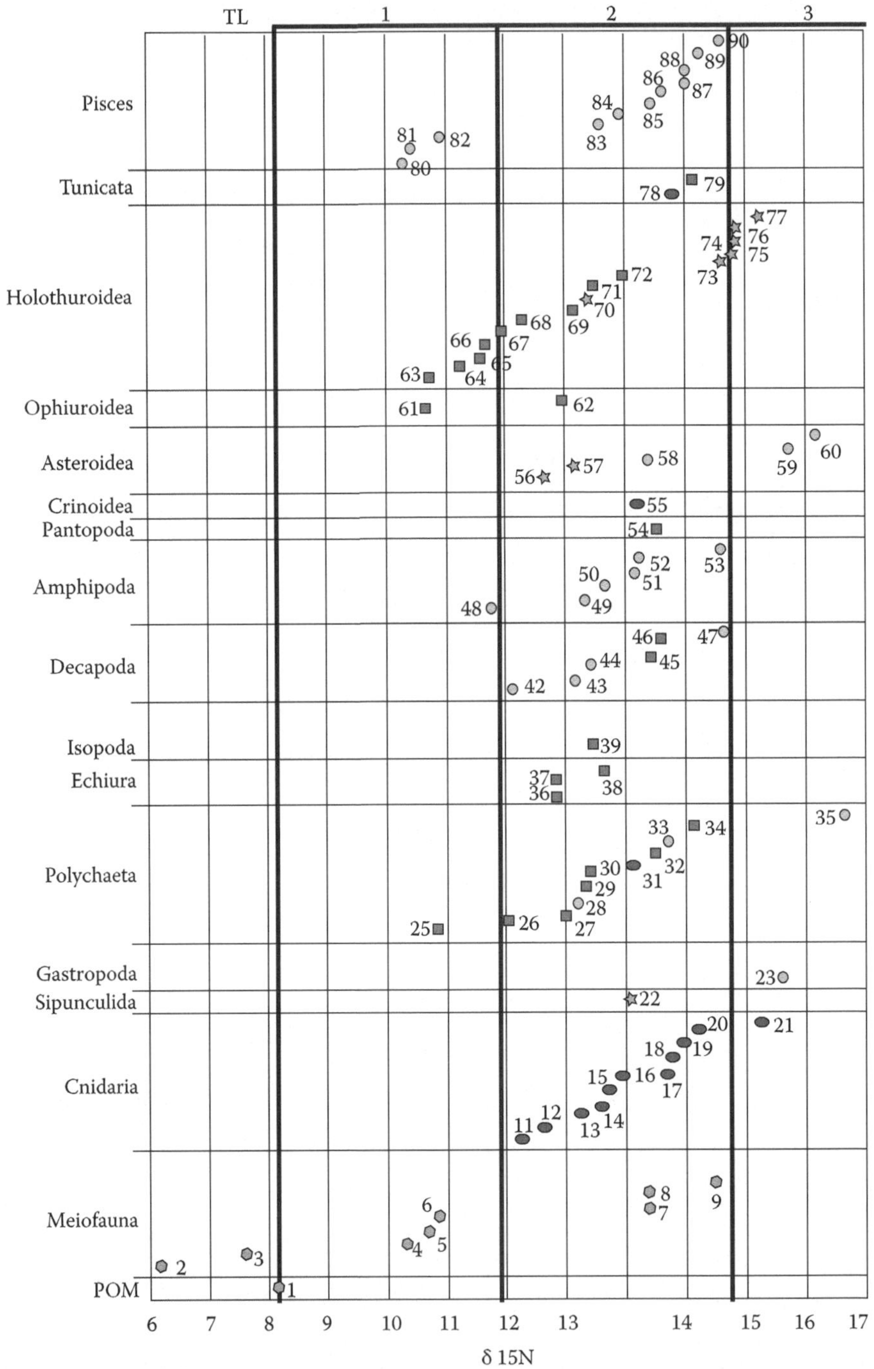

FIGURE 42.2 *(A color version of this figure follows page 234.)* ∂ ^{13}C Relative trophic position of benthic species/groups in deep-sea sediments based on ∂ ^{15}N. TL: relative trophic level. Different feeding types are designated with squares (suspension feeders), crosses (surface deposit feeders), filled circles (subsurface deposit feeders), triangles (predator/scavenger), and open circles (no feeding type associated). (Data from Iken, K., T. Brey, U. Wand, J. Voigt, and P. Junghans. 2001. *Progress in Oceanography* 50: 383–405.)

REFERENCES

Bett, B., A. Vanreusel, M. Vincx et al. 1994. Sampler bias in the quantitative study of deep-sea meiobenthos. *Marine Ecology Progress Series* 104: 197–203.

Bett, B. J. 2000. Comparative benthic ecology of the Rockall Trough and Faeroe-Shetland Channel. In *Atlantic margin environmental surveys of the seafloor 1996 and 1998 (Atlantic Frontier Environmental Network)* [CDROM], ed. Geotek Ltd, Section 4.3.2. London, UKOOA.

Bosley, K. L., J. W. Lavelle, R. D. Brodeur et al. 2004. Biological and physical processes in and around Astoria submarine Canyon, Oregon, USA. *Journal of Marine Systems* 50: 21–37.

Bosley, K. L. and S. C. Wainright. 1999. Effects of preservation and acidification on the stable isotope ratios (^{14}N: ^{15}N, ^{13}C: ^{12}C) of two species of marine animals. *Canadian Journal of Fisheries and Aquatic Science* 56: 2181–2185.

Chouch, C. A. 1989. Carbon and nitrogen stable isotopes of meiobenthos and their food resources. *Estuarine, Coastal, and Shelf Science* 28: 433–441.

Colaço, A., F. Dehairs, and D. Desbruyeres. 2002. Nutritional relations of deep-sea hydrothermal fields at the Mid-Atlantic Ridge: A stable isotope approach. *Deep-Sea Research Part I* 49: 395412.

Danovaro, R., D. Marrale, N. Della Croce, A. Dell'Anno, and M. Fabiano. 1999. Heterotrophic nanoflagellates, bacteria, and labile organic compounds in continental shelf and deep-sea sediments of the eastern Mediterranean. *Microbial Ecology* 35(3): 244–255.

DeNiro, M. J. and S. Epstein. 1978. Influence of diet on the distribution of carbon isotopes in animals. *Geochimica et Cosmochimica Acta* 42: 495–506.

Elmgren, R. and T. Radziejewska. 1989. Recommendations for quantitative benthic meiofauna studies in the Baltic. *Baltic Marine Biologists Publication* 12: 1–23.

Fleeger, J. W., W. B. Sikora, and J. P. Sikora. 1983. Spatial and long-term temporal variation of meiobenthic-hyperbenthic copepods in lake Pontchartrain, Lousiana. *Estuarine, Coastal, and Shelf Science* 16: 441–453.

Fry, B. 1988. Food web structure on George Bank from stable C, N, and S isotopic composition. *Limnology and Oceanography*.33: 1182–1190.

Fry B. and Sherr E. B. (1984) $\partial\ ^{13}C$ measurements as indicators of carbon flow in marine and freshwater ecosystems. *Contributions in Marine Science* 27: 13–47.

Griffin, M. P. A. and I. Valiela. 2001. $\partial^{15}N$ isotope studies of life history and trophic position Fundulus heteroclitus and Menidia menidia. *Marine Ecology Progress Series* 214: 299–305.

Harrigan, P., J. C. Zieman, and S. A. Macko. 1989. The base of nutritional support for the grey snapper (*Lutjanus griseus*): An evaluation based on combined stomach content and stable isotope analysis. *Bulletin of Marine Science* 44: 65–77

Hesslein, R. H., K. A. Hallard, and P. Ramal. 1993. Replacement of sulphur, carbon, and nitrogen of growing broad whitefish (*Coregonus nasus*) in response to a change in diet traced by ^{34}S, ^{13}C, and ^{15}N. *Canadian Journal of Fisheries and Aquatic Science* 50: 2071–2076.

Hyndes, G. A. and P. S. Lavery. 2005. Does transported seagrass provide an important trophic link in unvegetated, nearshore areas? *Estuarine, Coastal, and Shelf Science* 63: 633–643.

Hobson, K. A. and H. E. Welch. 1992. Determination of trophic relationships within a high Arctic marine food web using $\partial^{15}N$ and $\partial^{13}C$ analysis. *Marine Ecology Progress Series* 84: 9–18.

Iken, K., T. Brey, U. Wand, J. Voigt, and P. Junghans. 2001. Food web structure of the benthic community at the Porcupine Abyssal Plain (NE Atlantic): A stable isotope analysis. *Progress in Oceanography* 50: 383–405.

Kaehler, S. and E. A. Pakhomov. 2001. Effects of storage and preservation on the ∂15N and ∂13C signatures of selected marine organisms. *Marine Ecology Progress Series* 219: 299–304.

Levesque, C., S. K. Juniper, and H. Limen. 2006. Spatial organization of food webs along habitat gradients at deep-sea hydrothermal vents on Axial Volcano, Northeast Pacific. *Deep-Sea Research Part I* 53: 726–739.

Mann, K. H. 1988. Production and use of detritus in various freshwater, estuarine, and coastal marine ecosystems. *Limnology and Oceanography* 33: 910–930.

Michener, R. and K. Lajtha. 2007. *Stable isotopes in ecology and environmental science*. Blackwell Publishing.

Minagawa, M. and E. Wada. 1984. Stepwise enrichment of 15N along food chains: Further evidence and the relation between $\partial\ ^{15}N$ and animal age. *Geochimica et Cosmochimica Acta* 48: 1135–1140.

Mohtadi, M., D. Hebbeln, and M. Marchant. 2005. Upwelling and productivity along the Peru-Chile current derived from faunal and isotopic compositions of planktic foraminifera in surface sediments. *Marine Geology* 216: 107–126.

Monbet, Y. 1992. Control of phytoplankton biomass in estuaries—a comparative analysis of microtidal and macrotidal estuaries. *Estuaries* 15: 63–571.

Nieuwenhuize, J., Y. E. M. Mass, and J. J. Middelburg. 1994. Rapid analysis of organic carbon and nitrogen in particulate materials. *Marine Chemistry* 45: 217–224.

Pinnegar, J. K. and N. V. C. Polunin. 1999. Differential fractionation of $\partial^{13}C$ and $\partial^{15}N$ among fish tissues: Implications for the study of the trophic levels. *Functional Ecology* 13: 225–231.

Pinnegar, J. K. and N. V. C. Polunin. 2000. Contributions of stable-isotope data to elucidating food webs of Mediterranean rocky littoral fishes. *Oecologia* 122: 399–409.

Pocklington, R. and F. C. Tan. 1987. Seasonal and annual variations in the organic matter contributed by the St Lawrence river to the Gulf of St Lawrence. *Geochimica et Cosmochimica Acta* 51: 2579–2586.

Polunin, N. V. C., B. Morales-Nin, W. E. Pawsey, J. E. Cartes, J. K. Pinnegar, and J. Moranta. 2001. Feeding relationships in Mediterranean bathyal assemblages elucidated by stable nitrogen and carbon isotope data. *Marine Ecology Progress Series* 220: 13–23.

Rau, G. H. 1982. The relationship between trophic level and stable isotopes of carbon and nitrogen. In *Coastal water research project biennal report for the years 1981–1982*. Southern California Coastal Water Research Project Annual Report, ed. W. Bascom, 143–148. California, Long Beach.

Rau, G. H., A. J. Mearns., D. R. Young, R. J. Olson, H. A. Schafer, I. R. Kaplan. 1983. Animal $^{13}C/^{12}C$ correlates with trophic level in pelagic food webs. *Ecology* 64: 1314–1318.

Riera, P., L. J. Stal, J. Nieuwenhuize, P. Richard, G. Blanchard, and F. Gentil. 1999. Determination of food sources for benthic invertebrates in a salt marsh Aiguillon Bay, France by carbon and nitrogen stable isotopes: Importance of locally produced sources. *Marine Ecology Progress Series* 187: 301–307.

Schubert, C. and S. E. Calvert. 2001. Nitrogen and carbon isotopic composition of marine and terrestrial organic matter in Artic Ocean sediments: Implications for nutrient utilization and organic matter composition. *Deep-Sea Research Part I* 48: 789–810.

Schwinghamer, P., F. C. Tan, and D. C. Gordon. 1983. Stable carbon isotope studies on the pecks cove mudflat ecosystem in the Cumberland basin, Bay of Fundy. *Canadian Journal of Fisheries and Aquatic Science* 40: 262–273.

Shirayama, Y. and T. Fukushima. 1995. Comparison of deep-sea sediments and overlying water collected using multiple corer and box corer. *Journal of Oceanography* 51: 75–82.

Usui, T., S. Nagao, M. Yamamoto et al. 2006. Distribution and sources of organic matter in surficial sediments on the shelf and slope off Tokachi, western North Pacific, inferred from C and N stable isotopes and C/N ratios. *Marine Chemistry* 98: 241–259.

Vander Zanden, M. J., G. Cabana, and J. B. Rasmussen. 1997. Comparing trophic position of freshwater fish calculated using stable nitrogen isotope ratios ($\partial^{15}N$) and literature dietary data. *Canadian Journal of Fisheries and Aquatic Science* 54: 1142–1158.

Wada, E., M. Terazaki, Y. Kabaya, and T. Nemoto. 1987. 15N and 13C abundances in the Antartic ocean with emphasis on the biogeochemical structure of the food web. *Deep-Sea Research* 34: 829–841.

Section VI

Suggested Readings for the Study of Deep-Sea Biodiversity and Ecosystem Functioning

Suggested Readings

Aller, J. Y., R. C. Aller, and M. A. Green. 2002. Benthic faunal assemblages and carbon supply along the continental shelf/shelf break-slope off Cape Hatteras, North Carolina. *Deep-Sea Research Part II* 49: 4599–4625.

Andersson, J. H., J. W. M. Wijsman, P. M. J. Herman, J. J. Middelburg, K. Soetaert, and C. Heip. 2004. Respiration patterns in the deep ocean, *Geophysical Research Letters* 31: L03304.

Arico, S. and C. Salpin. 2005. *Bioprospecting of genetic resources in the deep seabed: Scientific, legal, and policy aspects.* Yokohama: United Nations University—Institute of Advanced Studies.

Arndt, H., K. Hausmann, and M. Wolf. 2003. Deep-sea heterotrophic nanoflagellates of the Eastern Mediterranean Sea: Qualitative and quantitative aspects of their pelagic and benthic occurrence. *Marine Ecology Progress Series* 256: 45–56

Arntz, W. E., S. Thatje, K. Linse et al. 2006. Missing link in the Southern Ocean: Sampling the marine benthic fauna of remote Bouvet Island. *Polar Biology* 29: 83–96.

Berge, J. and W. Vader. 2004. Two new Antarctic stegocephalid (Amphipoda: Stegocephalidae: Stegocephalinae) species, with implications for the phylogeny and classification of the two genera Pseudo and Schellenbergia. *Deep-Sea Research Part II* 51: 1709–1716.

Bernhard, J. M. 1992. Benthic foraminiferal distribution and biomass related to pore-water oxygen content: Central California continental slope and rise. *Deep-Sea Research* 39: 585–605.

Bett, B. J., A. Vanreusel, M. Vincx et al. 1994. Sampler bias in the quantitative study of deep-sea meiobenthos. *Marine Ecology Progress Series* 104: 197–203.

Bett, B. J., M. G. Malzone, B. E. Narayanaswamy, and B. D. Wigham. 2001. Temporal variability in phytodetritus and megabenthic activity at the seabed in the deep Northeast Atlantic. *Progress in Oceanography* 50: 349–368.

Billett, D. S. M. 1991. Deep-sea holothurians. *Oceanography and Marine Biology—An Annual Review* 29, 259–317.

Billett, D. S. M., R. S. Lampitt, A. L. Rice, and R. F. C. Montoura. 1983. Seasonal sedimentation of phytoplankton to the deep-sea benthos. *Nature* 302: 520–522.

Billett, D. S. M., B. J. Bett, A. L. Rice et al. 2001. Long-term change in the megabenthos of the Porcupine Abyssal Plain (NE Atlantic). *Progress in Oceanography* 50: 325–348.

Billett, D. S. M. and A. L. Rice. 2001. The BENGAL programme: Introduction and overview. *Progress in Oceanography* 50: 13–25.

Blake, J. A. and B. E. Narayanaswamy. 2004. Benthic infaunal communities across the Weddell Sea Basin and South Sandwich Slope, Antarctica. *Deep-Sea Research Part II* 51: 1797–1816.

Boetius, A. and K. Lochte. 1994. Regulation of microbial enzymatic degradation of organic matter in deep-sea sediments. *Marine Ecology Progress Series* 104: 299–307.

Boetius, A., K. Ravenschlag, C. J. Schubert et al. 2000. A marine microbial consortium apparently mediating anaerobic oxidation of methane *Nature* 407: 623–627.

Böggemann, M. and G. Purschke. 2005. Abyssal benthic Syllidae (Annelida: Polychaeta) from the Angola Basin. *Organisms, Diversity, and Evolution* 5(Suppl. 1): 221–226.

Bolam, S. G., T. F. Fernandes, and M. Huxham. 2002. Diversity, biomass, and ecosystem processes in the marine benthos. *Ecological Monographs* 72(4): 599–615.

Bongers T., R. Alkemade, and G. W. Yeates. 1991. Interpretation of disturbance induced maturity decrease in marine nematode assemblages by means of the Maturity Index. *Marine Ecology Progress Series* 76: 135–1472.

Bongers, T. and H. Ferris. 1999. Nematode community structure as a bioindicator in environmental monitoring. *Trends in Ecology and Evolution* 14: 224–228.

Bouchet, P. and M. Taviani. 1992. The Mediterranean deep-sea fauna: Pseudopopulations of Atlantic species? *Deep-Sea Research Part A.* 39: 169–184.

Bouchet, P. and B. A. Marshall. 2001. *Tropical Deep-Sea Benthos, Vol. 22. Mémoires du Muséum National d'Histoire Naturelle, 185.* Paris: Publications Scientifiques du Muséum.

Braker, G., H. L. Ayala-del-Rio, A. H. Devol, A. Fesefeldt, and J. M. Tiedje. 2001. Community structure of denitrifiers, bacteria, and archaea along redox gradients in Pacific Northwest marine sediments by terminal restriction fragment analysis of amplified nitrite reductase (nirS) and 16S rRNA genes. *Applied and Environmental Microbiology* 67: 1893–1901.

Brandt, A., W. Brökeland, S. Brix, and M. Malyutina. 2004. Diversity of Southern Ocean deep-sea Isopoda (Crustacea, Malacostraca)—a comparison with shelf data. *Deep-Sea Research Part II* 51 (14–16): 1753–1768.

Brandt, A., S. Brix, W. Brökeland et al. 2007. The Southern Ocean deep sea: First insights into biodiversity and biogeography. *Nature* 447: 307–311.

Brandt, A., C. De Broyer, I. De Mesel et al. 2007. The biodiversity of the deep Southern Ocean benthos. In *Antarctic ecology: From genes to ecosystems*, ed. A. Rogers, 362: 39–66. London: Philosophical Transactions of the Royal Society B. 1477: 39–66.

Brix, S. 2007. Four new species of Desmosomatidae Sars, 1897 (Crustacea: Isopoda) from the deep sea of the Angola Basin. *Marine Biology Research* 3 (4): 205–230.

Bröhldick, K. 2005. A new taxon of Idyanthidae (Copepoda, Harpacticoida) from the deep sea of the Angola Basin. *Organisms, Diversity, and Evolution* 5(Suppl. 1): 43–57.

Brökeland, W. and A. Brandt. 2004. Two new species of Ischnomesidae (Crustacea: Isopoda) from the Southern Ocean displaying neoteny. *Deep-Sea Research Part II* 51: 1769–1785.

Brökeland, W. 2006. Three species of the isopod crustacean genus *Antennuloniscus* Menzies, 1962 (Asellota: Haploniscidae) from the Southern Ocean. *Zootaxa* 1115: 1–29.

Butman, C. A., J. T. Carlton, and S. R. Palumbi. 1995. Whaling effects on deep-sea biodiversity. *Conservation Biology* 9: 462–464.

Canals, M., G. Lastras, R. Urgeles et al. 2004. Slope failure dynamics and impacts from seafloor and shallow sub-seafloor geophysical data: Case studies from the COSTA project; *Marine Geology* 213 (1–4): 9–72.

Canals, M., P. Puig, X. Durrieu de Madron et al. 2006. Flushing submarine canyons. *Nature* 444: 354–357.

Carmen, K. R., K. M. Sherman, and D. Thistle. 1987. Evidence that sediment type influences the horizontal and vertical distribution of nematodes at a deep-sea site. *Deep-Sea Research Part I* 34: 45–53.

Carney, R. S. 2005. Zonation of deep biota on continental margins. *Oceanography and Marine Biology: An Annual Review* 43: 211–278.

Cartes, J. E. 1998. Dynamics of the bathyal benthic boundary layer in the north-western Mediterranean: Depth and temporal variations in macrofaunal–megafaunal communities and their possible connections within deep-sea trophic webs. *Progress in Oceanography* 41: 11–139.

Cartes J. E., V. Papiol, and B. Guijarro. 2008. The feeding and diet of the deep-sea shrimp *Aristeus antennatus* off the Balearic Islands (Western Mediterranean): Influence of environmental factors and relationship with the biological cycle. *Progress in Oceanography* 79: 37–54.

Chase, J. M. and M. A. Leibold. 2002. Spatial scale dictates the productivity–biodiversity relationship. *Nature* 416: 427–430.

Clarke, A. and A. Crame. 1997. Diversity, latitude, and time: Patterns in the shallow sea. In *Marine diversity: Patterns and processes,* ed. R. F. Ormond, J. D. Gage, and M. V. Angel, 122–147. Cambridge: Cambridge University Press.

Clarke, A. and S. Lidgard. 2000. Spatial patterns of diversity in the sea: Bryozoan species richness in the North Atlantic. *Journal of Animal Ecology* 69: 799–814.

Clarke, K. R. 1993. Non parametric multivariate analyses of changes in community structure. *Australian Journal of Ecology* 18: 117–143.

Cochonat, P., S. Durr, V. Gunn et al. 2007. The deep-sea frontier: Science challenges for a sustainable future. European Commission Directorate-General for Research Publication EUR 22812.

Company, J. B., F. Sardà, P. Puig, J. Cartes, and A. Planques. 2003. Duration and timing of reproduction in decapod crustaceans of the NW Mediterranean continental margins: is there a general pattern? *Marine Ecology Progress Series* 261: 201–216.

Company, J. B., P. Puig, F. Sardà, A. Palanques, M. Latasa, and R. Scharek. 2008. Climate influence on deep sea populations. *PLoS ONE* 3 (1): e1431. doi: 10.1371/journal.pone.0001431.

Cook, A. A, P. J. D. Lambshead, L. E. Hawkins, N. Mitchell, and L. A. Levin. 2000. Nematode abundance at the oxygen minimum zone in the Arabian Sea. *Deep-Sea Research Part II* 47: 75–85.

Corinaldesi, C., A. Dell'Anno, and R. Danovaro. 2007. Early diagenesis and trophic role of extracellular DNA in different benthic ecosystems. *Limnology and Oceanography* 52 (4): 1710–1717.

Corliss, B. H. and T. C. E. Van Weering. 1993. Living (stained) benthic Foraminifera within surficial sediments of the Skagerrak. *Marine Geology* 111: 325–335.

Cornelius, N. and A. J. Gooday. 2004. "Live" (stained) deep-sea benthic foraminiferans in the western Weddell Sea: Trends in abundance, diversity, and taxonomic composition in relation to water depth. *Deep-Sea Research Part II* 51: 1571–1602.

Cornell, H. V. and R. H. Karlson. 1996. Species richness of reef building corals determined by local and regional processes. *Journal of Animal Ecology* 65: 233–241.

Corselli, C. and D. Basso. 1996. First evidence of benthic communities based on chemosynthesis on the Napoli mud volcano (Eastern Mediterranean). *Marine Geology* 132: 227–239.

Cosson, N., M. Sibuet, and J. Galéron. 1997. Community structure and spatial heterogeneity of the deep-sea macrofauna at 3 contrasting stations in the tropical northeast Atlantic. *Deep-Sea Research Part I* 44: 247–269.

Cosson-Sarradin, N., M. Sibuet, G. L. J. Paterson, and A. Vangriesheim. 1998. Polychaete diversity as tropical Atlantic deep-sea sites: Environmental effects. *Marine Ecology Progress Series* 165: 173–185.

Cronin, T. M. and Raymo M. E. Orbital forcing of deep-sea species diversity. *Nature* 385: 624–627.

Culver, S. J. and M. A. Buzas. 2000. Global latitudinal species diversity gradient in deep-sea benthic foraminifera. *Deep-Sea Research Part I* 47: 259–275.

D'Onghia G., C. Y. Politou, A. Bozzano et al. 2004. Deep-water fish assemblages in the Mediterranean Sea. *Scientia Marina* 68: 87–99.

D'Onghia, G., R. Carlucci, P. Maiorano, and M. Panza. 2003. Discards from deep-water bottom trawling in the eastern-central Mediterranean Sea and effects of mesh size changes. *Journal of Northwest Atlantic Fishery Science* 31: 245–261.

Danovaro, R., M. Fabiano, and N. Della Croce. 1993. Labile organic matter and microbial biomasses in deep sea sediments (Eastern Mediterranean Sea). *Deep-Sea Research Part I* 40: 953–965.

Danovaro, R., N. Della Croce, A. Eleftheriou et al. 1995. Meiofauna of deep Eastern Mediterranean Sea: Distribution and abundance in relation to bacterial biomass, organic matter composition, and other environmental factors. *Progress in Oceanography* 36: 329–34.

Danovaro, R., D. Marrale, N. Della Croce, A. Dell'Anno, and M. Fabiano. 1998. Heterotrophic nano-flagellates, bacteria, and labile organic compounds in continental shelf and deep-sea sediments of the Eastern Mediterranean. *Microbial Ecology* 35: 244–255.

Danovaro, R., A. Dinet, G. Duineveld, and A. Tselepides. 1999. Benthic response to particulate fluxes in different trophic environments: A comparison between Gulf of Lions–Catalan Sea (western-Mediterranean) and the Cretan Sea (eastern-Mediterranean). *Progress in Oceanography* 44: 287–312.

Danovaro, R., A. Tselepides, A. Otegui, and N. Della Croce. 2000. Dynamics of meiofaunal assemblages on the continental shelf and deep-sea sediments of the Cretan Sea (NE Mediterranean): Relationships with seasonal changes in food supply. *Progress in Oceanography* 46: 367–400.

Danovaro, R., A. Dell'Anno, M. Fabiano, A. Pusceddu, and A. Tselepides. 2001. Deep-Sea ecosystem response to climate changes: The Eastern Mediterranean case study. *Trends in Ecology and Evolution* 16: 505–510.

Danovaro, R., E. Manini, and A. Dell'Anno. 2002. Higher abundance of bacteria than of viruses in deep Mediterranean sediments. *Applied and Environmental Microbiology* 68 (3): 1468–1472.

Danovaro, R. 2003. Organic inputs and ecosystem efficiency in the Mediterranean Sea. *Chemistry and Ecology* 19: 391–398.

Danovaro, R., A. Dell'Anno, and A. Pusceddu. 2004. Biodiversity response to climate change in a warm deep sea. *Ecology Letters* 7: 821–828.

Danovaro, R., G. M. Luna, A. Dell'Anno, and B. Pietrangeli. 2006. Comparison of two fingerprinting techniques, terminal restriction fragment length polymorphism, and automated ribosomal intergenic spacer analysis, for determination of bacterial diversity in aquatic environments. *Applied and Environmental Microbiology*. 72: 5982–5989.

Danovaro, R., A. Dell'Anno, C. Corinaldesi et al. 2008. Major viral impact on the functioning of benthic deep-sea ecosystems. *Nature* 454: 1084–1088.

Danovaro, R., C. Gambi, A. Dell'Anno et al. 2008. Exponential decline of deep-sea ecosystem functioning linked to benthic biodiversity loss. *Current Biology* 18: 1–8.

Dell'Anno, A. and R. Danovaro. 2005. Extra-cellular DNA plays a key role in deep-sea ecosystem functioning. *Science* 309: 2179.

Deming, J. W. and R. R. Colwell, 1982. Barophilic bacteria associated with digestive tracts of abyssal holothurians. *Applied Environmental Microbiology* 44: 1222–1230.

Deming, J. W. 1985. Bacterial growth in deep sediment trap and boxcore samples. *Marine Ecology Progress Series* 25: 305–312.

Deming, J. W. and P.L. Yager. 1992. Natural bacterial assemblages in deep-sea sediments: Toward a global view. In: *Deep-sea food chains and the global carbon cycle* ed. G. T. Rowe, and V. Pariente, 11–27. Dordrecht, Netherlands: Kluwer.

Deming, J. W. and J. A. Baross. 1993. The early diagenesis of organic matter: Bacterial activity. In *Organic geochemistry*, ed. M. H. Engel and S. A. Macko, 119–144. New York: Plenum Press.

Deming, J. W. and J. A. Baross 2000. Survival, dormancy, and nonculturable cells in extreme deep-sea environments. In *Nonculturable microorganisms in the environment*, ed. R. R. Colwell and D. J. Grimes, 147–197. Washington, DC: ASM Press.

Deming, J. W. and S. D. Carpenter. 2003. Factors influencing benthic bacterial abundance and activity on the northern continental slope of the Gulf of Mexico. *Oceanography* 16: 35.

Dinet, A. and M. H. Vivier. 1979. Le meiobenthos abyssal du Golfe de Gascogne. II. Les peuplements de Nematodes et Leur divesite specifique. *Cahiers de Biologie Marine* 20: 109–123.

Dixon, J. L. and C. M. Turley. 2000. The effect of water depth on bacterial numbers, thymidine incorporation rates, and C:N ratios in northeast Atlantic surficial sediments. *Hydrobiologia* 440: 217–225.

Drazen, J. C., R. J. Baldwin, and K. L. Smith. 1998. Sediment community response to a temporally varying food supply at an abyssal station in the North East Pacific. *Deep-Sea Research Part II* 45: 893–913.

Druffel, E. R. M. and B. H. Robison. 1999. Is the deep sea on a diet?. *Science* 284: 1139–1140.

Duineveld, G. C. A., A. Tselepides, R. Witbaard et al. 2000. Benthic–pelagic coupling in the oligotrophic Cretan Sea. *Progress in Oceanography* 46: 457–481.

Eardly, D. C., M. Gallagher, and J. Patching. 2001. Bacterial abundance and activity in deep-sea sediments from the eastern North Atlantic. *Progress in Oceanography* 50: 245–259.

Epping, E., C. van der Zee, K. Soetaert, and W. Helder. 2002. On the oxidation and burial of organic carbon in sediments of the Iberian margin and Nazaré Canyon (NE Atlantic). *Progress in Oceanography* 52: 399–431.

Etter, R. J., M. A. Rex, M. C. Chase, and J. M. Quattro. 1999. A genetic dimension to deep-sea biodiversity. *Deep-Sea Research Part I* 46: 1095–1099.

Etter, R. J., M. A. Rex, M. R. Chase, and J. M. Quattro. 2005. Population differentiation decreases with depth in deep-sea bivalves. *Evolution* 59: 1479–1491.

Flach, E., A. Muthumbi, and C. Heip. 2002. Meiofauna and macrofauna community structure in relation to sediment composition at the Iberian margin compared to the Goban Spur (NE Atlantic). *Progress in Oceanography* 52: 433–457.

Floyd, R. M., A. D. Rogers, P. J. D. Lambshead, and C. R. Smith. 2005. Nematode-specific PCR primers for the 18S small subunit rRNA gene. *Molecular Ecology Notes* 5: 611–612.

Fonseca, G., W. Decraemer, and A. Vanreusel. 2006. Taxonomy and species distribution of the genus Manganonema (Nematoda: Monhysterida; Bussau, 1993). *Cahiers de Biologie Marine* 47: 189–204.

Fosså, J. H., B. Lindberg, O. Christensen et al. 2006. Mapping of Lophelia reefs in Norway: Experiences and survey methods. In *Cold-Water Corals and Ecosystems*, ed. A. Freiwald and J. M. Roberts, 359–391. Berlin: Springer.

Freiwald, A., J. H. Fosså, A. Grehan, T. Koslow, and J. M. Roberts. 2004. *Cold-water coral reefs*. Cambridge: UNEP-WCMCì.

Gad, G. 2004. Giant Higgins-larvae with paedogenetic reproduction from the deep sea of the Angola Basin—evidence for a new life cycle and for abyssal gigantism in Loricifera? *Organisms Diversity and Evolution* 5(Suppl. 1): 59–76.

Gage, J. D. and P. A. Tyler. 1991. *Deep sea biology: A natural history of organisms at the deep-sea floor*. Cambridge: Cambridge University Press.

Gage, J. D., P. A. Lamont, and P. A. Tyler. 1995. Deep-sea macrobenthic communities at contrasting sites off Portugal, preliminary results: I. Introduction and diversity comparisons. *Internationale Revue der gesamten Hydrobiologie* 80: 235–250.

Gage, J. D. 1996. Why are there so many species in deep-sea sediments? *Journal of Experimental Marine Biology and Ecology* 200: 257–286.

Gage, J. D. 1997. High benthic species diversity in deep-sea sediments: The importance of hydrodynamics. In *Marine biodiversity*, ed. R. F. G. Ormond, J. D. Gage, and M. V. Angel, 148–177. Cambridge: Cambridge University Press.

Gage, J. D., D. J. Hughes, and J. L. Gonzalez Vecino. 2002. Sieve size influence in estimating biomass, abundance, and diversity in samples of deep-sea macrobenthos. *Marine Ecology Progress Series* 225: 97–107.

Gage, J. D. 2004. Diversity in deep-sea benthic macrofauna: The importance of local ecology, the larger scale, history and the Antarctic. *Deep-Sea Research Part II* 51: 1689–1708.

Galéron, J., M. Sibuet, M. L. Mahaut, and A. Dinet. 2000. Variation in structure and biomass of the benthic communities at three contrasting sites in the tropical Northeast Atlantic. *Marine Ecology Progress Series* 197: 121–137.

Galéron, J., M. Sibuet, A. Vanreusel et al. 2001. Temporal patterns among meiofauna and macrofauna taxa related to changes in sediment geochemistry at an abyssal NE Atlantic site. *Progress in Oceanography* 50: 303–324.

Gambi, C., A. Vanreusel, and R. Danovaro. 2003. Biodiversity of nematode assemblages from deep-sea sediments of the Atacama Slope and Trench (South Pacific Ocean). *Deep-Sea Research Part I* 50: 103–117.

Genin A., P. K. Dayton, P. F. Lonsdale, and F. N. Spiess. 1986. Corals on seamount peaks provide evidence of current acceleration over deep-sea topography. *Nature* 322: 59–61.

Gili, J.-M., W. E. Arntz, A. Palanques et al. 2006. A unique assemblage of epibenthic sessile suspension feeders with archaic features in the high-Antarctic. *Deep-Sea Research Part II* 53: 1029–1052.

Gili, J.-M., S. Rossi, S., Pagès et al. 2006. A new trophic link between the pelagic and benthic systems on the Antarctic shelf. *Marine Ecology Progress Series* 322: 43–49.

Giribet, G., A. Okusu, A. R. Lindgren et al. 2006. Evidence for a clade composed of mollusks with serially repeated structures—monopolacophorans are related to chitons. *Proceedings of the National Academy of Sciences of the United States* 103: 7723–7728.

Gooday, A. J. 1988. A response by benthic Foraminifera to the deposition of phytodetritus in the deep sea. *Nature* 332: 70–73.

Gooday, A. J. and C. M. Turley. 1990. Responses by benthic organisms to inputs of organic material to the ocean floor: A review. *Philosophical Transactions of the Royal Society of London* Series A 331: 119–138.

Gooday, A. J. 1996. Epifaunal and shallow infaunal foraminiferal communities at three abyssal NE Atlantic sites subject to differing phytodetritus input regimes. *Deep-Sea Research Part I* 43: 1395–1421.

Gooday, A. J., O. Pfannkuche, and P. J. D. Lambshead. 1996. An apparent lack of response by metazoan meiofauna to phytodetritus deposition in the bathyal north-eastern Atlantic. *Journal of the Marine Biological Association of the United Kingdom* 76: 297–310.

Gooday, A. J., B. J. Bett, R. Shires, and P. J. D. Lambshead. 1998. Deep sea benthic foraminiferal species diversity in the NE Atlantic and NW Arabian Sea: A synthesis. *Deep-Sea Research Part I* 45: 165–201.

Gooday, A. J. and E. Alve. 2001. Morphological and ecological parallels between sublittoral and abyssal foraminiferal species in the NE Atlantic: A comparison of *Stainforthia fusiformis* and *Stainforthia* sp. *Progress in Oceanography* 50: 261–283.

Gooday, A. J., D. W. Pond, and S. S. Bowser. 2002. Ecology and nutrition of the large agglutinated foraminiferan *Bathysiphon capillare* in the bathyal NE Atlantic: Distribution within the sediment profile and lipid biomarker composition. *Marine Ecology Progress Series* 245: 69–82.

Gooday, A. J., M. Holzmann, J. Guiard, N. Cornelius, and J. Pawlowski. 2004. A new monothalamous foraminiferan from 1000–6300 m water depth in the Weddell Sea: Morphological and molecular characterisation. *Deep-Sea Research Part II* 51: 1603–1616.

Gooday, A. J., S. Hori, Y. Todo, T. Okamoto, H. Kitazato, and A. Sabbatini. 2004. Soft-walled, monothalamous benthic foraminiferans in the Pacific, Indian, and Atlantic Oceans: Aspects of biodiversity and biogeography. *Deep-Sea Research Part I* 51: 33–53.

Gooday, A. J., O. E. Kamenskaya, and T. Cedhagen. 2007. New and little-known Komokiacea (Foraminifera) from the bathyal and abyssal Weddell Sea and adjacent areas. *Zoological Journal of the Linnean Society* 151: 219–251.

Graf, G. 1989. Benthic–pelagic coupling in a deep-sea benthic community. *Nature* 341: 437–439.

Grassle, J. F. and L. S. Morse-Porteous. 1987. Macrofaunal colonization of disturbed deep-sea environments and the structure of deep-sea benthic communities. *Deep-Sea Research Part A*. 34: 1911–1950.

Grassle, J. F. 1991. Deep-sea benthic biodiversity. *Bioscience* 41: 464–469.

Grassle, J. F. and N. J. Maciolek. 1992. Deep-sea species richness: Regional and local diversity estimates from quantitative bottom samples. *American Naturalist* 139: 313–341.

Grassle, J. F. 1989. Species diversity in deep-sea communities. *Trends in Ecology and Evolution* 4: 12–15.

Grehan A. J. and S. K. Juniper. 1996. Clam distribution and subsurface hydrothermal processes at Chowder Hill (Middle Valley), Juan de Fuca Ridge. *Marine Ecology Progress Series* 130: 105–115.

Guerrero-Kommritz, J. and M. Błażewicz-Paskowycz. 2004. New species of Tanaella Norman et Stebbing, 1886 (Tanaidacea: Crustacea) from the deep-sea of the Antarctic and the Angola Basin. *Zootaxa* 459: 1–20.

Guerrero-Kommritz, J. 2005. Notes on the Tanaidacea (Crustacea: Peracarida) from the Angola Basin Organisms. *Diversity and Evolution* 5: 171–178.

Guinotte, J., M., J. Orr, S. Cairns, L. Morgan, and R. George. 2006. Will human-induced changes in seawater chemistry alter the distribution of deep-sea scleractinian corals? *Frontiers in Ecology and the Environment* 4: 141–146.

Haedrich, R. L. and G. T. Rowe. 1977. Megafaunal biomass in the deep sea. *Nature* 269: 141–142.

Haedrich, R. L., G. T. Rowe, and P. T. Polloni. 1980. The megabenthic fauna in the deep sea south of New England, USA. *Marine Biology* 57: 165–179.

Heinz, P. and C. Hemleben. 2003. Regional and seasonal variations of recent benthic deep-sea foraminifera in the Arabian Sea. *Deep-Sea Research Part I* 50: 435–447.

Heip, C. H. R., G. Duineveld, E. Flach et al. 2001. The role of the benthic biota in sedimentary metabolism and sediment–water exchange processes in the Goban Spur area (NE Atlantic). *Deep-Sea Research Part II* 48: 3223–3243.

Heip, C., Vincx, M., and G. Vranken. 1985. The ecology of marine nematodes. *Oceanography and Marine Biology Annual Review* 23: 399–489.

Hemleben, C. and H. Kitazato. 1995. Deep-sea foraminifera under long time observation in the laboratory. *Deep-Sea Research Part I* 42: 827–832.

Henry, L.-A. and J. M. Roberts. 2007. Biodiversity and ecological composition of macrobenthos on cold-water coral mounds and adjacent off-mound habitat in the bathyal Porcupine Seabight, NE Atlantic. *Deep-Sea Research Part I* 54 (4): 654–672.

Herman, P. M. J., K. Soetaert, J. J. Middelburg et al. 2001. The seafloor as the ultimate sediment trap—using sediment properties to constrain benthic–pelagic exchange processes at the Goban Spur. *Deep-Sea Research Part II* 48: 3245–3264.

Hessler, R. R., J. D. Isaacs, and E. L. Mills. 1972. Giant amphipod from the abyssal Pacific Ocean. *Science* 175: 636–637.

Heussner, S., A. Monaco, and J. Carbonne. 1990. The PPS 3 time-series sediment trap and the sample processing technique used during the ECOMARGE experiment. *Continental Shelf Research* 10: 943–958.

Hilário, A., P. A. Tyler, and D. W. Pond. 2008. A new method to determine the reproductive condition in female tubeworms tested in *Seepiophila jonesi* (Polychaeta: Siboglinidae: Vestimentifera). *Journal of the Marine Biological Association of the United Kingdom* 88: 909–912.

Hoppe, H. G. 1993. Use fluorogenic model substrates for extracellular enzyme activity (EEA) measurements of bacteria. In *Handbook of methods in aquatic microbial ecology*, ed. P. F. Kemp, B. F. Sherr, E. B. Sherr, and J. J. Cole, 509–512. London: Lewis, 1993.

Howe, J. A., T. M. Shimmield, and R. Diaz. 2004. Deep-water sedimentary environments of the northwestern Weddell Sea and South Sandwich Islands, Antarctica. *Deep-Sea Research Part II* 54: 1489–1514.

Howell, K., D. S. M. Billett, and P. A. Tyler. 2002. Depth-related distribution and abundance of seastars (Echinodermata: Asteroidea) in the Porcupine Seabight and Porcupine Abyssal Plain, N.E. Atlantic. *Deep-Sea Research Part I* 49: 1901–1920.

Hudson, I. R., B. D. Wigham, D. S. M. Billett, and P. A. Tyler. 2003. Seasonality and selectivity in the feeding ecology and reproductive biology of deep-sea bathyal holothurians. *Progress in Oceanography* 59: 381–407.

Huthnancea, J. M., H. M. Van Aken, M. White et al. 2002. Ocean margin exchange—water flux estimates. *Journal of Marine Systems* 32(1–3): 107–137.

Iken, K., T. Brey, U. Wand, I. Voight, and P. Junghans. 2001. Food web structure of the benthic community at Porcupine Abyssal Plain (N. Atlantic): A stable isotope analysis. *Progress in Oceanography* 50: 383–405.

Ingole, B. S., Z. A. Ansari, V. Rathod, and N. Rodrigues. 2001 Response of deep-sea macrobenthos to a small-scale environmental disturbance. *Deep-Sea Research Part II* 48 (16): 3401–3410.

Ingole B. S., Z. A. Ansari, V. Rathod, and N. Rodrigues. 2000. Response of meiofauna to immediate benthic disturbance in the Central Indian Ocean Basin. *Marine Georesources and Geotechnology* 18(3): 263–272.

Jahnke, R. A. 1996. The global ocean flux of particulate organic carbon: Areal distribution and magnitude. *Global Biogeochemical Cycles* 10: 71–88.

Jensen, P. 1988. Nematode assemblages in the deep-sea benthos of Norwegian Sea. *Deep-Sea Research Part I* 35: 1173–1184.

Jones, D. O. B., B. J. Bett, and P. A. Tyler. 2007. Depth-related changes in the Arctic epibenthic megafaunal assemblages of Kangerdlugssuaq, East Greenland. *Marine Biology Research* 3 (4): 191–204.

Joergensen, B. B. and A. Boetius. 2007. Feast and famine–microbial life in the deep-sea bed. *Nature Reviews: Microbiology* 5: 770–781.

Kaiser, S. and S. Brix. 2007. Two new species of *Pseudomesus* Hansen, 1916 (Isopoda: Asellota) from the Southern hemisphere: *Pseudomesu pitombo* sp. nov and *Pseudomesus satanus* sp. nov. *Zootaxa*: 1658: 21–38.

Kaiser, S. and D. K. A. Barnes. 2008. Southern Ocean deep-sea biodiversity: Sampling strategies and predicting responses to climate. *Climate Research* 37: 165–179.

Kallianiotis, A., K. Sophronidis, P. Vidoris, and A. Tselepides. 2000. Demersal fish and megafaunal assemblages on the Cretan continental shelf and slope (NE Mediterranean): Seasonal variation in species density, biomass, and diversity. *Progress in Oceanography* 46: 429–455.

Khripounoff, A., A. Vangriesheim, N. Babonneau, P. Crassous, B. Dennielou, and B. Savoye. 2003. Direct observation of intense turbidity current activity in the Zaire submarine valley at 4000 m water depth. *Marine Geology* 194: 151–158.

Klages, M., K. Vopel, H. Bluhm, T. Brey, T. Soltwedel, and W. E. Arntz. 2001. Deep-sea food falls: First observation of a natural event in the Arctic Ocean. *Polar Biology* 24: 292–295.

Klages, M., S. Muyakshin, T. Soltwedel, and W. E. Arntz. 2002. Mechanoreception, a possible mechanism for food fall detection in deep-sea scavengers. *Deep-Sea Research Part I* 49: 143–155.

Klein, B., W. Roether, B. B. Manca et al. 1999. The large deep water transient in the Eastern Mediterranean. *Deep-Sea Research Part I* 46: 371–414.

Kröncke, I. and M. Türkay. 2003. Structural and functional aspects of the macrofauna communities in the deep Angola-Basin. *Marine Ecology Progress Series* 260: 43–53.

Lambshead, P. J. D., H. M. Platt, and K. M. Shaw. 1983. The detection of differences among assemblages of marine benthic species based on an assessment of dominance and diversity. *Journal of Natural History* 17: 859–874.

Lambshead, P. J. D., B. J. Elce, D. Thistle, J. E. Eckman, and P. R. O. Barnett. 1994. A comparison of the biodiversity of deep-sea marine nematodes from three in the Rockall Trough, Northeast Atlantic, and one station in the San Diego Trough, Northeast Pacific. *Biodiversity Letters* 2: 95–107.

Lambshead, P. J. D., J. Tietjen, T. Ferrero, and P. Jensen. 2000. Latitudinal diversity gradients in the deep-sea with special reference to North Atlantic nematodes. *Marine Ecology Progress Series* 194: 159–167.

Lambshead, P. J. D., J. Tietjen, C. B. Moncrieff, and T. Ferrero. 2001. North Atlantic latitudinal diversity patterns in deep-sea marine nematode data: A reply to Rex et al. *Marine Ecology Progress Series* 210: 299–301.

Lambshead, P. J. D., C. J. Brown, T. Ferrero et al. 2002. Latitudinal diversity patterns of deep-sea marine nematodes and organic fluxes: A test from the central equatorial Pacific. *Marine Ecology Progress Series* 236: 129–135.

Lambshead, P. J. D., C. Brown, T. Ferrero, L. E. Hawkins, C. R. Smith, and N. J. Mitchell. 2003. Biodiversity of nematode assemblages from the region of the Clarion–Clipperton Fracture Zone, an area of commercial mining interest. *Biomed Central Ecology* 3: 1.

Lambshead, P. J. D. and G. Boucher. 2003. Marine nematode deep-sea biodiversity—hyperdiverse or hype? *Journal of Biogeography* 30 (4): 475–485.

Lampadariou N. and A. Tselepides. 2006. Spatial variability of meiofaunal communities at areas of contrasting depth and productivity in the Aegean Sea (NE Mediterranean). *Progress in Oceanography* 69: 19–36.

Lampitt R. S., D. S. M. Billett, and A. L. Rice. 1986. Biomass of the invertebrate megabenthos from 500 to 4100 m in the northeast Atlantic Ocean. *Marine Biology* 93: 69–81.

Lampitt, R. S., B. J. Bett, K. Kiriakoulakis, E. E. Popova, O. Ragueneau, A. Vangriesheim, and G. A. Wolff. 2001. Material supply to the abyssal seafloor in the Northeast Atlantic. *Progress in Oceanography* 50: 27–63.

Lee, R. W. 2003 Thermal tolerances of deep-sea hydrothermal vent animals from the Northeast Pacific. *Biology Bulletin* 205: 98–101.

Levin, L., S. E. Childers, and C. R. Smith. 1991. Epibenthic, agglutinating foraminiferans in the Santa Catalina Basin and their response to disturbance. *Deep-Sea Research* 38: 465–483.

Levin, L. A. and R. H. Michener. 2002. Isotopic evidence for chemosynthesis-based nutrition of macrobenthos: The lightness of being at Pacific methane seeps. *Limnology Oceanography* 47: 1336–1345.

Levin, L. A., E. L. Leithold, T. F. Gross, C. L. Huggett, and C. Di Bacco. 1994. Contrasting effects of substrate mobility on infaunal assemblages inhabiting two high-energy settings on Fieberling Guyot. *Journal of Marine Research* 52: 489–522.

Li, C., C. Kato, and K. Horikoshi. 1999. Bacterial diversity in deep-sea sediments from different depths. *Biodiversity and Conservation* 8: 659–677.

Lochte, K. and Turley C. M. 1988. Bacteria and cyanobacteria associated with phytodetritus in the deep sea. *Nature* 333: 67–69.

Loreau, M. 2000. Are community saturated? On the relationship between α, β, and γ diversity. *Ecology Letters* 3: 73–76.

Lykousis, V., G. Chronis, A. Tselepides et al. 2002. Major outputs of the recent multidisciplinary biogeochemical researches undertaken in the Aegean Sea. *Journal of Marine Systems* 33–34: 313–334.

Martín-Cuadrado, A. B., P. López-García, G. Gottschalk, and F. Rodríguez-Valera. 2007. Metagenomics of the deep Mediterranean, a warm bathypelagic habitat. *PLoS ONE* 2: 914.

McClain, C. R., N. A. Johnson, and M. A. Rex. 2004. Morphological disparity as a biodiversity metric in lower bathyal and abyssal gastropod assemblages. *Evolution* 58: 338–348.

McClain, C. R., M. A. Rex, and R. J. Etter. 2009. Patterns in deep-sea macroecology. In *Marine macroecology*, ed. K. Roy and J. Witman. Chicago: University of Chicago Press. in press.

Middelburg, J. J. 1989. A simple rate model for organic matter decomposition in marine sediments. *Geochimica et Cosmochimica Acta* 53 (7): 1577–1581.

Moodley L., J. J. Middelburg, K. Soetaert, H. T. S. Boschker, P. M. J. Herman, and C. H. R. Heip. 2005. *Journal of Marine Research* 63: 457–469.

Mienert, J., K. Andreassen, J. Posewang, and D. Lukas. 2000. Changes of the hydrate stability zone of the Norwegian margin from glacial to interglacial times. *Annals of the New York Academy of Sciences* 912: 200–210.

Moodley, L., J. J. Middelburg, H. T. S. Boschker et al. 2002. Bacteria and Foraminifera: Key players in a short-term deep-sea benthic response to phytodetritus. *Marine Ecology Progress Series* 236: 23–29.

Neira, C., J. Sellanes, L. A. Levin, and W. E. Arntzd. Meiofaunal distributions on the Peru margin: Relationship to oxygen and organic matter availability. *Deep-Sea Research Part I* 48 (11): 2453–2472.

Nozawa, F., H. Kitazato, M. Tsuchiya, and A. J. Gooday. 2006. "Live" benthic foraminifera at an abyssal site in the equatorial Pacific nodule province: Abundance, diversity, and taxonomic composition. *Deep-Sea Research Part I*: 1406–1422.

Olu-Le Roy, K., M. Sibuet, A. Fiala-Médioni et al. 2004. Cold seep communities in the deep eastern Mediterranean Sea: Composition, symbiosis, and spatial distribution on mud volcanoes. *Deep-Sea Research Part I* 51: 1915–1936.

Paterson, G. L. J. and P. J. D. Lambshead. 1995. Bathymetric patterns of polychaete diversity in the Rockall Trough, northeast Atlantic. *Deep-Sea Research Part I* 42: 1199–1214.

Paterson, G. L. J., G. D. F. Wilson, N. Cosson, and P. A. Lamont. 1998. Hessler and Jumars (1974) revisited: Abyssal polychaete assemblages from the Atlantic and Pacific. *Deep-Sea Research Part II* 45 (1–3): 225–251.

Parkes R. J., G. Webster, B. A. Cragg et al. 2005. Deep sub-seafloor prokaryotes stimulated at interfaces over geological time. *Nature* 436: 390–394.

Pawlowski, J., M. Holzmann, C. Berney et al. 2003. The evolution of early Foraminifera. *Proceedings of the National Academy of Sciences USA* 100 (20): 11494–11498.

Pfannkuche, O., A. Boetius, K. Lochte, U. Lundgreen, and H. Thiel. 1999. Responses of deep-sea benthos to sedimentation patterns in the North-East Atlantic in 1992. *Deep-Sea Research Part I* 46 (4): 573–596.

Price, A. R. G. 2002. Simultaneous "hotspots" and "coldspots" of marine biodiversity and implications for global conservation. *Marine Ecology Progress Series* 241: 23–27.

Priede, I. G., R. Froese, D. M. Bailey et al. 2006. The absence of sharks from abyssal regions of the world's oceans. *Proceedings of Royal Society B* 273: 1435–41.

Puig, P., J. B. Company, F. Sardà, and A. Palanques. 2001. Responses of deep-water shrimp populations to the presence of intermediate nepheloid layers on continental margins. *Deep-Sea Research Part I* 48: 2195–2207.

Quéric, N. V., T. Soltwedel, and W. E. Arntz. 2004. Application of a rapid direct viable count method to deep-sea sediment bacteria. *Journal of Microbiological Methods* 57: 351–367.

Raghukumar, C., P. A. L. Bharathi, Z. A. Ansari et al. 2001. Bacterial standing stock, meiofauna, and sediment–nutrient characteristics: Indicators of benthic disturbance in the Central Indian Basin. *Deep-Sea Research Part II* 48: 3381–3399.

Ramirez-Llodra, E. 2002. Fecundity: Characteristics and role in life-history strategies of marine invertebrates. *Advances in Marine Biology* 43: 87–170.

Ramirez-Llodra, E., W. D. K. Reid, and D. S. M. Billett. 2005. Long-term changes in reproductive patterns of the holothurian *Oneirophanta mutabilis* from the Porcupine Abyssal Plain. *Marine Biology* 146: 683–693.

Ramirez-Llodra, E. and M. Segonzac. 2006. Reproductive biology of *Alvinocaris muricola* (Decapoda: Caridea: Alvinocarididae) from cold seeps in the Congo Basin. *Journal Marine Biology Association United Kingdom* 86: 1347–1356.

Ramirez-Llodra E., J. B. Company, M. Camps, and G. Rotllant. 2007. Spatio-temporal variations in reproductive patterns and population structure of *Pasiphaea multidentata* (Decapoda: Caridea) in the Blanes canyon and adjacent margin, North-western Mediterranean Sea. *Marine Ecology* 28: 470–479.

Ramirez-Llodra, E., M. Ballesteros, J. B. Company, L. Dantart, and F. Sardà. 2008. Spatio-temporal variations of biomass and abundance in bathyal non-crustacean megafauna in the Catalan Sea (North-western Mediterranean). *Marine Biology* 153: 297–309.

Ramirez-Llodra, E., J. B. Company, F. Sardà, G. Rotllant. 2009. Megabenthic diversity patterns and community structure of the Blanes submarine canyon and adjacent slope in the Northwestern Mediterranean: a human overprint? *Marine Ecology* in press.

Renaud-Mornant, J. and N. Gourbault. 1990. Evaluation of abyssal meiobenthos in the eastern central Pacific (Clarion-Clipperton Fracture Zone). *Progress in Oceanography* 24: 317–329.

Rex, M. A. 1973. Deep-sea species diversity: Decreased gastropod diversity at abyssal depths. *Science* 181: 1051–1053.

Rex, M. A. 1981. Community structure in the deep-sea benthos. *Annual Review of Ecological Systems* 12: 331–353.

Rex, M. A., C. R. McClain, N. A. Johnson et al. 2005. A source-sink hypothesis for abyssal biodiversity. *The American Naturalist* 165: 163–178.

Rex, M. A., C. T. Stuart, and R. J. Etter. 2001. Do deep-sea nematodes show a positive latitudinal gradient of species diversity? The potential role of depth. *Marine Ecology Progress Series* 210: 297–298.

Rice, A. L., R. G. Aldred, E. Darlington, and R. A. Wild. 1982. The quantitative estimation of the deep-sea megabenthos—a new approach to an old probem. *Oceanologica Acta* 5: 63–72.

Rice, A. L., D. S. M. Billett, J. Fry, A. W. G. John, R. S. Lampitt, R. F. C. Montoura, and R. J. Morris. 1986. Seasonal deposition of phytodetritus to the deep-sea floor. *Proceedings of the Royal Society of Edinburg* 88B: 265–279.

Ricklefs, R. E. 2000. The relationship between local and regional species richness in birds of the Caribbean basin. *Journal of Animal Ecology* 69: 1111–1116.

Roberts, D. and H. M. Moore. 1997. Tentacular diversity in deep-sea deposit-feeding holothurians: Implications for biodiversity in the deep sea. *Biodiversity and Conservation* 6: 1487–1505.

Roberts, D., A. V. Gebruk, V. Levin, and B. A. D. Manship. 2000. Feeding and digestive strategies in deposit-feeding holothurians. *Oceanography and Marine Biology: An Annual Review* 38: 257–310.

Roberts, D., H. M. Moore, J. Berges, J. W. Patching, M. W. Carton, and D. F. Eardly, 2001. Sediment distribution, hydrolytic enzyme profiles, and bacterial activities in the guts of *Oneirophanta mutabilis, Psychropotes longicauda,* and *Pseudostichopus villosus*: What do they tell us about digestive strategies of abyssal holothurians? *Progress in Oceanography* 50: 443–458.

Roberts, J. M., A. J. Wheeler, and A. Freiwald. 2006. Reefs of the deep: the biology and geology of cold-water coral ecosystems. *Science* 312: 543–547.

Rosenzweig, M. L. 1995. *Species diversity in space and time*. Cambridge, U.K.: Cambridge University Press.

Rowe, G. T. 1983. Biomass and production of the deep-sea macrobenthos. In *Deep-sea biology. The sea,* Vol. 8, ed. G. T. Rowe, 97–122. New York: Wiley.

Rowe, G. T., A. Lohse, F. Hubbard, G. S. Boland, E. E. Briones, and J. Deming. 2003. Preliminary trophodynamic carbon budget for the Sigsbee Deep benthos, Northern Gulf of Mexico. *American Fisheries Society Symposium* 36: 225–238.

Rowe, G. T., M. Sibuet, J. Deming et al. 1991. "Total" sediment biomass and preliminary estimates of organic carbon residence time in deep-sea benthos. *Marine Ecology Progress Series* 79: 99–114.

Roy, K., D. Jablonski, J. W. Valentine, and G. Rosemberg. 1998. Marine latitudinal diversity gradients: Tests of causal hypotheses. *Proceeding of the National Academy of Sciences USA* 95: 3699–3702.

Ruhl, H. A. and K. L. Smith Jr. 2004. Shifts in deep-sea community structure linked to climate and food supply. *Science* 305: 513–515.

Sanders, H. L. 1968. Marine benthic diversity: A comparative study. *The American Naturalist* 102: 243–282.

Sardà, F., G. D'Onghia, C.-Y. Politou, A. Tselepides. 2004. Mediterranean deep-sea biology. *Scientia Marina* 68 (3):1–204.

Sardà, F., J. B. Company, G. Rotllant, M. Coll. 2009. Biological patterns and ecological indicators for Mediterranean fish and crustaceans below 1,000 m: a review. *Rev Fish Biol Fisheries* 19: 329–347.

Schewe, I. and T. Soltwedel. 1998. Deep-sea meiobenthos of the central Arctic Ocean: Distribution patterns and size-structure under extreme oligotrophic conditions. *Vie Milieu* 49: 79–92.

Schippers, A., L. N. Neretin, J. Kallmeyer et al. 2005. Prokaryotic cells of the deep sub-seafloor biosphere identified as living bacteria *Nature* 433: 861–4.

Schmidt, J. L., J. W. Deming, P. A. Jumars, and R. G. Keil. 1998. Constancy of bacterial abundance in surficial marine sediments. *Limnology and Oceanography* 43: 976–982.

Schmidt, S., H. C. deStigter, and T. C. E. Van Weering. 2001. Enhanced short-term sediment deposition within the Nazare Canyon, North-East Atlantic. *Marine Geology* 173: 55–67.

Schwabe, E., J. M. Bohn, W. Engl, K. Linse, and M. Schrödl. 2007. Rich and rare—first insights into species diversity and abundance of Antarctic abyssal Gastropoda (Mollusca). *Deep-Sea Research Part II* 54: 1831–1847.

Schwinghamer, P. 1985. Observations on size–structure and pelagic coupling of some shelf and abyssal benthic communities. In *Proceedings of the 19th European Marine Biology Symposium,* Plymouth, Devon, UK, September 16–21, 1984, ed. P. E. Gibbs, 347–359. Cambridge: Cambridge University Press.

Seiter, K., C. Hensen, J. Schröter, and M. Zabel. 2004. Organic carbon content in surface sediments-defining regional provinces. *Deep-Sea Research Part I* 51: 2001–2026.

Sellanes, J., E. Quiroga, and V. A. Gallardo. 2004. First direct evidence of methane seepage and associated chemosynthetic communities in the bathyal zone off Chile. *Journal of the Marine Biological Association of the UK* 84: 1065–1066.

Sicot, F. X., M. Mesnage, M. Masselot et al. 2000. Molecular adaptation to an extreme environment: Origin of the thermal stability of the pompeii worm collagen. *Journal of Marine Biology* 302: 811–820.

Sibuet, M., A. Khripounoff, J. Deming, R. Colwell, and A. Dinet 1982. Modification of the gut contents in the digestive tract of abyssal holothurians In *International Echinoderms Conference, Tampa Bay*, ed. J. M. Lawrence, 421–428. Rotterdam, Netherlands: A. A. Balkema.

Sibuet, M. 1985. Quantitative distribution of echinoderms (Holothuroidea, Asteroidea, Ophiouroidea, Echinoidea) in relation to organic matter in the sediment, in deep sea basins of the Atlantic Ocean. In *Echinodermata. Proceedings of the Fifth International Echinoderm Conference Galway 24–29 September 1984,* Galway, Ireland, ed. B. F. Keegan, and B. D. S. O'Connor, 99–108. Rotterdam, Netherlands: A. A. Balkema.

Sibuet, M., C. E. Lambert, R. Chesselet, and L. Laubier. 1989. Density of the major size groups of benthic fauna and trophic input in deep basins of the Atlantic Ocean. *Journal of Marine Research* 47: 851–867.

Sibuet, M. and K. Olu. 1998. Biogeography, biodiversity, and fluid dependence of deep-sea cold-seep communities at active and passive margin. *Deep-Sea Research Part II* 45: 517–567.

Smith C. R., F. C. De Leo, A. F. Bernardino, A. K. Sweetman, and P. M. Arbizu. 2008. Abyssal food limitation, ecosystem structure, and climate change. *Trends in Ecology and Evolution* 23 (9): 518–528.

Smith, C. R., W. Berelson, D. J. DeMaster et al. 1997. Latitudinal variations in benthic processes in the abyssal equatorial Pacific: Control by biogenic particle flux. *Deep-Sea Research Part II* 44: 2295–2317.

Smith, C. R. and A. W. J. Demopoulos. 2003. The deep Pacific Ocean floor. In *Ecosystems of the deep oceans*, ed. P. A. Tyler, 179–218. Amsterdam: Elsevier.

Smith, K. L. Jr., R. S. Kaufmann, R. J. Baldwin, and A. F. Carlucci. 2001. Pelagic–benthic coupling in the abyssal eastern North Pacific: An 8-year time-series study of food supply and demand. *Limnology and Oceanography* 46: 543–556.

Smith, K. L., Jr., R. J. Baldwin, H. A. Ruhl, M. Kahru, B. G. Mitchell, and R. S. Kaufman. 2006. Climate effect on food supply to depths greater than 4,000 meters in the northeast Pacific. *Limnology and Oceanography* 51: 166–176.

Smith, C. R., L. A. Levin, A. Koslow, P. A. Tyler, and A. G. Glover. 2008. The near future of deep seafloor ecosystems. In *Aquatic ecosystems: Trends and global prospects*, ed. N. Polunin. 334–351. Cambridge: Cambridge University Press.

Smith, K. L., R. J. Baldwin, D. M. Karl, and A. Boetius. 2002. Benthic community responses to pulses in pelagic food supply: North Pacific subtropical gyre. *Deep-Sea Research Part I* 49: 971–990.

Smith, K. L., R. J. Baldwin, and H. A. Ruhl. 2006. Climate effect on food supply to depths greater than 4,000 meters in the Northeast Pacific. *Limnology and Oceanography* 51(1): 166–176.

Smith., K. L. and R. S. Kaufmann. 1999. Long-term discrepancy between food supply and demand in the deep Eastern North Pacific. *Science* 284: 1174.

Soetaert, K., C. Heip, and M. Vincx. 1991. Diversity of nematode assemblages along a Mediterranean deep-sea transect. *Marine Ecology Progress Series* 75: 275–282.

Soetaert, K. and C. Heip. 1995. Nematode assemblages of deep-sea and shelf break sites in the North Atlantic and Mediterranean Sea. *Marine Ecology Progress Series* 125: 171–183.

Soetaert K., P. M. J. Herman, and J. J. Middelburg 1996. Dynamic response of deep-sea sediments to seasonal variations: a model. *Limnology and Oceanography* 41: 1651–1668.

Sogin, M. L., H. G. Morrison, J. A. Huber et al. 2006. Microbial diversity in the deep sea and the underexplored "rare biosphere." *Proceedings of the National Academy of Sciences USA* 103: 12115–12120.

Soltwedel, T. 1997. Meiobenthos distribution pattern in the tropical East Atlantic: Indication for fractionated sedimentation of organic matter to the sea floor? *Marine Biology* 129: 747–756.

Soltwedel, T. 1997. Temporal variabilities in benthic activity and biomass on the western European continental margin. *Oceanologica Acta* 20: 871–879.

Soltwedel, T. 2000. Metazoan meiobenthos along continental margins: A review. *Progress in Oceanography* 46: 59–84.

Soyer, J. 1985. Mediterranean Sea meiobenthos. In *Mediterranean marine ecosystems*, ed. M. Moraitou-Apostoloupolou and V. Kiortsis, 85–108. New York: Plenum.

Thiel, H. 1975. The size structure of the deep-sea benthos. *Intermediate Revue dergesamten Hydrobiologie* 60: 576–606.

Thiel, H. 1979. Structural aspects of the deep-sea benthos. *Ambio Special Report* 6: 25–31.

Thistle, D. and K. Sherman. 1985. The nematode fauna of a deep-sea site exposed to strong near bottom currents. *Deep-Sea Research Part I* 31: 1077–1088.

Thistle, D., L. A. Levin, A. J. Gooday, O. Pfannkuche, and P. J. D. Lambshead. 1999. Physical reworking by near-bottom flow alters the metazoan meiofauna of Fieberling Guyot (northeast Pacific). *Deep-Sea Research Part I* 46: 2041–2052.

Thistle, D., P. J. D. Lambshead, and K. Sherman. 1995. Nematode tail-shape groups respond to environmental differences in the deep sea. *Vie Milieu* 45: 107–115.

Tietjen, J. H. 1976. Distribution and species diversity of the deep-sea nematodes off north Carolina. *Deep-Sea Research Part I* 23: 755–768.

Tietjen, J. H. 1984. Distribution and species diversity of deep-sea nematodes in the Venezuela basin. *Deep-Sea Research Part I* 31: 119–132.

Tietjen, J. H. 1989. Ecology of deep-sea nematodes from the Puerto Rico Trench area and Hatteras Abyssal Plain. *Deep-Sea Research Part I* 36: 1579–1594.

Trincardi, F. and A. Argnani. 2006 Gela submarine slide: A major basin-wide event in the plio-quaternary foredeep of Sicily. *Geo-Marine Letters* 10: 13–21.

Trincardi, F., A. Correggiari, M. E. Field, and W. R. Normark. 1995. Turbidite deposition from multiple sources: Quaternary Paola Basin (Eeastern Tyrrhenian Sea). *Journal of Sedimentary Research* 65 (4): 469–483.

Tselepides, A. and A. Eleftheriou. 1992. South Aegean (Eastern Mediterranean) continental slope benthos: Macrofaunal–environmental relationships. In *Deep-sea food chain and the global carbon cycle,* ed. G. T. Rowe and V. Pariente 360: 139–156. Dordrecht, Netherlands: Kluwer.

Tselepides, A., K. N. Papadopoulou, D. Podaras, W. Plati, D. Koutsoubas, and F. Pantazoglou. 1997. Macrobenthic community structure in the Cretan Sea. *5th Hellenic Symposium on Oceanography and Fisheries–Kawala, Greece, April 15–18, 1997. Proceedings in Oceanography, NCMR Association of Employers* 1: 233–236.

Tunnicliffe, V., S. K. Juniper, and M. Sibuet. 2003. Reducing environments of the deep-sea floor. In *Ecosystems of the deep oceans*, ed. P. A. Tyler, 81–110. Amsterdam: Elsevier.

Tursi, A., F. Mastrototaro, A. Matarrese, P. Maiorano, and G. D'onghia. 2004. Biodiversity of the white coral reefs in the Ionian Sea (Central Mediterranean). *Chemistry and Ecology* 20: 107–116.

Tyler, P. A. 1988. Seasonality in the deep sea. *Oceanography and Marine Biology Annual Review* 26: 227–258.

Tyler, P. A. 2003. *Ecosystems of the Deep Oceans Ecosystems of the World.* Elsevier: Amsterdam.

Tyler, P. A., C. M. Young, and F. Dove. 2007. Settlement, growth, and reproduction in the deep-sea wood-boring bivalve mollusc *Xylophaga depalmai. Marine Ecology Progress Series* 343: 151–159.

Tyler, P. A., C. M. Young, E. Dolan, S. M. Arellano, S. D. Brooke, and M. Baker. 2007. Gametogenic periodicity in the chemosynthetic cold mussel "*Bathymodiolus*" *childressi. Marine Biology* 150 (5): 829–840.

Tyler, P. A., S. Pendlebury, S. W. Mills et al. 2008. Reproduction of gastropods from vents on the East Pacific Rise and the Mid-Atlantic Ridge. *Journal of Shellfish Research* 27(1): 107–118.

Van Den Hove, S. and V. Moreau. 2007. Deep-sea biodiversity and ecosystems: A scoping report on their socio-economy, management, and governance. Cambridge: United Nations Environment Programme.

Van Dover, C. L., J. F. Grassle, B. Fry, R. H. Garritt, and V. R. Starczak. 1992. Stable isotope evidence for entry of sewage-derived organic material into a deep-sea food web. *Nature* 360: 153–156.

Van Dover, C. L. 2000. *The ecology of deep-sea hydrothermal vents*. Princeton: Princeton University Press.

Vanhove, S., H. Vermeeren, and A. Vanreusel. 2004. Meiofauna towards the South Sandwich Trench (750–6300 m), focus on nematodes. *Deep-Sea Research Part II* 51(14–16): 1665–1688.

Vanreusel, A., N. Cosson-Sarradin, A. J. Gooday et al. 2001. Evidence for episodic recruitment in a small opheliid polychaete species from the abyssal NE Atlantic. *Progress in Oceanography* 50: 285–301.

Vanreusel, A., I. Van-den-Bossche, and F. Thiermann. 1997. Free-living marine nematodes from hydrothermal sediments: Similarities with communities from diverse reduced habitats. *Marine Ecology Progress Series* 157: 207–219.

Vanreusel, A., L. Clough, K. Jacobsen et al. 2000. Meiobenthos of the Central Arctic ocean with special emphasis on the nematode community structure. *Deep-Sea Research Part I* 47: 1855–1879.

Vetter, E. W. and P. K. Dayton. 1998. Macrofaunal communities within and adjacent to a detritus-rich submarine Canyon system. *Deep-Sea Research Part II* 45: 25–54.

Vopel, K. and H. Thiel. 2001. Abyssal nematode assemblages in physically disturbed and adjacent sites of the eastern equatorial Pacific. *Deep-Sea Research Part II* 48: 3795–3808.

Waide, R. B., M. R. Willig, C. F. Steiner et al. 1999. The relationship between productivity and species richness. *Annual Review Ecological Systematic* 30: 257–300.

Walsh, J. J. 1991. Importance of continental margins in the marine biogeochemical cycling of carbon and nitrogen. *Nature* 350: 53–55.

Weaver, P. E. 2005. *The geobiology of the Nazare and Setubal Canyons, Portuguese Continental Margin. Cruise report No. 1 RRS Discovery cruise D297 27 July–16 August 2005.* Southampton: National Oceanography Centre.

Weaver, P. E., D. S. M. Billett, A. Boetius, R. Danovaro, A. Freiwald, and M. Sibuet. 2004. Hotspot ecosystems research on Europe's deep-ocean margins. *Oceanography I* 17: 132–143.

Wigham, B. D., I. R. Hudson, D. S. M. Billett, and G. A. Wolff. 2003. Is long-term change in the abyssal Northeast Atlantic driven by qualitative changes in export flux? Evidence from selective feeding in deep-sea holothurians. *Progress in Oceanography* 59: 409–441.

Witbaard, R., A. Duineveld, A. Kok, J. van der Weele, and E. M. Berghuis. 2000. The benthic response to the seasonal deposition of phytopigments at the Porcupine Abyssal Plain in the northeast Atlantic. *Journal of Sea Research* 43: 15–31.

Witbaard, R., A. Duineveld, A. Kok, J. van der Weele, and E. M. Berghuis. 2001. The response of *Oneirophanta mutabilis* (Holothuroidea) to the seasonal deposition of phytopigments at the Porcupine Abyssal Plain in the Northeast Atlantic. *Progress in Oceanography* 50: 423–441.

Yakimov, M. M., S. Cappello, E. Crisafi et al. 2006. Phylogenetic survey of metabolically active microbial communities associated with the deep-sea coral *Lophelia pertusa* from the Apulian plateau, Central Mediterranean Sea. *Deep-Sea Research Part I* 53: 62–75.

Young, C. M. 2003. Reproduction, development, and life history traits. In *Ecosystems of the deep oceans*, ed. P. A. Tyler, 381–426. Amsterdam: Elsevier.

Index

"f" indicates material in figures. "t" indicates material in tables.

G

H

N

O

P

R

Non-Homogeneous Markov Chains and Systems

Non-Homogeneous Markov Chains and Systems: Theory and Applications fulfills two principal goals. It is devoted to the study of non-homogeneous Markov chains in the first part, and to the evolution of the theory and applications of non-homogeneous Markov systems (populations) in the second. The book is self-contained, requiring a moderate background in basic probability theory and linear algebra, common to most undergraduate programs in mathematics, statistics, and applied probability. There are some advanced parts, which need measure theory and other advanced mathematics, but the readers are alerted to these so they may focus on the basic results.

Features

- A broad and accessible overview of non-homogeneous Markov chains and systems
- Fills a significant gap in the current literature
- A good balance of theory and applications, with advanced mathematical details separated from the main results
- Many illustrative examples of potential applications from a variety of fields
- Suitable for use as a course text for postgraduate students of applied probability, or for self-study
- Potential applications included could lead to other quantitative areas

The book is primarily aimed at postgraduate students, researchers, and practitioners in applied probability and statistics, and the presentation has been planned and structured in a way to provide flexibility in topic selection so that the text can be adapted to meet the demands of different course outlines. The text could be used to teach a course to students studying applied probability at a postgraduate level or for self-study. It includes many illustrative examples of potential applications in order to be useful to researchers from a variety of fields.

P.-C.G. Vassiliou is an Honorary Professor in the Department of Statistical Sciences, University College London. He has been affiliated with UCL for almost a decade.

Non-Homogeneous Markov Chains and Systems

Theory and Applications

P.-C.G. Vassiliou
University College London, UK

CRC Press
Taylor & Francis Group
Boca Raton London New York

CRC Press is an imprint of the
Taylor & Francis Group, an **informa** business
A CHAPMAN & HALL BOOK

Front cover image: Photo of pond lilies in the garden of author's house, created by author's wife, Fevronia Vassiliou. See Section 14.2.

First edition published 2023
by CRC Press
6000 Broken Sound Parkway NW, Suite 300, Boca Raton, FL 33487-2742

and by CRC Press
4 Park Square, Milton Park, Abingdon, Oxon, OX14 4RN

Library of Congress Cataloging-in-Publication Data

Names: Vassiliou, P-C. G. (Panos-C. G.), author.
Title: Non-homogeneous Markov chains and systems : theory and applications / P.-C.G. Vassiliou, University College London, UK.
Description: First edition. | Boca Raton : C&H/CRC Press, 2023. | Includes bibliographical references and index.
Identifiers: LCCN 2022023854 (print) | LCCN 2022023855 (ebook) | ISBN 9781138034525 (hardback) | ISBN 9781032378046 (paperback) | ISBN 9781315269986 (ebook)
Subjects: LCSH: Markov processes.
Classification: LCC QA274.7 .V37 2023 (print) | LCC QA274.7 (ebook) | DDC 519.2/33--dc23/eng20221006
LC record available at https://lccn.loc.gov/2022023854
LC ebook record available at https://lccn.loc.gov/2022023855

ISBN: 978-1-138-03452-5 (hbk)
ISBN: 978-1-032-37804-6 (pbk)
ISBN: 978-1-315-26998-6 (ebk)

DOI: 10.1201/b23204

Typeset in Computer Modern Font
by KnowledgeWorks Global Ltd.

Publisher's note: This book has been prepared from camera-ready copy provided by the authors.